Surface Preparation and Cleaning Conference (SPCC 2017)

Austin, Texas, USA
28 - 29 March 2017

ISBN: 978-1-5108-4814-6

Printed from e-media with permission by:

Curran Associates, Inc.
57 Morehouse Lane
Red Hook, NY 12571

Some format issues inherent in the e-media version may also appear in this print version.

Printed by Curran Associates, Inc. (2017)

Surface Preparation and Cleaning Conference (SPCC)
c/o Mark Thirsk
Linx Consulting-PO Box 384
Mendon, MA 01756-0384
USA

Phone: 617-273-8837

mthirsk@linx-consulting.com

Additional copies of this publication are available from:

Curran Associates, Inc.
57 Morehouse Lane
Red Hook, NY 12571 USA
Phone: 845-758-0400
Fax: 845-758-2633
Email: curran@proceedings.com
Web: www.proceedings.com

TABLE OF CONTENTS

SESSION 1: KEYNOTE/BEOL

KEYNOTE—Wafer Cleanliness: Challenges from an Increasingly Complex Fab Process – 1
Ben Eynon

INVITED—Chemical Bonding Transformation Mapping to Optimize Low-k Dielectric Nanostructure Fabrication and Post-etch Residue Clean 15
Oliver Chyan, Sirish Rimal, Tamal Mukherjee, Muthappan Asokan, Seare A. Berhe, Kanwaljit Singh, Ian Brown, Jacques Faguet

SESSION 1: PART 2, BEOL

Wet Clean Transfer Challenges in 22 NM ½ Pitch and 16 NM ½ Pitch Structures 37
Els Kesters, Q. Le, C. Lorant, X. Xu, F. Holsteyns

High Temperature Water Clean and Etch Reactions with Low-k and SiO_2 Films: Experiments and Simulations 50
Rick Reidy, Joshua Barclay, Lu Deng, Oseoghaghare Okobiah, Tina Sengphanlaya, Jincheng Du

Contact Cleaning Opportunities on Single Wafer Tool 59
Lucile Broussous, S. Zoll, H. Ishikawa, F. Buisine, A. Lamaury

Removal of Edge Cluster Defects by Improving Recipe and Hardware for Backside Polysilicon Wet Etch Process 67
Hong-Ying Zhai, Colin Weidner, Dhiman Bhattacharyya, Norberto DeOliveria, Vincent Sih

SESSION 2: CONTAMINATION CONTROL

INVITED:The Future of Micro-Contamination Control in Chemical Delivery Systems for Advanced Lithography & Wet Etch and Clean Semiconductor Processes 77
Archita Sengupta, Thomas Phely-Bobin, Paul Trio, Robert McIntosh

Effect of Dilute Hydrogen Peroxide in Ultrapure Water 91
Yuichi Ogawa, Toru Masaoka, Nobuko Gan, Yu Fujimura, Yasuharu Minato, Yoichi Miyazaki, Kurt Wostyn, Antoine Pacco, Frank Holsteyns, Yukifumi Yoshida

Effect of Sulfuric Acid Manufacturing Process on Semiconductor Inline Defects 102
Dhiman Bhattacharyya, Sushil Patil, Mark Conrad, Hayley Manning, Fauzia Khatkhatay, Alexander Mena, Norberto DeOliveira

Blisters Formation Mechanism During High Dose Implant Stripping 112
Marion Croisy, Cecile Jenny, Claire Richard, Denis Guiheux, Sylvain Joblot, Alain Campo, Erwine Pargon, Nicolas Posseme

Gas Purge or Wet Cleaning? Decontamination Solutions to Control AMCs in FOUPs 121
Paola González-Aguirre, Herve Fontaine, Carlos Beitia, Jim Ohlsen, Jorgen Lundgren

Invited: Non-uniform Contamination Results in Sub-ppm Fail Rates: Analytical Challenges 129
Martin Knotter

SESSION 4: ADVANCED SURFACE PREP FOR FEOL

INVITED: —Nitride Device Surface Processes: A Review of Surface Science for Power and RF Technologies with Nitrides 142
H. Rusty Harris

Aqueous Ammonium Sulfide Passivation and $Si_{1-x}Ge_x$ (100) MOSCaps 159
Lauren Peckler, Stacy Heslop, Anthony Muscat

Surface Preparation and Wet Cleaning for Germanium Surface 170
Hajime Shirakawa, J. Snow, M. Otsuji, Y. Yoshida, H. Takahashi, F. Sebaai, F. Holsteyns, M. Sato

Study and Development of Advanced Surface Preparations for the Integration of New Materials in Micro/Nanoelectronics 181
Raynal Pierre-Edouard

SESSION 4, PART 2: ADVANCED SURFACE PREP FOR FEOL AND FINFET

A Wet Clean Solution to Reduce Unwanted eSiGe Growth Defect in FinFET ... 191
Jian Li, Vincent Sih, Talapady Bhat

Wet Etching of TiN in 1-D and 2-D confined nano-spaces of FinFET Transistors ... 199
Guy Vereecke, Hanne De Coster, Patrick Carolan, Hugo Bender, Kherim Willems, Lars-Ake Ragnarsson, Pol Van Dorpe, Naoto Horiguchi, Frank Holsteyns

SESSION 5: ETCHING

INVITED: Shifting Global Economic Forces Shaping the Semiconductor Outlook ... 210
Duncan Meldrum

Thermal SiO_2 Atomic Layer Etching by a "Conversion-Etch" Mechanism Using Sequential Reactions of Trimethylaluminum and Hydrogen Fluoride ... 227
Steven George, Jaime DuMont, Amy Marquardt, Austin Cano

Effect of Additives in Diluted HF Solutions on Removal of Metal Contaminants and Particles on Silicon Wafer ... 239
Jin-Goo Park, Hyun-Tae Kim, Dong-Hwan Lee, Jae-Hwan Lee, Eun-Suck Choi, Jin-Goo Park

SESSION 6: POST CMP CLEANING

Invited: Post CMP In-situ Cleaning for 14/7nm Transistor Scaling: A Crucial Process for Yield Enhancement at Advanced Node Device Fabrication ... 252
TaeHoon Lee

High Performance Ceria Post-CMP Cleaning Formulations for STI Dielectric Substrates ... 256
Daniela White, Atanu Das, Thomas Parson, Michael White

Mechanistic and Electrochemical Aspects of Copper and Cobalt Post CMP Cleaners for 5-7 nm Nodes ... 267
Mike White, Daniela White, Volley Wang, Jun Liu, Elizabeth Thomas, Don Frye, Ruben Lieten, Thomas Parson, Atanu Das

CMP Stack-Trek ... 275
Viorel Balan

POSTERS

Chemical/Mechanical Balance Management through Pad Microstructure in CMP ... 289
Ratanak Yim, C. Perrot, V. Balan, P.-Y. Friot, B. Qian, N. Chiou, G. Jacob, E. Gourvest, F. Salvatore, S. Valette

Process Model Identification for Optimizing Particle-to-Part Selectivity during Ultrasonic Parts Cleaning ... 290
Osama Khalil, John Deem, Iordan Iordanov, Ardeshir Sidhwa, Matthew House, David Zuck

Advantages of Using RGA as Part of Atomically Clean Surface (ACS™) for Sub-10nm Parts Cleaning Process ... 291
Matthew House, Osama Khalil, John Deem, Ardeshir Sidhwa, David Zuck

Sulfate Ion Removal by Combined UV and Bake Process ... 292
Fei Xu, Eric Guo, Crystal Wang, Sandy Qian, Mars Wang, Harry Zhang, Keanu Wu, Wei Jiang, Jun Jin, Jian Shen

Particle Clean-Up of Various Filter Media in WEC Chemistry via 20 nm Particle Counting ... 302
Patrick Connor, Lawrence Johnson, Rao Varanasi, Tony Shucosky

Effective Point of Use Filtration of Ultrapure Water for Critical Tools and Processes ... 308
Glen Sundstrom, Alan Knapp

Continuous Monitoring of Particles at 20 nm in Critical Semiconductor Process Chemicals ... 316
Dan Rodier

Implementing Chemical Cost Saving in 14 nm High Volume Manufacturing Line ... 324
Tejasvi Sadasivuni, Norberto DeOliveira

Improved Missing Pattern Defects Using Megasonic for Hard Mask Clean at 14 nm and Beyond ... 325
Sangita Kumari, SherJang Singh, F. Devendar, Silas Scott

Dummy Poly Gate Etch Residue Removal ... 326
Wen Dar Liu, Yi Chia Lee, John G. Langan, Tianniu Rick Chen, Aiping Wu, Andrew J. Adamczyk

Indium Bump Liftoff Using the MEI FluidJet ... 333
Scott Tice

Author Index

Wafer Cleanliness: Challenges from an Increasingly-Complex Fab Process

Benjamin Eynon

Assistant Director, NASCENT Center
University of Texas at Austin

March 28, 2017

2017 SPCC – Austin, TX

Tech Leadership = Following 'Moore's Law'

Gordon Moore
ca 1965

ca 2010

Moore's famous prediction:

"The number of transistors on an integrated circuit will double every ~18 months."

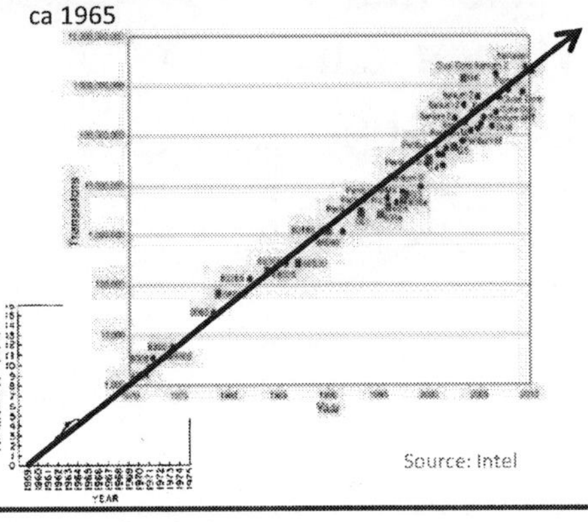

Source: Intel

Cost may end Moore's Law before physics will:

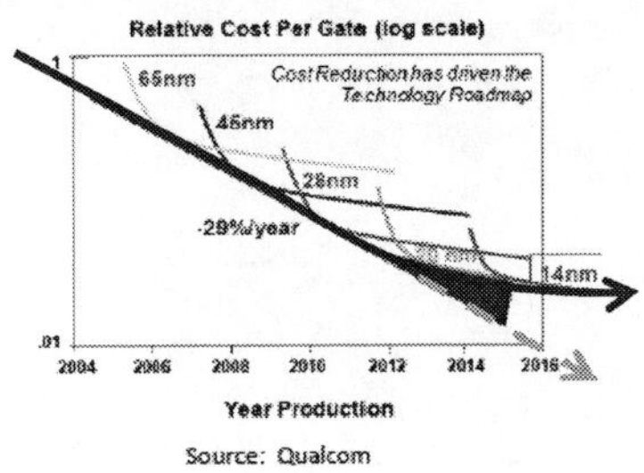

Source: Qualcom

BGE – March 28th, 2017

SPCC – Austin, TX

Moore's Law in Perspective

- Airplane
  - 1903: Wright Bro's of NC
  - 7mph @ 0.7mpg
  - After 56 years (1959)
  - 224 billion mph
- Integrated Circuit
  - 1959: Jack Kilby of TI
  - 1 transistor
  - After 56 years (2015)
  - 32 billion transistors

Jack would now be able to travel from ***Dallas, TX to NYC*** in ***6.9 nanoseconds*** on ***just the fumes*** from ***a drop of gas!***

Challenges in IC fabrication

- Photomask and Wafer Defects
- Integration Complexity
- Data Comprehension
- Process Limitations

So many things can go wrong!

Fab Defects on Photomasks

The materials and ambient trapped between the pellicle film and the mask surface create a reactant-rich atmosphere.

This "reaction chamber" produces organic and inorganic precipitates by photochemical means and called progressive defects.

A similar mechanism can trigger backside glass progressive defect formation (haze) simultaneously.

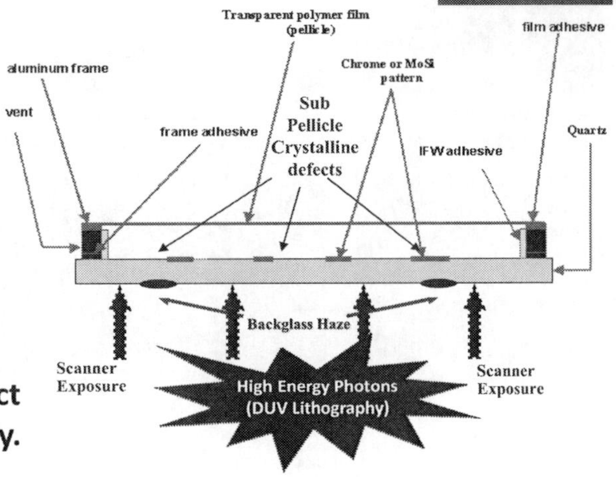

Photomask Defect Composition Analysis

Most common haze crystals are made of:

- **ammonium sulfate** [$(NH_4)_2SO_4$]
- **ammonium oxalate** [$(NH_4)_2C_2O_4.H_2O^+$]

Cleaning / Environment

Many other compounds were also identified on photomask surfaces:

• ammonium carbonate • ammonium carbamate • acrylonitrile butadiene styrene co-polymer • dimethyl phthalate • diethyl phthalate • dibutyl phthalate • dioctyl phthalate • palmitic acid • 2,4-di-t-butyl phenyl phosphite • 2,6-di-t-butyl phenyl phosphite • cellulose nitrate • methyl palmitate • potassium phosphate • cyanuric acid • stearic acid • t-butyl benzene • Tinuvin® • oleic acid amide • brassidic acid amide • cyanoacrylate adhesive • poly(dimethyl siloxane) • silicic acid • ammonium silicate • poly(methyl methacrylate) • poly(butyl methacrylate) • poly(methacrylic acid) • poly(vinyl chloride) • poly(vinylidene chloride) • poly(tetrafluoroethylene) • sodium ions • potassium ions • calcium ions • ceric ammonium salts • organosilanes • organosiloxanes • alkyl fluorosulfonic acids • aryl fluorosulfonic acids • langmuir-blodgett films of long chain polar aliphatic compounds • sodium chloride • potassium chloride • glycine • protein residues • santovar® • nickel acetate • nickel sulfate • chromium sulfate • nylon • organoamines • nitrile compounds

Source: Grenon Consulting, Inc.

Fab Defects on Wafers

Near-perfect tool maintenance and wafer cleaning capabilities required !

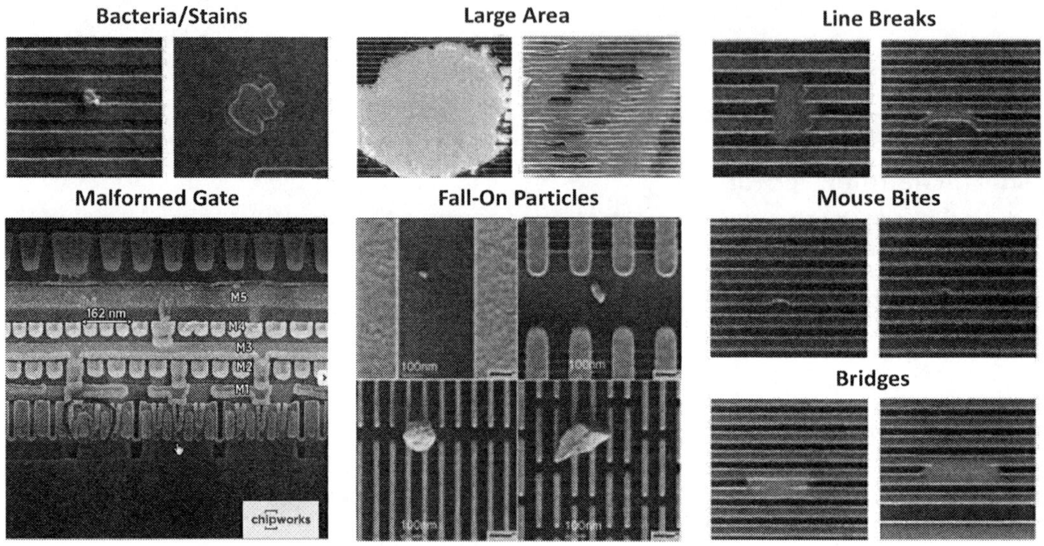

Integration Complexity - Structure

- Planar to Vertical Transistor Architecture

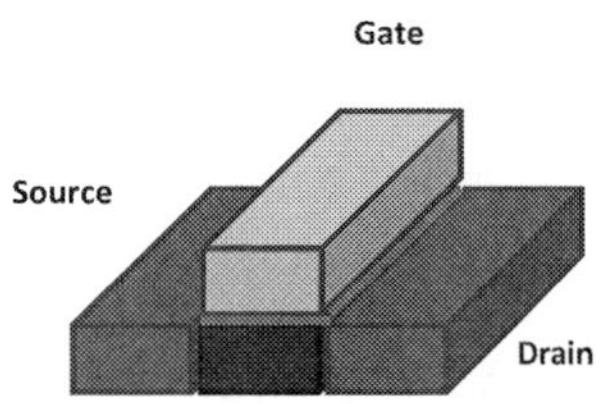

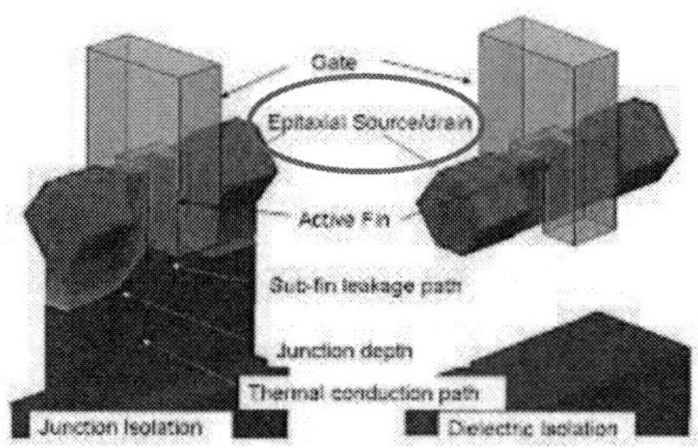

Planar MOSFET

Vertical FinFET

Integration Complexity – IC Periodic Table

(after John Robertson, CUED)

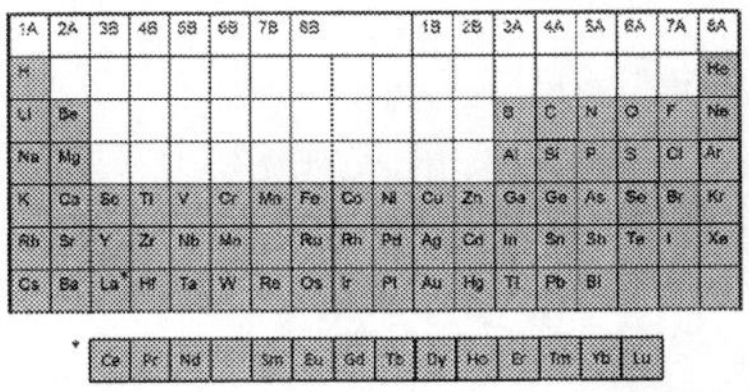

(after John Robertson, CUED)

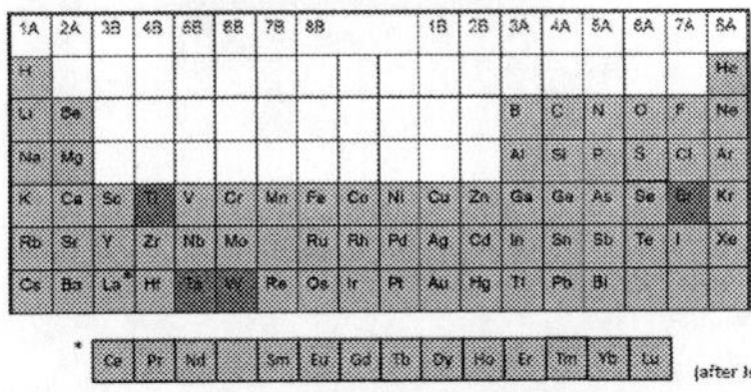

(after John Robertson, CUED)

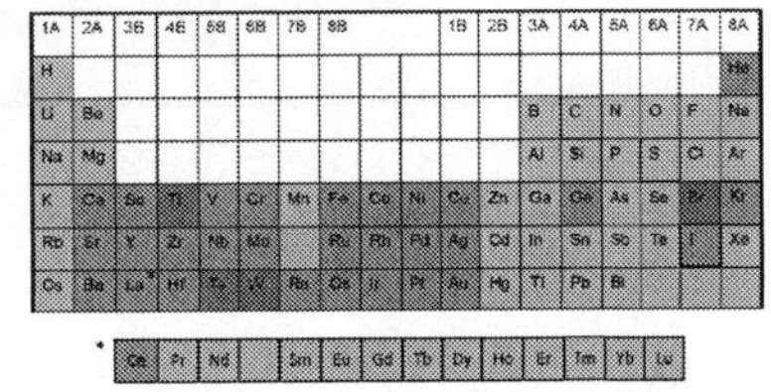

Today, nearly two thirds of the non-radioactive elements are used in every chip!

Data Comprehension: Sensory Overload!

>100,000 sensors

More information is ignored than is used!

Too Much For Humans To Digest!

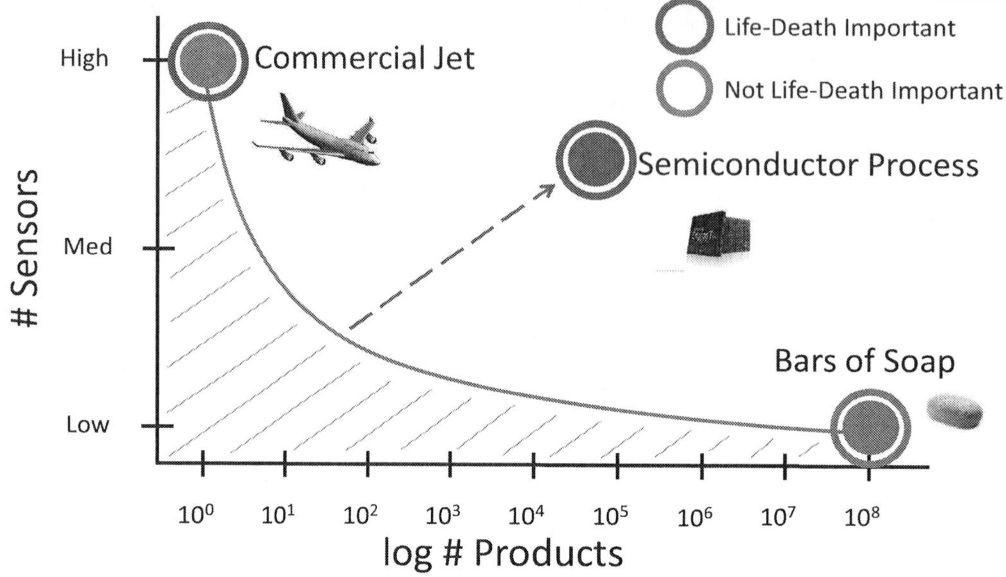

- Toward high end of Sensors <u>AND</u> Products!

Process Limitations: The Lithographer's Paradox

Lord Rayleigh

$$\text{Min. Feature} = \frac{k_1\lambda\downarrow}{NA\uparrow} \;\mathbf{X}\; \frac{\uparrow k_2\lambda}{\downarrow(NA)^2} = \text{Depth of Focus}$$

Wavelength and NA Evolution

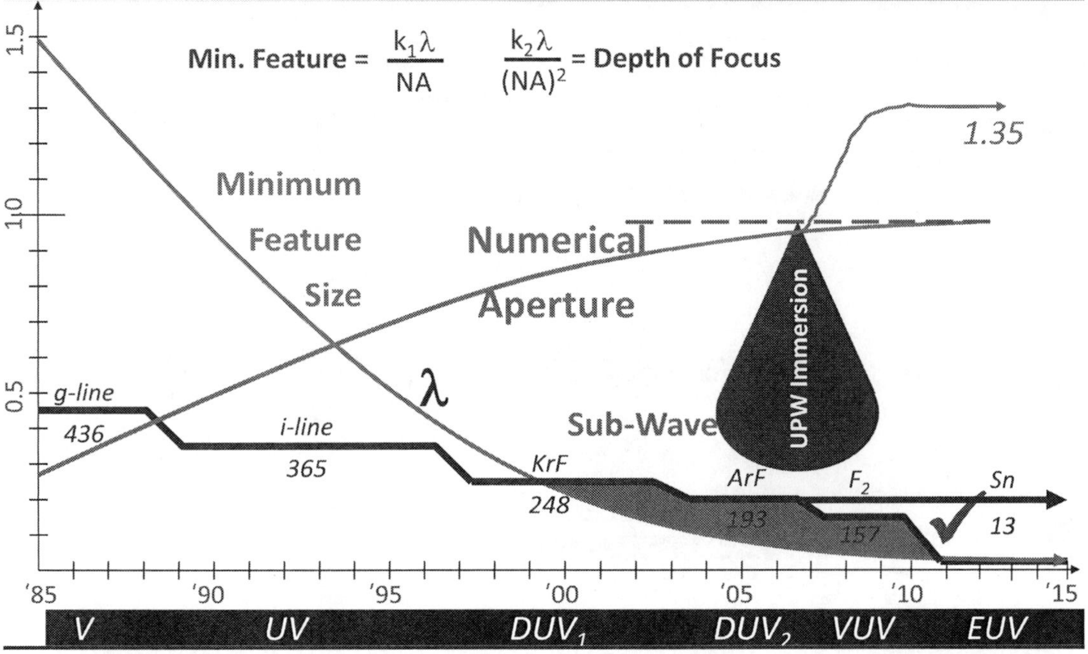

NA Game Changer: Immersion Lithography (UPW)

Snell's law: $NA = \eta_0 \sin\theta_0 = \eta_f \sin\theta_f = \eta_r \sin\theta_r$

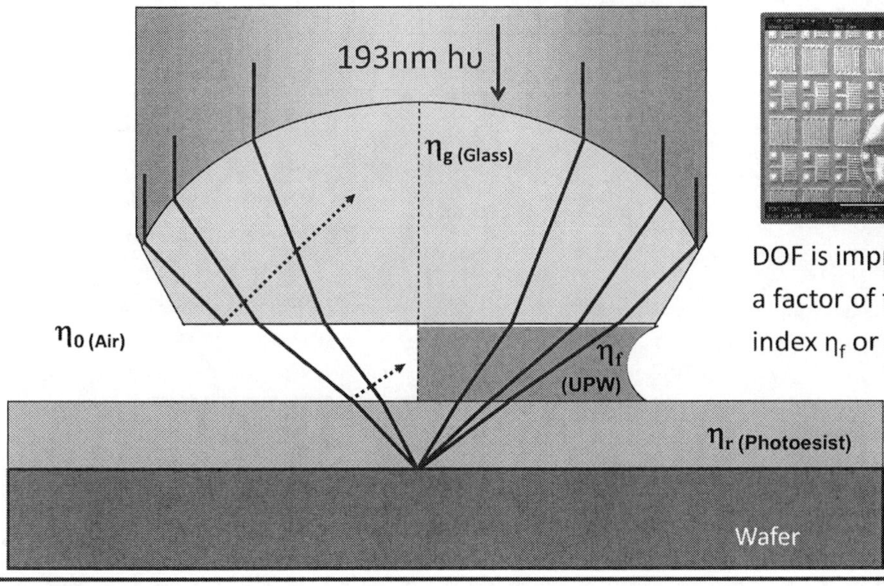

DOF is improved by a factor of the UPW index η_f or more.

Optical Proximity Correction - k_1

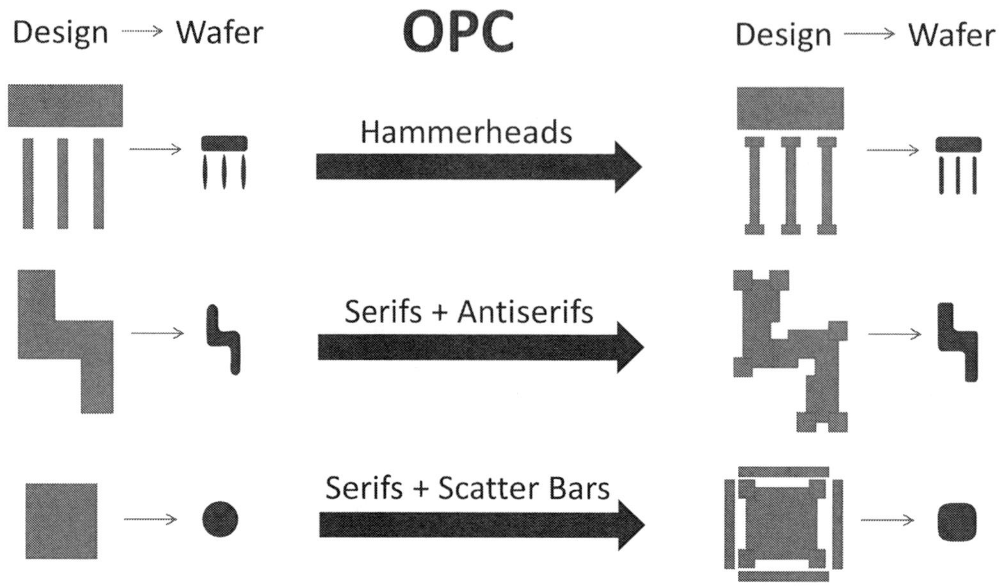

Another k_1 Innovation: Double Patterning

Spacer Type:

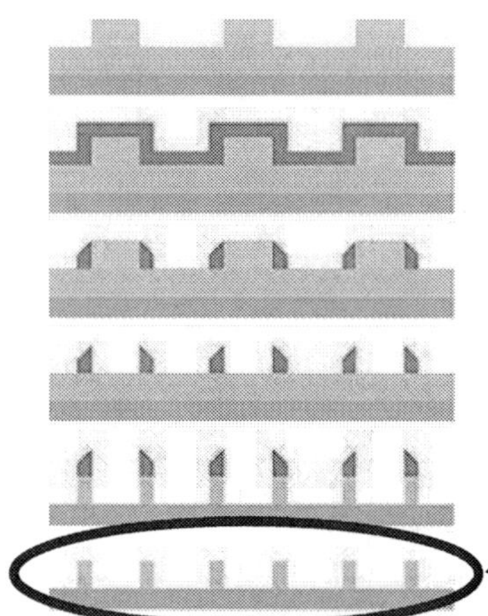

Ref: Paul Zimmerman

Litho-Etch-Litho-Etch:

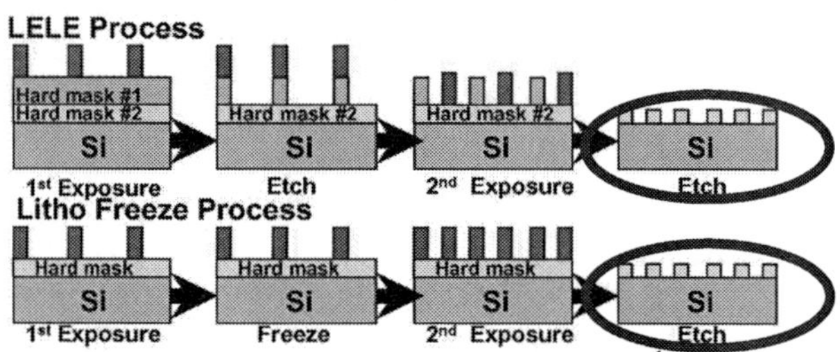

Ref: Brian Wang

Previously took only one lithography step

Too Much Innovation? Quadruple Patterning

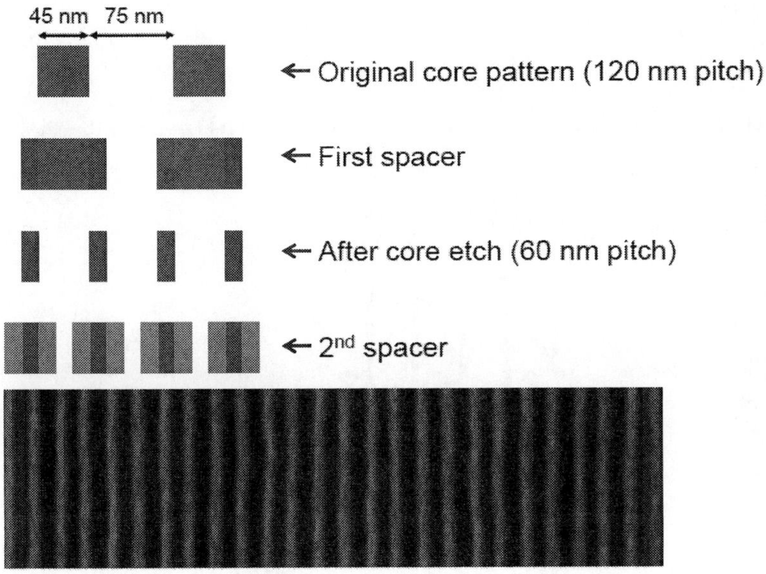

Source: Nikon LithoVision 2015

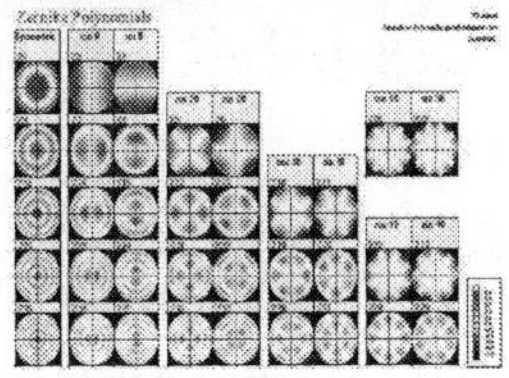

Future Options

- Octuple
- LELELE
- LELELELE
- ...
- EUV (Single)?

Limiting Factors

1. Overlay
2. Yield
3. Cost

What IC Challenges Lie Ahead?

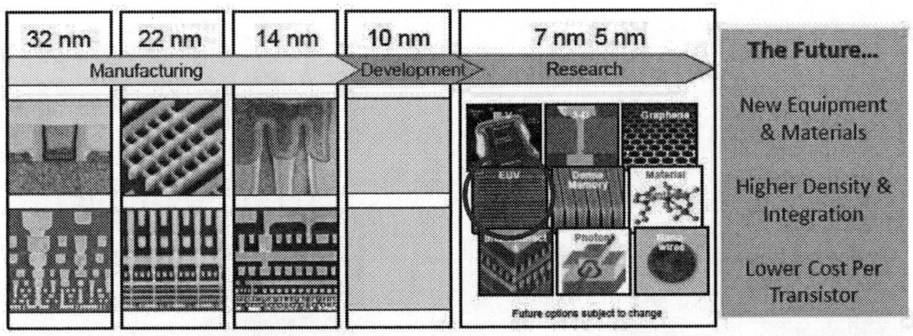

Source: LithoVision 2015

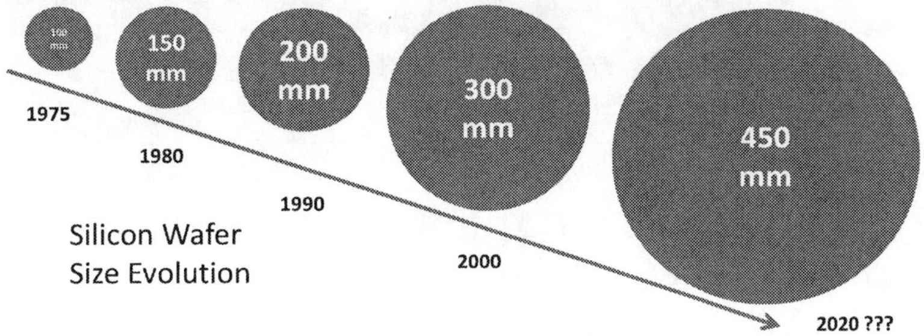

Moore's Law and Contaminant Size

- Particulates Example
  - UPW spec <200pcs/liter @ >**50nm**
  - Max defect size on mask 7-10% of wafer CD (~1-2 nm)
  - Size spec needs to track with node shrinks
  - 14nm: **1.0nm** → 10nm: **0.7nm(7Å)** → ...

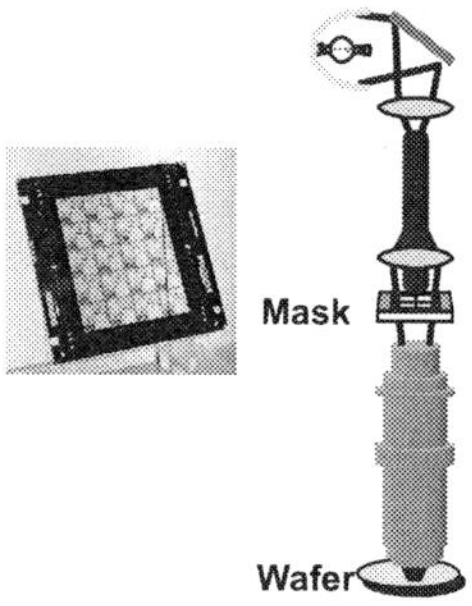

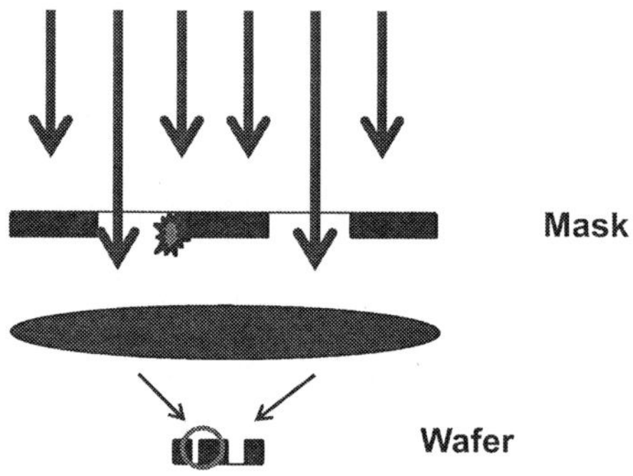

Contaminants Can Destroy Wafer Yield

- Bacteria Colony-Forming Units (CFUs)

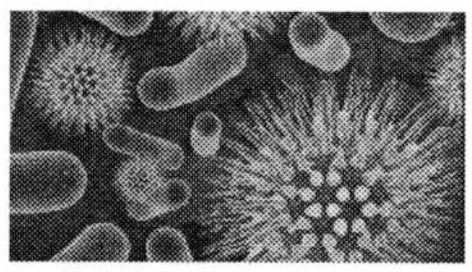

  - Typical size is 1-8μm in length
  - One CFU/liter could easily destroy dozens of logic die
    - UPW is used dozens of times in Cleans, CMP, & Photo steps

  - If it kills one die on a photomask, then all wafers shot with that mask will suffer yield loss for that die

A Mathematical Perspective

>5 billion *non-redundant* transistors in a logic chip

x >600 chips per wafer

>**3 trillion** sites on every wafer needing to be perfect!

There are also ~**$3.35x10^{25}$** molecules/liter in UPW

Today's UPW impurity specs:	"Allowed Bad"
ppt for metals, ions, boron	→ **33.5 trillion**
ppb for TOC, DO, Total SiO_2, H_2O_2	→ **33,500 trillion**
pcs/liter for particles and bacteria	→ **Dozens**

BGE – March 28th, 2017 SPCC – Austin, TX

Yield-Up Speed is Critical

- Typical Yield Improvement Dynamics

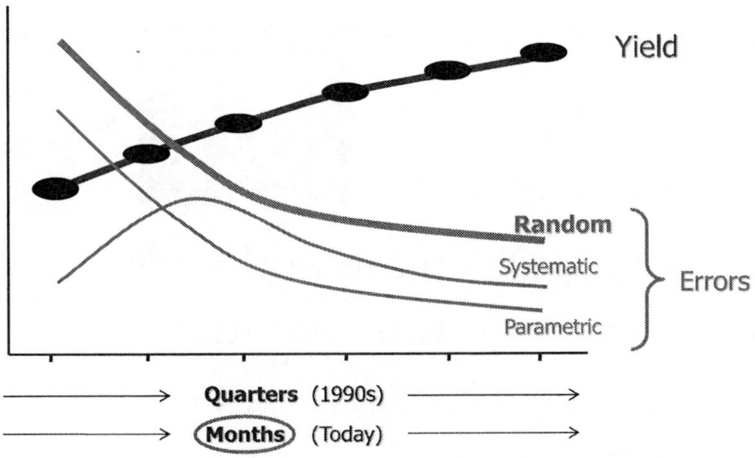

Now vulnerable to picoscale differences and perturbations!

BGE – March 28th, 2017 SPCC – Austin, TX

What Do Fabs Need of Suppliers?

- Excel in the Four C's
  - Capability
  - Collaboration
  - Consistency
  - Communication

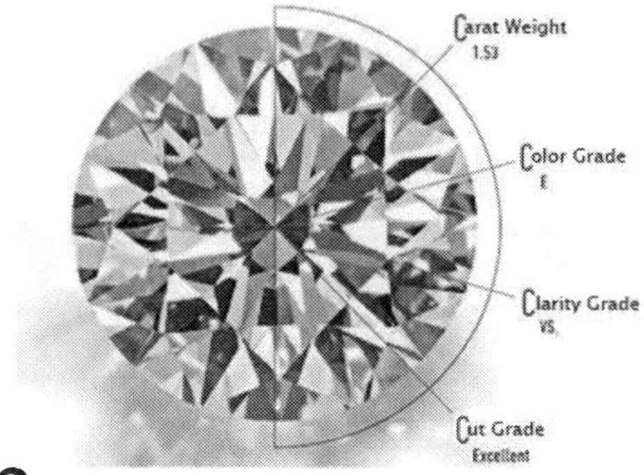

Capability – Key to *getting* yield

- Anticipate & explore new formulations and methods of synthesis
- Utilize most advanced metrology & inspection
- Innovate better particulate and ionic contamination elimination solutions
- Outgoing quality testing sampling to guarantee 100% of shipment

Collaboration – Key to *improving* yield

- Share roadmaps and needs from the current node down to n+2

- Share metrology, inspection, & analytical tool lists and cross-verify with fab
- Share and test ideas with fab and fab tool suppliers (e.g. layer-specific formulations)
- Share in enabling green-ness (packaging, recycling, etc.)

BGE – March 28th, 2017 SPCC – Austin, TX

Consistency – Key to *stable* yield

- Structured risk elimination
  - Assume all changes matter
- Sub-supplier management by fab QA personnel
- Batch-to-batch uniformity beyond the CofA
- Reduce variability post manufacture through installation and use
  - Shipment and storage conditions
  - Filter prep time

BGE – March 28th, 2017 SPCC – Austin, TX

Communication – Key to *not 'tanking'* yield

- Advise fab on filtration and blending strategies
- Report manufacture, delivery, and storage conditions

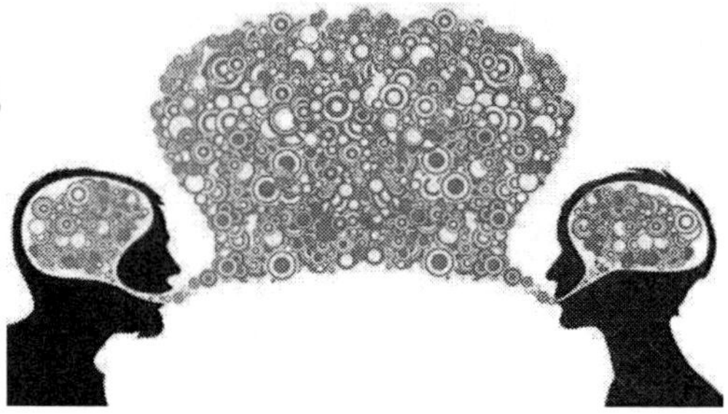

- Proactively report or recall excursions of any kind once discovered
- Be open to fab feedback (Avoid "I told you so" moments.)

Summary

- **Making Advanced Semiconductors is HARD!**
  - Defectivity
  - Integration complexity
  - Data/Sensor comprehension
  - Process limitations

- **Four C's to Success in this Business**
  - Capability is key to getting yield.
  - Collaboration is key to improving yield.
  - Consistency is key to stable yield.
  - Communication is key to not "tanking" yield.

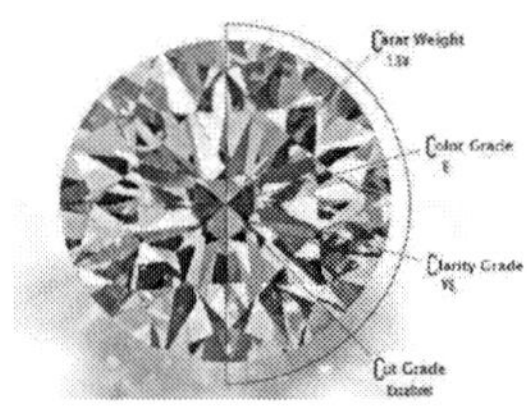

Thank you, any questions?

Chemical Bonding Transformation Mapping to Optimize Low-k Dielectric Nanostructure Fabrication and Post-etch Residue Clean

Oliver Chyan*, Sirish Rimal, Tamal Mukherjee,

Muthappan Asokan, Seare A. Berhe

Interfacial Electrochemistry and Materials Research Lab,
University of North Texas

Kanwaljit Singh

Components Research, Intel Corporation

Ian Brown, Jacques Faguet

US-Technology Development Center,
TEL Technology Center

1

Present to SPCC-2017, **chyan@unt.edu*

Interfacial Electrochemistry & Material Research Lab UNT

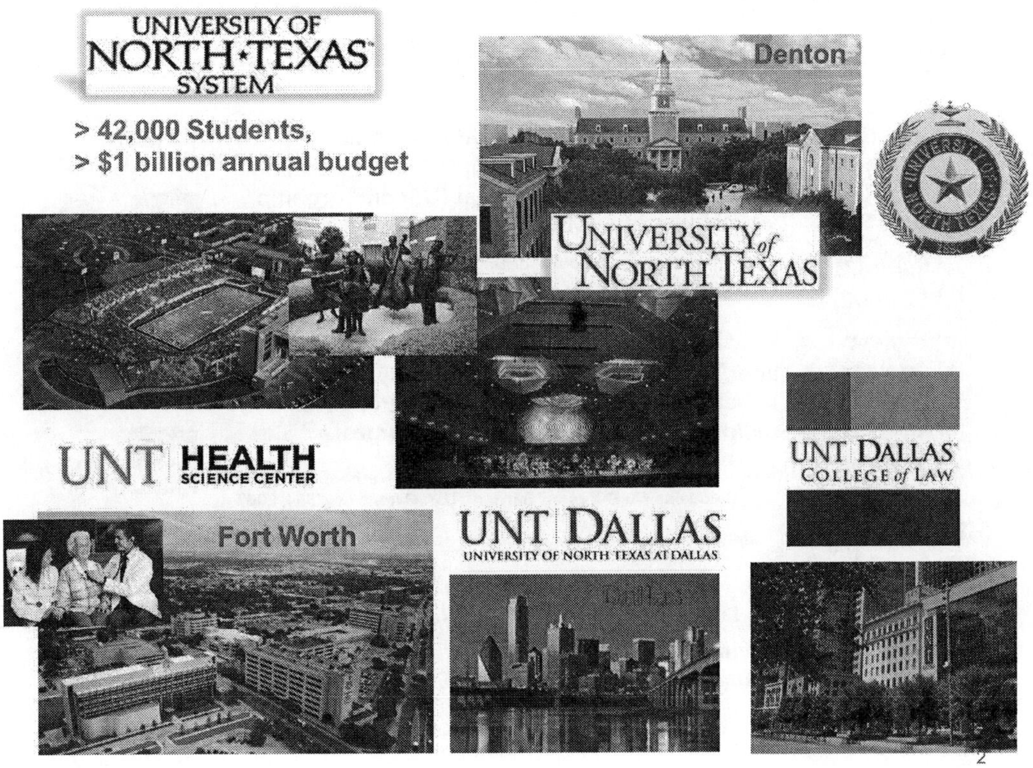

2

Goal: Explore the underlying fundamental science that leads to real engineering solutions for important interfacial issues in microelectronic fabrication.

Interfacial Electrochemistry & Material Research Lab
UNT

Group members:

Alex Lambert, Issac Goutham, Muthappan Asokan, Seare Berhe

Alumni :

Dr. Sirish Rimal (*JSR-Micro*)

Dr. Tamal Mukherjee (*LAM*)

Dr. Nick Ross (*Intel*)

Dr. Kyle Yu (*TEL*)

Dr. Karthik Pillai (*TEL*)

Dr. Simon Koskey (*Intel*)

3

Interfacial Electrochemistry & Materials Research Lab

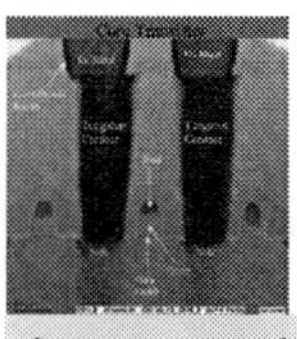

(1992-2002)

Front end (FEOL): Silicon surface preparation, H-passivation, SC1/SC2 cleaning, Silicon oxide etching, metal (Cu) and organic contamination control/detection.

"An Apparatus and Method for Detecting Impurities in Wet Chemicals" *U.S. Patent 6145372*, 2000.

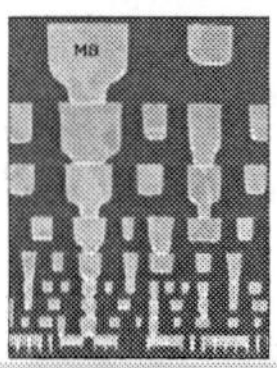

(2000 - Date)

Back end (BEOL): Cu ECD, Ru-based liner/barrier, Cu cleaning chemistry, Cu bimetallic corrosion mechanism, Post etch residue, Porous Low-k damage control.

"Method of Making Integrated Circuits Using Ruthenium and Its Oxides as Cu Diffusion Barrier", *U.S. Patent 7247554*, 2007.

"Method for Wafer Characterization", *U.S. Patent 9366601*, 2016.

(2015 - Date)

IC Packaging: Cu vs. Au wire bonded device, Al pad corrosion prevention, molding compounds, Subtractive Cu Etching.

"Systems and Methods for Copper Etch Monitoring and Control", *U.S. Patent Application*, 2016.

Requirements for BEOL Cleans

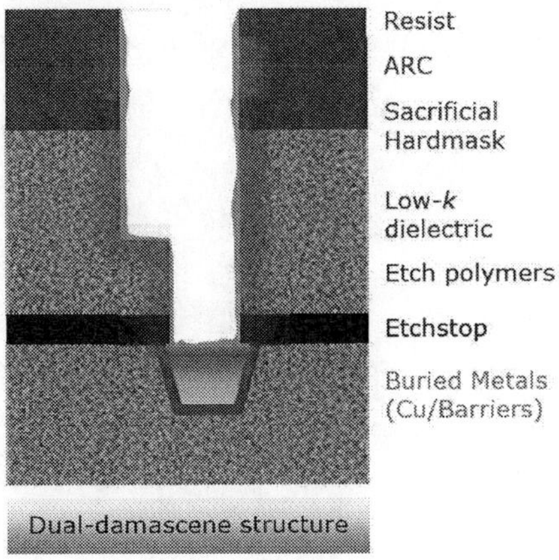

Removal of:
- Litho materials
- Etch polymers
- Sacrificial hardmasks

Compatibility with:
- Low-*k* dielectric
- Lower-layer metals

Complex interactions:
- Plasma-etch modified materials
- Galvanic corrosion
- Interfacial adhesion

Kanwal J. Singh, *"Process and Environmental Challenges Facing BEOL Semiconductor Cleans Beyond the 32-nm node"*, Electrochemical Society 220th Meeting, Oct, 2011

BEOL Patterning & Damascene Processing

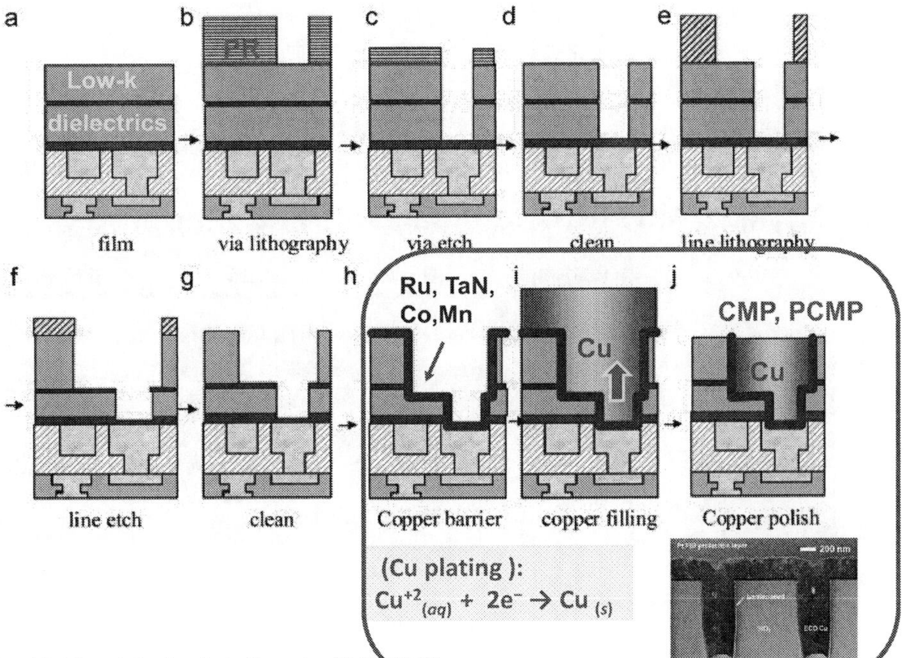

C. Weng, *Materials Science in Semiconductor Processing*, **13**, 376 (2010)

Invent Ru Barrier/liner

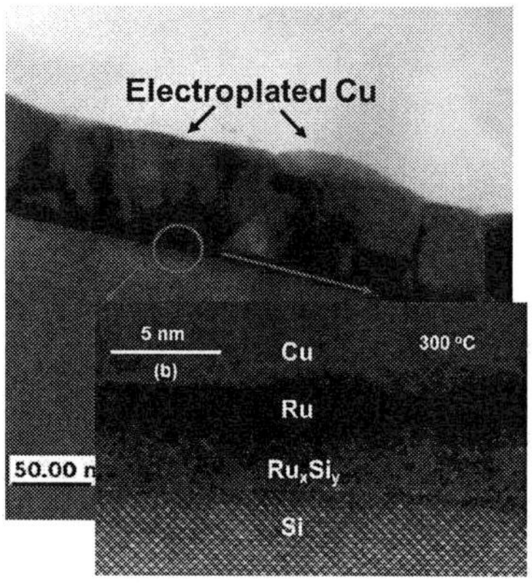

J. Electrochem. Soc., 2003, 150, 347.
J. Electrochem. Solid-State Lett., 2004, 7, G154.
J. Electrochem. Soc., 2005, *152*, G808
Applied Physics Letter, 2005, 86, 083104
U.S. Patent, 7,247,554, 2007.

Micro-pattern Corrosion Screening Cu/Ru, Co/Cu

Co on Cu **without inhibitor**

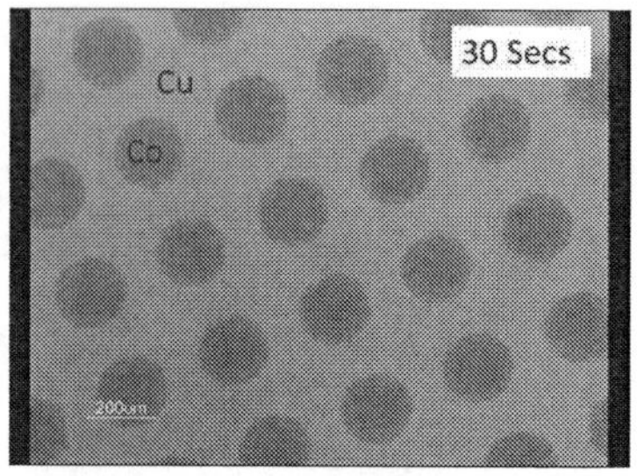

Co on Cu **with inhibitor**

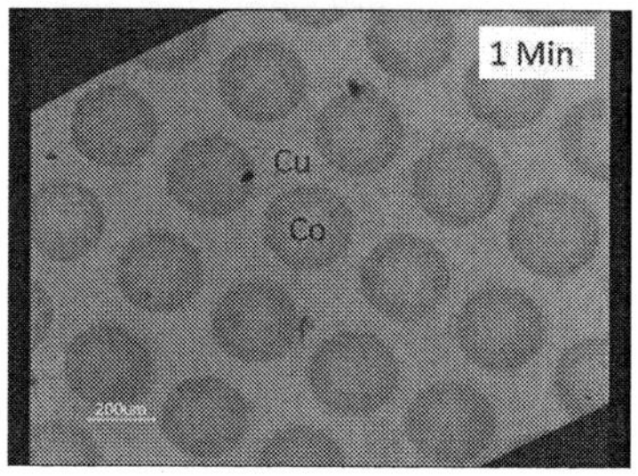

BEOL Patterning & Damascene Processing

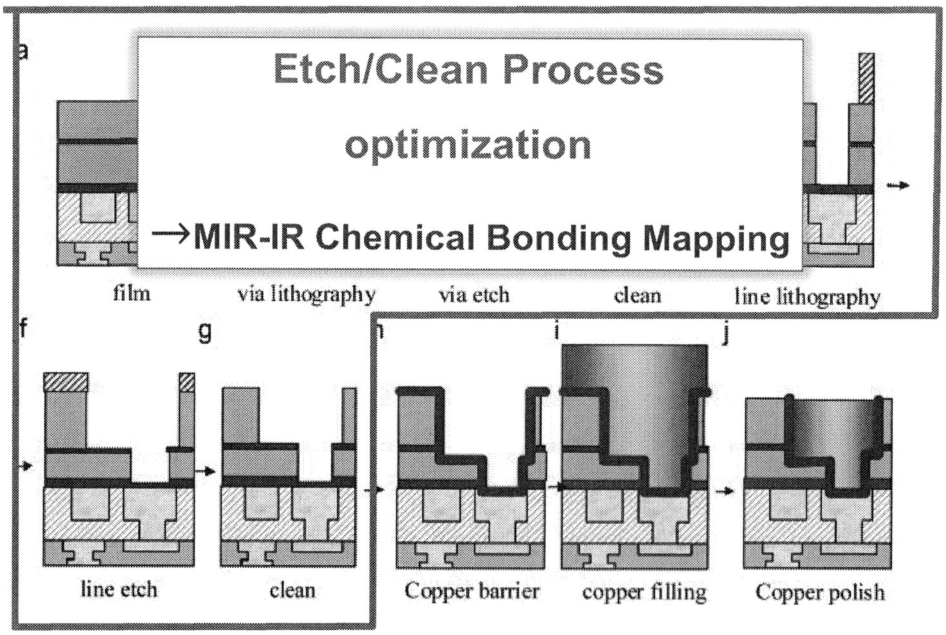

C. Weng, *Materials Science in Semiconductor Processing*, **13**, 376 (2010)

8

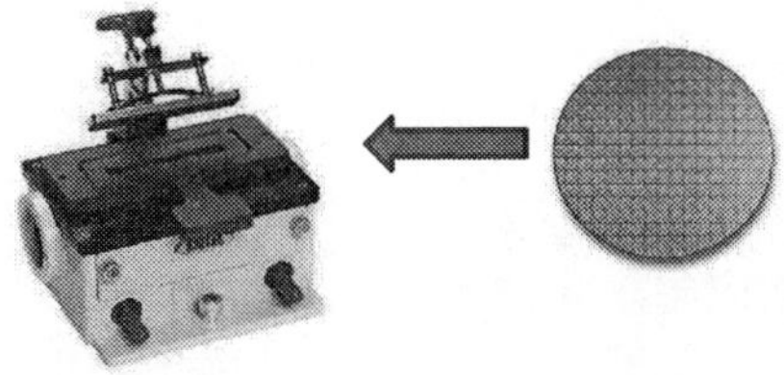

Patterned Si wafer

Attenuated total reflection (ATR) Infrared Spectroscopy (ATR-IR)

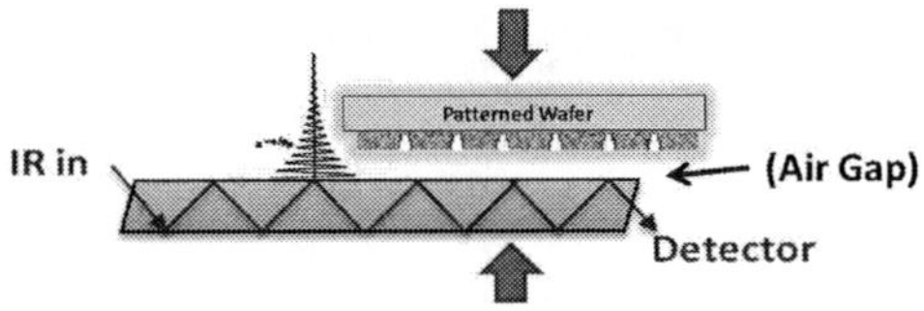

External ATR-IR

- ➢ Not- reproducible
- sample elasticity
- flatness, position
- fragile ILD damage
- ➢ Low sensitivity
- miss max evanescent wave interaction

Interfacial Electrochemistry & Material Research Lab
UNT

Multiple Internal Reflection Infrared Spectroscopy (MIR-IR)

Patterned Si wafer = IR waveguide

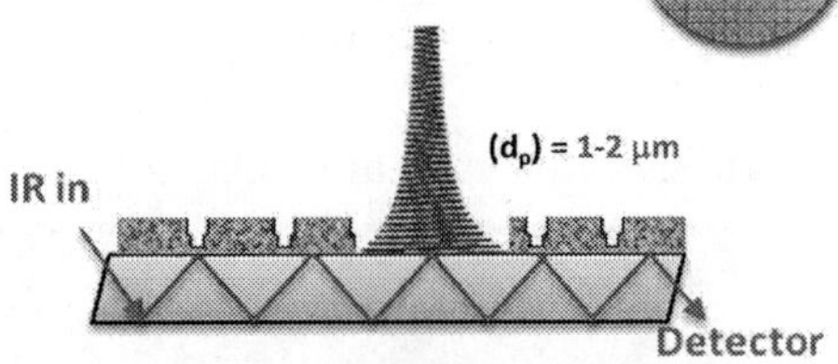

MIR-IR

MIR-IR Advantages

- No sample-contact problem
- Reproducible
- Max Evanescent Wave Interaction
- Superior Sensitivity (<10% monolayer)

"Method for Wafer Characterization", *U.S. Patent 9366601*, 2016.

External ATR-IR

- ➢ Not- reproducible
- sample elasticity
- flatness, position
- fragile ILD damage
- ➢ Low sensitivity
- miss max evanescent wave interaction

Interfacial Electrochemistry & Material Research Lab
UNT

MIR-IR Application:
Ultra-clean Silicon Wafer Surface Preparation

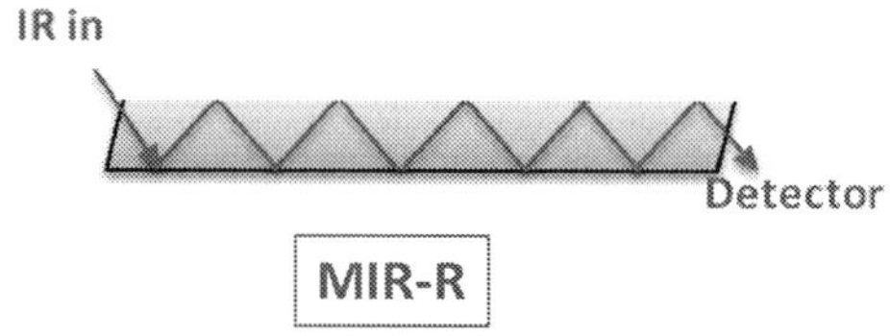

TEXAS INSTRUMENTS

Interfacial Electrochemistry & Material Research Lab UNT

Surface Bonding Transformation on Si(100) by Trace Cu^{+2} impurity

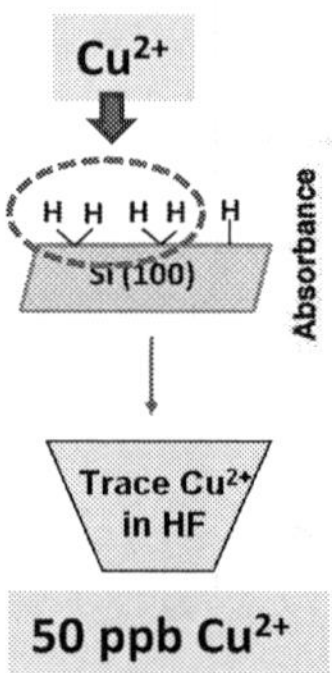

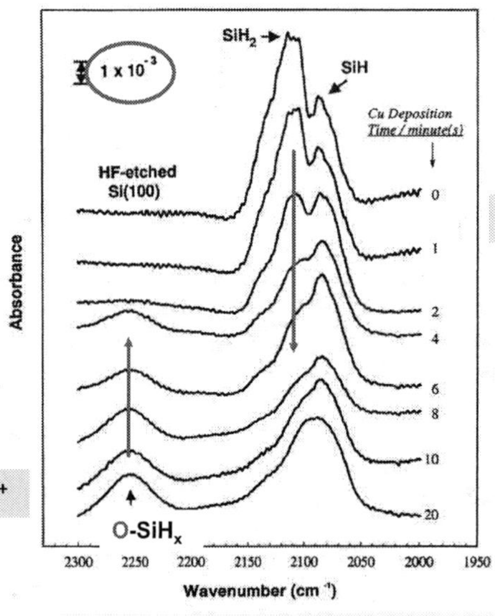

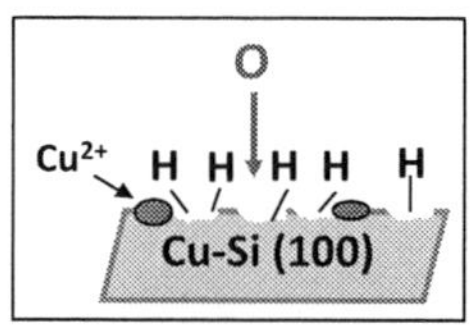

Trace Cu^{2+} induced early oxide and pitting

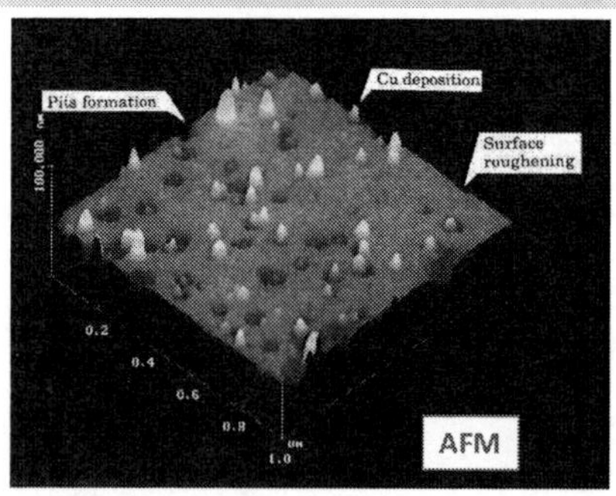

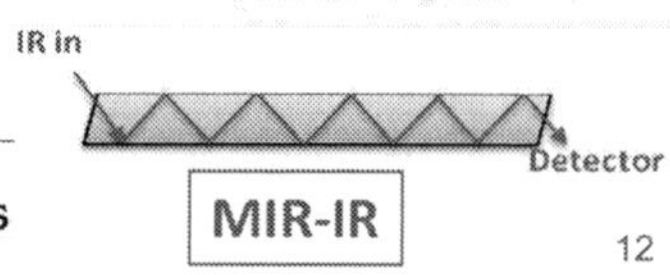

TEXAS INSTRUMENTS

12

J. Electrochem. Soc., 1996, ***143***, 92
J. Electrochem. Soc., 1996, ***143***, L235
J. Electrochem. Soc., 1997, ***144***, L17
U.S. Patent, 6,145,372, 2000

Interfacial Electrochemistry & Material Research Lab UNT

MIR-IR : Optimize a-Si:H Microbolometer for Night Vision Camera

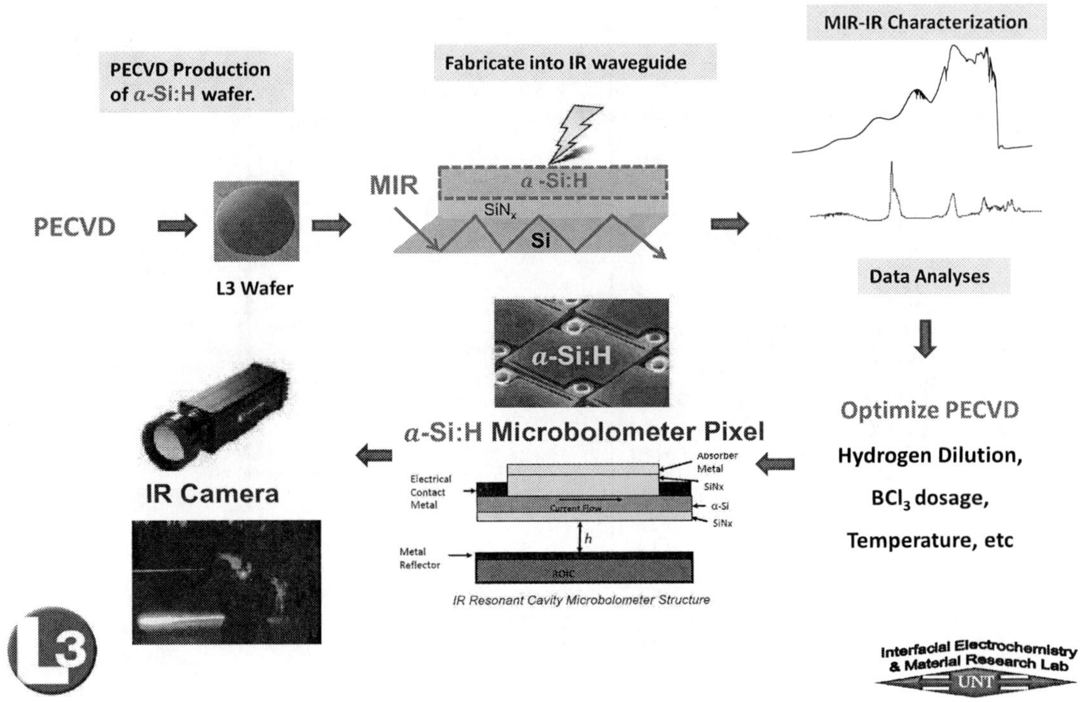

Comparing MIR vs Transmission (TIR) Sensitivity on a-Si:H

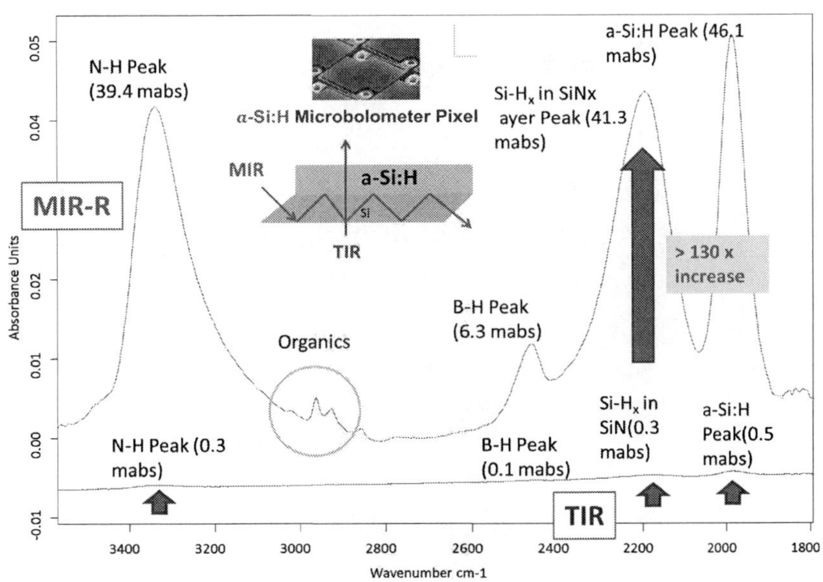

N. Ross, K. Shrestha, O. Chyan, C. L. Littler, V. C. Lopes, and A. J. Syllaios, *Mater. Res. Soc. Symp. Proc.*, **1536**, 127 (2013).

Interfacial Electrochemistry & Material Research Lab
UNT

MIR-IR : Optimizes *α*-Si:H Thin Film Quality

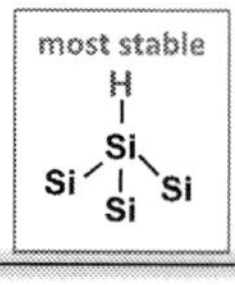

Bonding to Hydrogen is the key

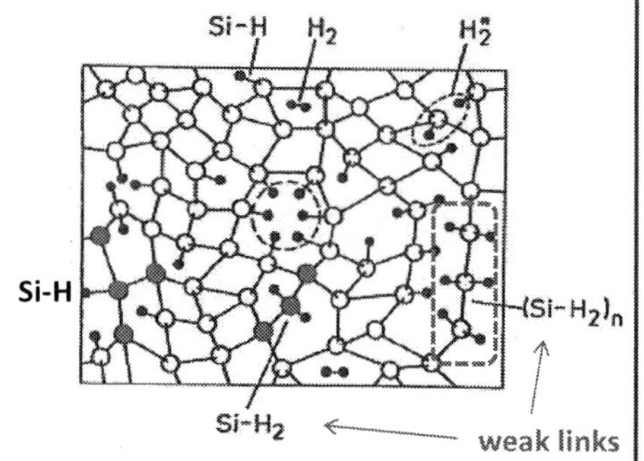

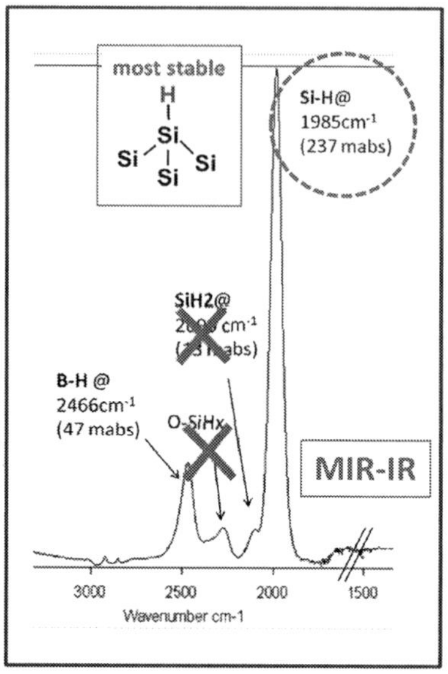

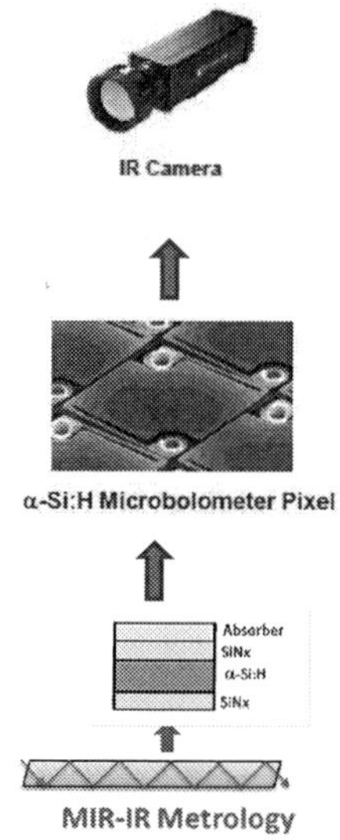

Chemical Bonding Data Accelerates PECVD Process Optimization

Interfacial Electrochemistry & Material Research Lab UNT

MIR-IR : Apply to BEOL Processing

Objective: Utilize MIR-IR to optimize RIE etching and post RIE cleaning with minimum dielectric damages

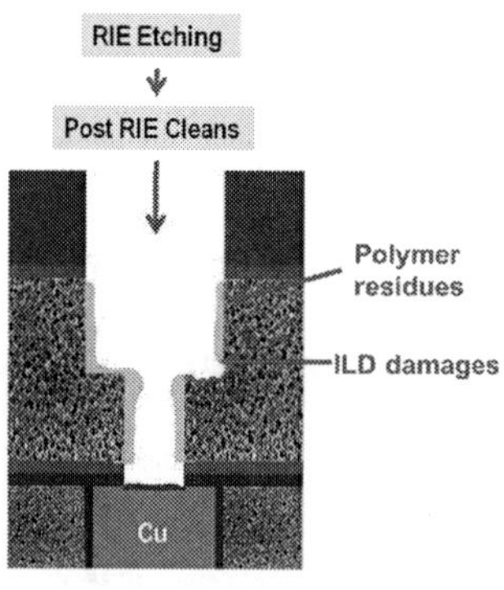

Interfacial Electrochemistry & Material Research Lab UNT

Current Metrology for Trench ILD Nano-structures

SEM and TEM
- Electron beam based metrology
- Cross sectional profiling
- Destructive analysis, limited chemical bonding info
- Laborious and time consuming

XPS and TOF-SIMS
- Evaluate depth of carbon depletion
- Provides elemental composition
- Based on x-ray, ion bombardment
- Can't reach deep-sidewalls and bottoms

Spectroscopic Ellipsometry
- For unpatterned films
- Technique based on refractive index
- Requires modeling

IR – traditional
- Provide chemical bonding structure
- Weaker signal and limited resolution

XPS, TOF-SIMS, SEM

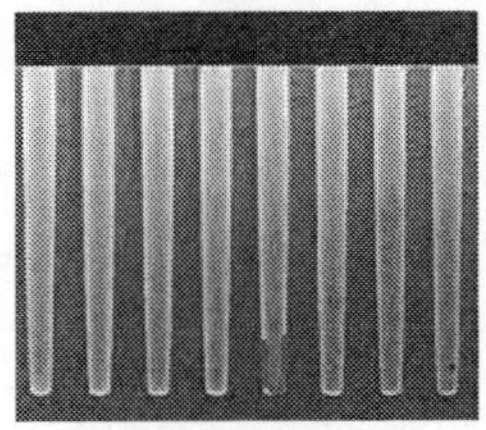

Trench bottom

New MIR-IR Metrology

Interfacial Electrochemistry & Material Research Lab UNT

17

MIR-IR is a Powerful Substantive Too -- Isolate Low-*k* Film Stack only Spectra

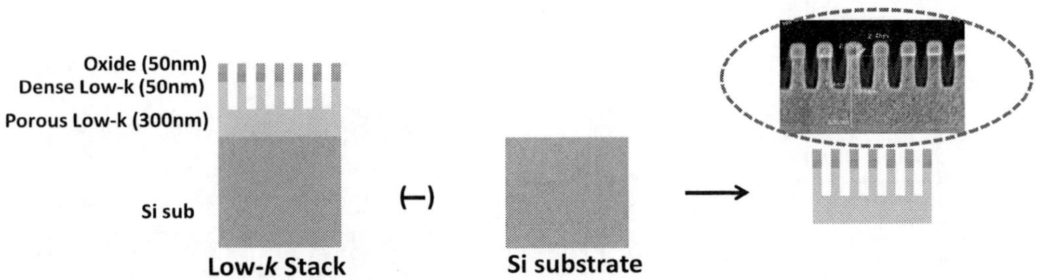

Sample Spectrum

Vs.

Background Spectrum

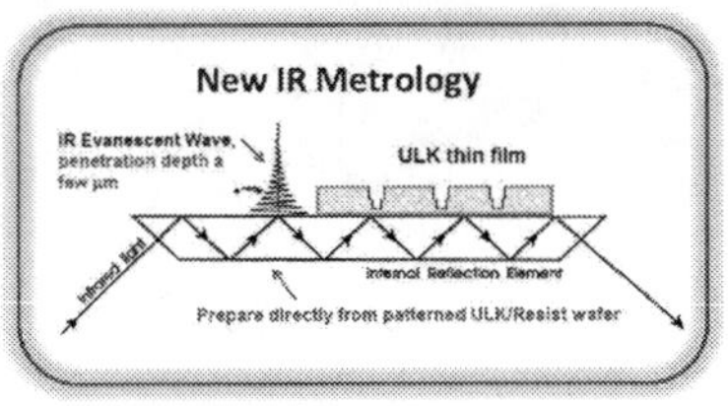

Interfacial Electrochemistry & Material Research Lab UNT

18

Chemical Bonding Transformation Map for Porous Low-k NanoStructure

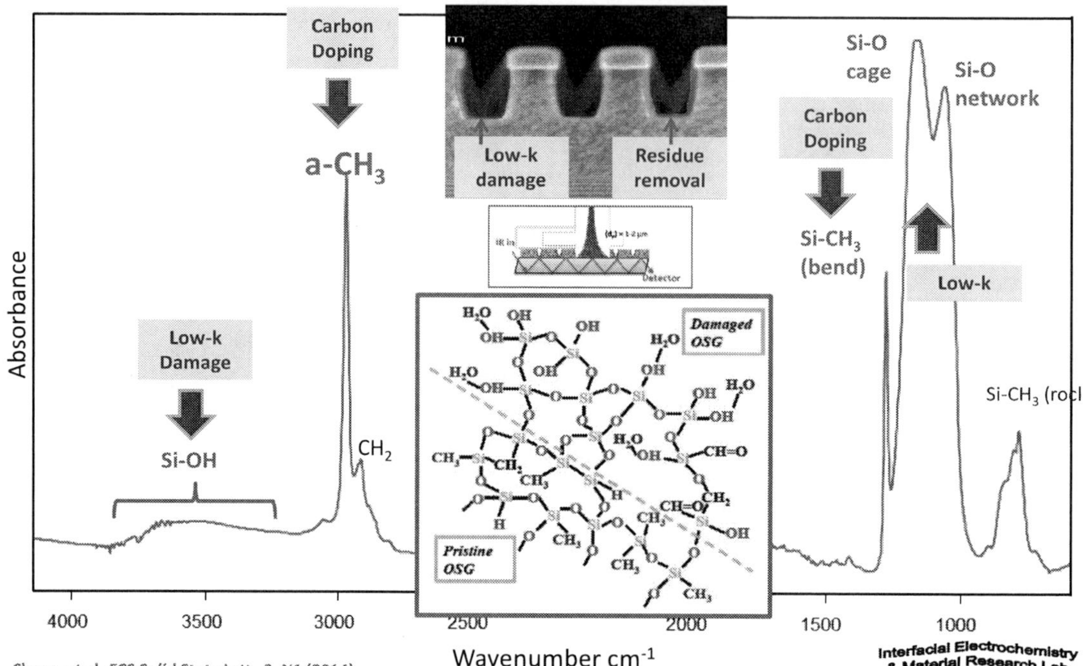

Chyan, et al, *ECS Solid State Lett.*, 3, N1 (2014)

MIR-IR: Evaluating Plasma-induced Damages on Trench ILD

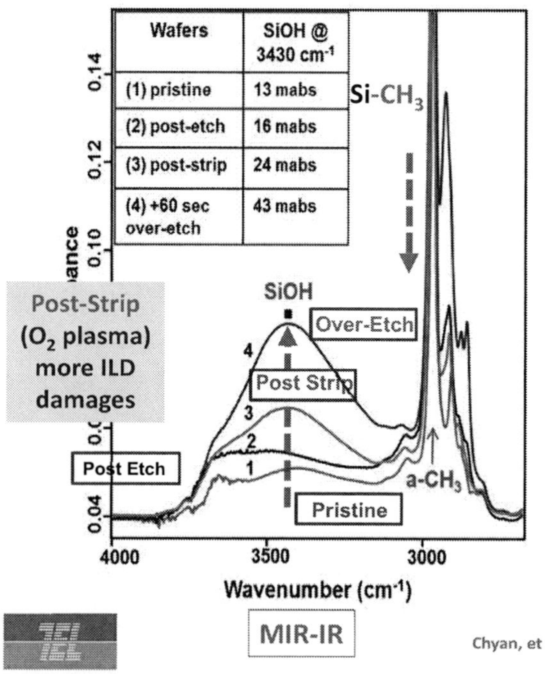

Wafers	SiOH @ 3430 cm^{-1}
(1) pristine	13 mabs
(2) post-etch	16 mabs
(3) post-strip	24 mabs
(4) +60 sec over-etch	43 mabs

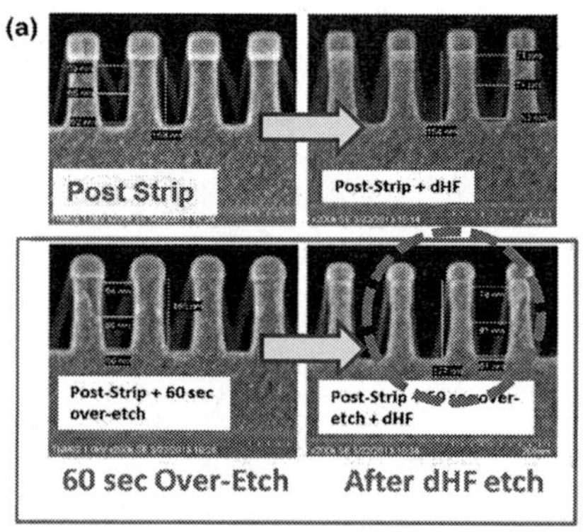

Confirmed by DHF etch

Chyan, et al, *ECS Solid State Lett.*, 3, N1 (2014)

MIR-IR: Minimize Low-*k* Damage of Manufacture Strip Processes

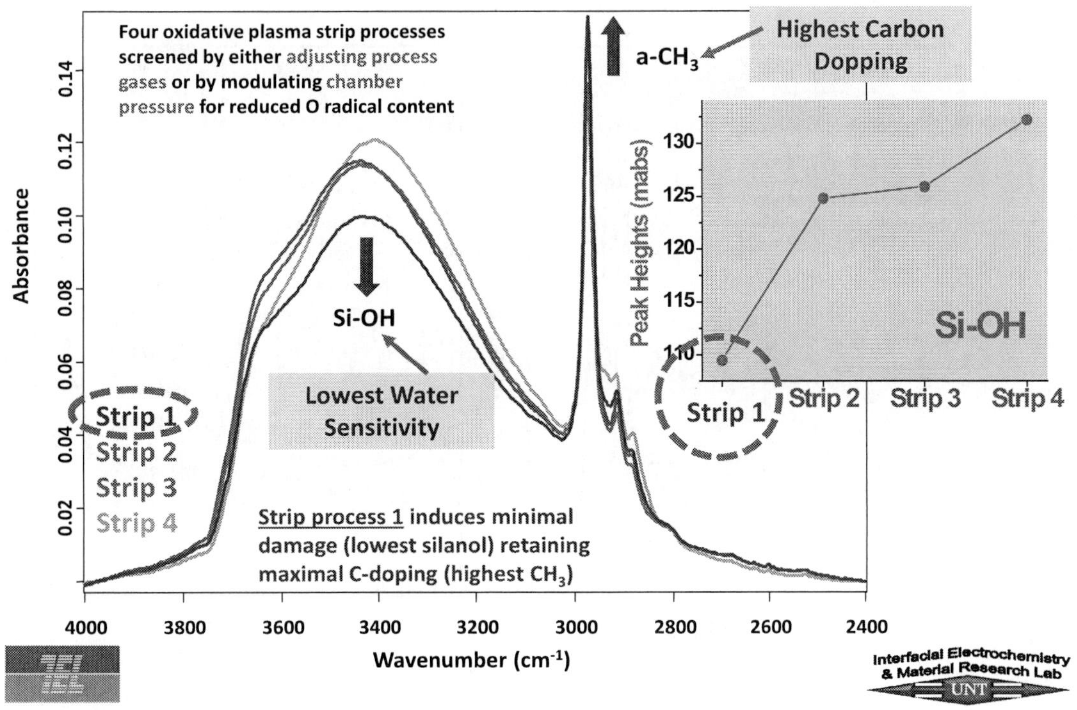

Chemical Bonding Transformation Map for Porous Low-k NanoStructure

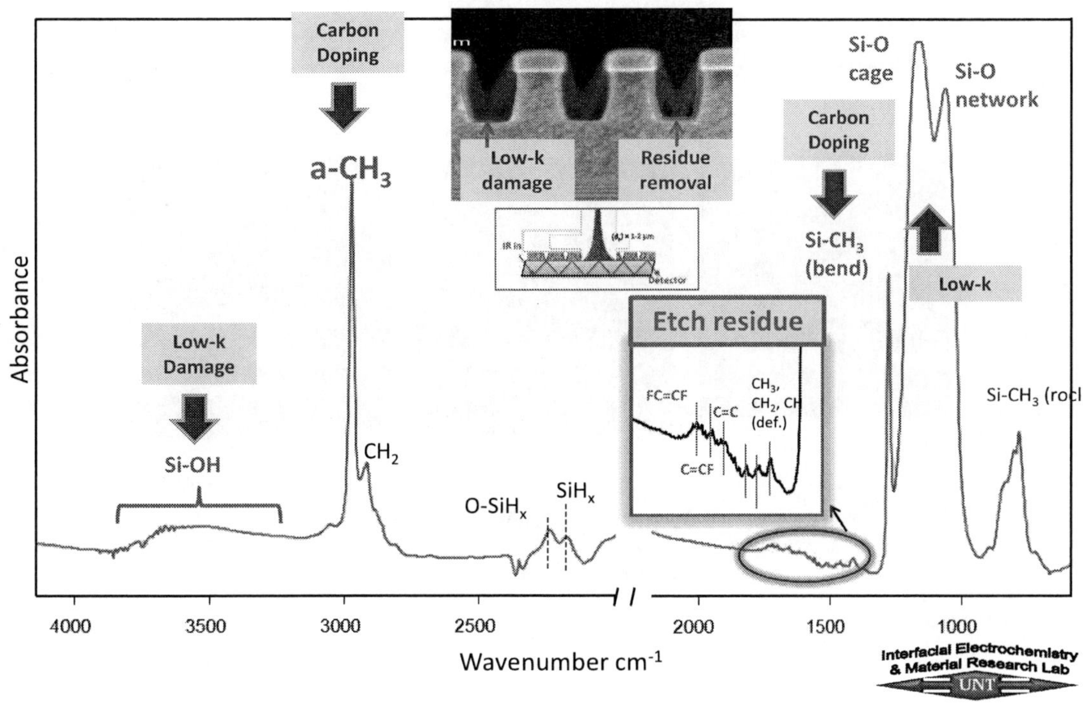

MIR-IR: Evaluating Plasma Damages and Etch Residue Removal

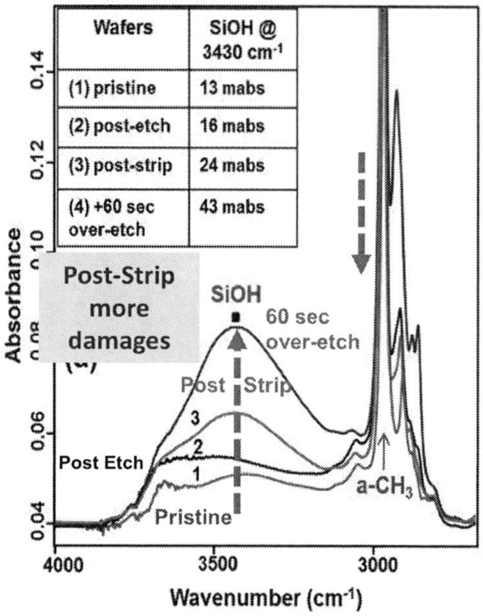

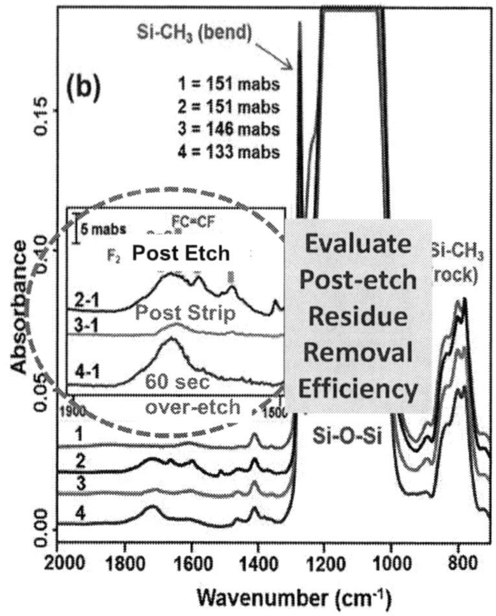

Chyan, et al, *ECS Solid State Lett.*, 3, N1 (2014)

Interfacial Electrochemistry & Material Research Lab UNT

MIR-IR : Identify Chemical Bonding Structure for Post-Etch Residues

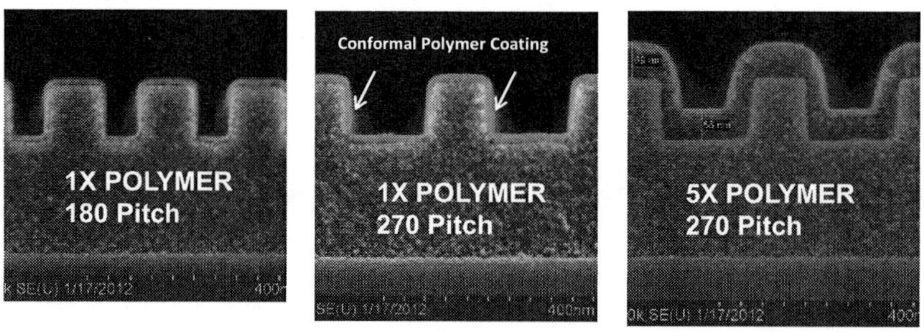

Kanwaljit Singh *Intel*

Alan M Myers *Intel*

Interfacial Electrochemistry & Material Research Lab UNT

24

XPS Data: Difficult to Assign Chemical Bonding

Curve Fitting

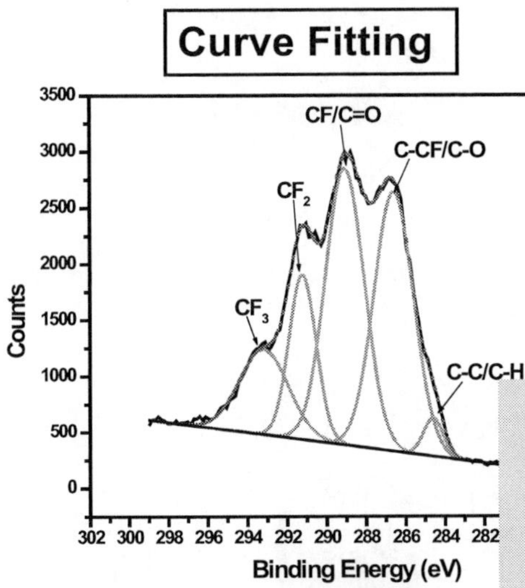

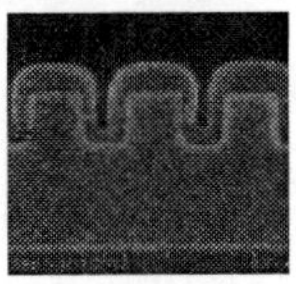

ECS Solid State Lett., **2**, N11 (2013)

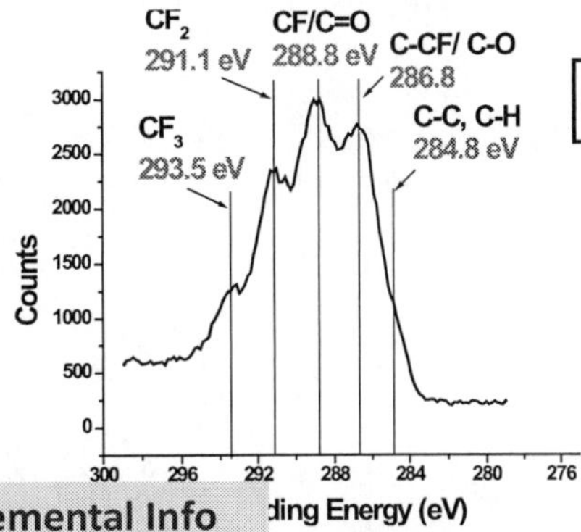

C 1s

XPS: Elemental Info

(C, F, O, H)

C-F, C-H, C-O, C-C

But, no reliable chemical bonding structure info

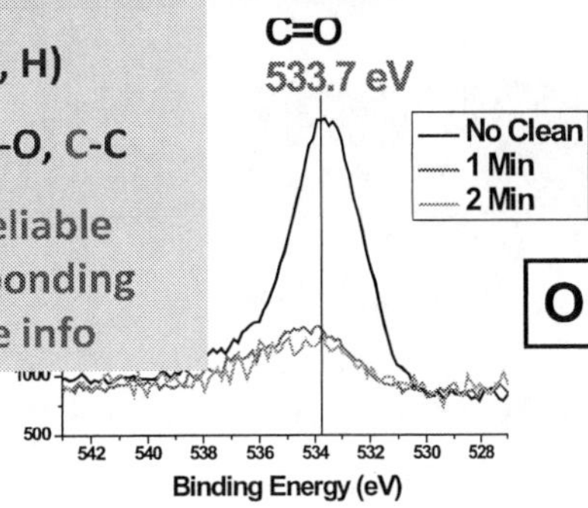

O 1s

Chemical Bonding Structure of Etch Residues

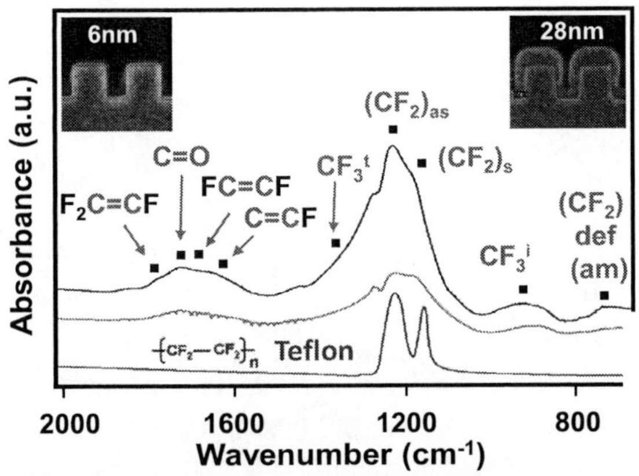

- Mainly fluoropolymer backbone
- Significant branching/cross-linking
- Olefinic unsaturation (fluorinated)
- Carbonyl functional groups

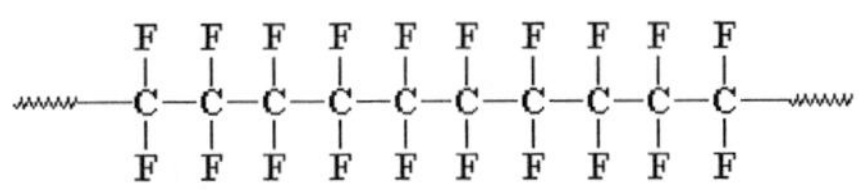

Polytetrafluoroethylene (Teflon)

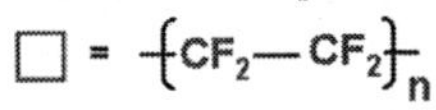

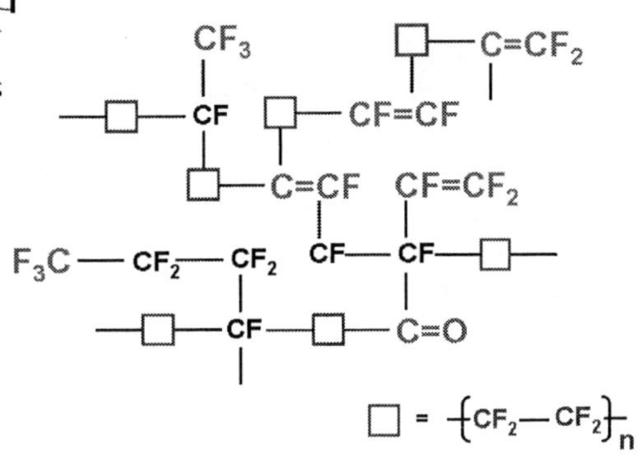

ECS Solid State Lett., 2, N11 (2013)

Identify Chemical Bonding Structure by Functional Group Specific Surface Derivatization

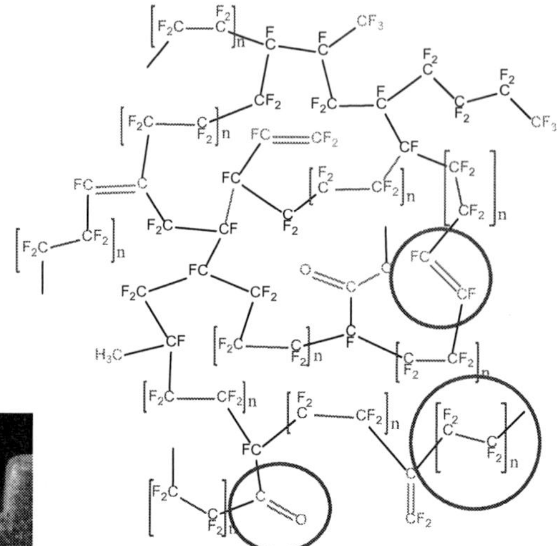

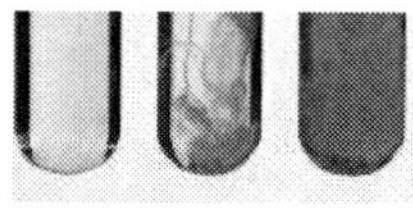

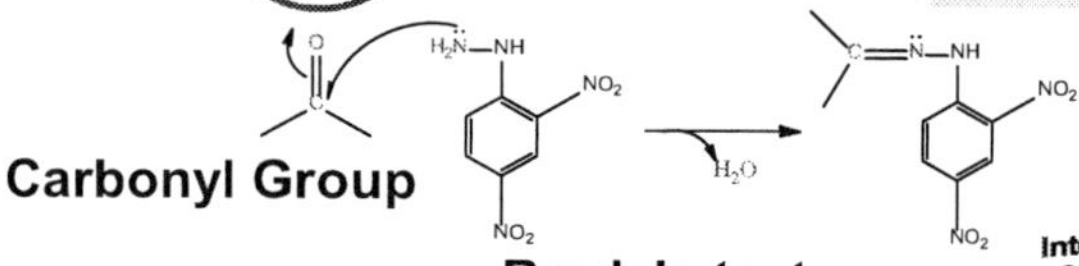

Brady's test

Interfacial Electrochemistry & Material Research Lab UNT

Confirmation of C=O by DNPH Reaction

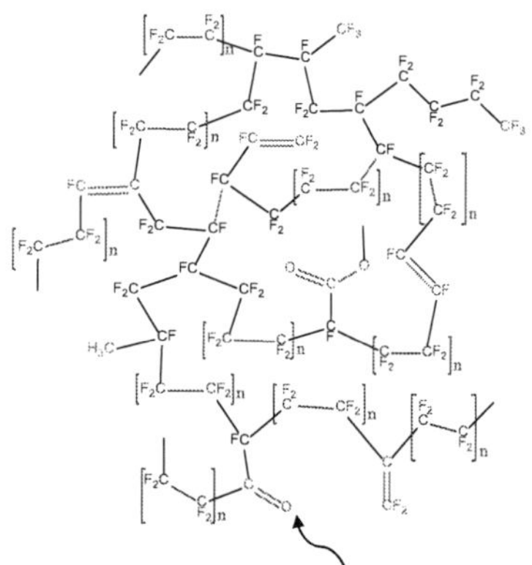

DNPH
2,4-Dinitrophenyl Hydrazine

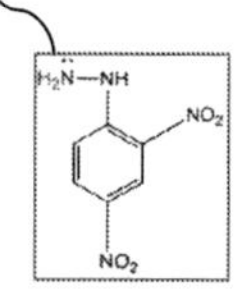

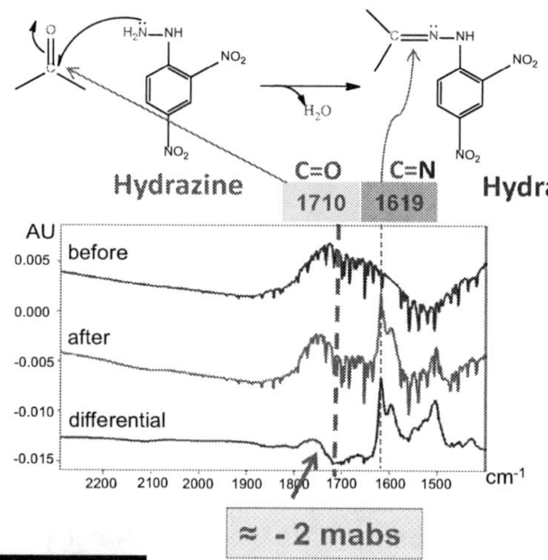

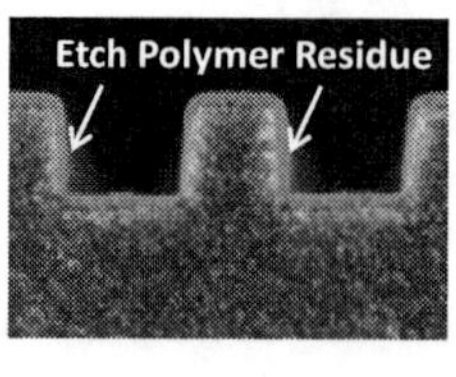

Interfacial Electrochemistry & Material Research Lab UNT

Assignment	Lit* (cm^{-1})	Exp (cm^{-1})
C-H stretching (aromatic)	3000-3100	3108
C-H wagging (aromatic)	795-880	870-910
C-H scissoring (aromatic)	1000-1600	1000-1230
Aromatic ring sextant out of plane deformation (ring pucker)	680-720	746
NO_2 stretching (sym)	1318-1357	1341
NO_2 stretching (asym)	1485-1555	1505
C=C stretching (aromatic ring)	1500-1555	1510-1555
N-H stretching	3300-3500	3314
N-H bending	1580-1650	1596
C=N stretching	1610-1680	1619

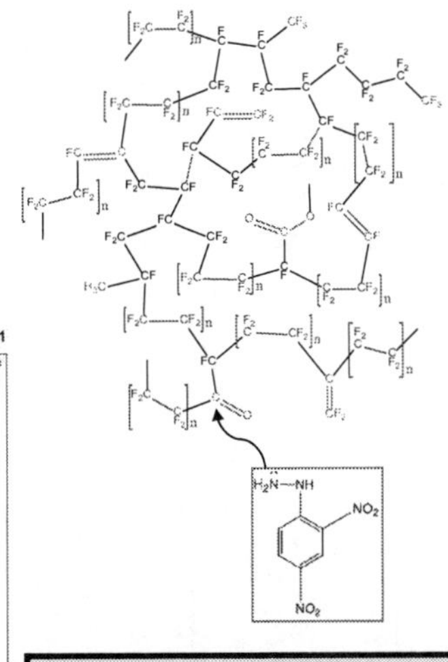

Hydrazone absorption peak table: Experimental (exp) and reference (ref) data in cm^{-1}

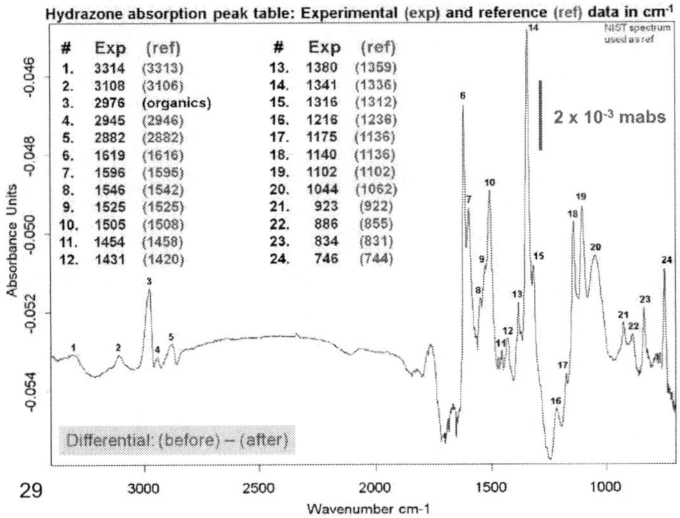

#	Exp	(ref)	#	Exp	(ref)
1.	3314	(3313)	13.	1380	(1359)
2.	3108	(3106)	14.	1341	(1336)
3.	2976	(organics)	15.	1316	(1312)
4.	2945	(2946)	16.	1216	(1236)
5.	2882	(2882)	17.	1175	(1136)
6.	1619	(1616)	18.	1140	(1136)
7.	1596	(1595)	19.	1102	(1102)
8.	1546	(1542)	20.	1044	(1062)
9.	1525	(1525)	21.	923	(922)
10.	1505	(1508)	22.	886	(855)
11.	1454	(1458)	23.	834	(831)
12.	1431	(1420)	24.	746	(744)

Confirm all 24 spectroscopic features of Hydrazone modified Post-Etch Residues.

29

Identification of C=C by Br_2 Reaction

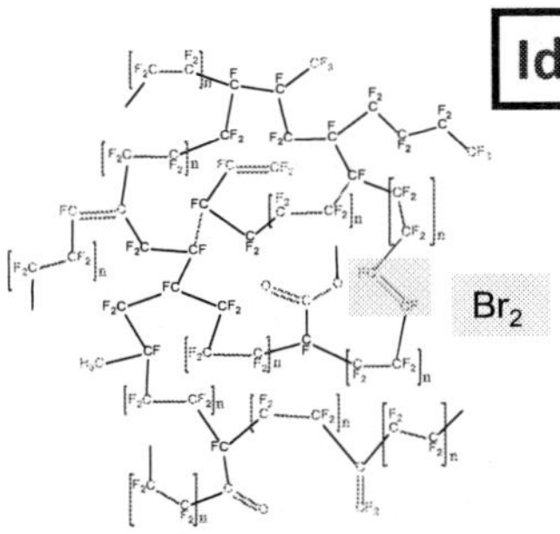

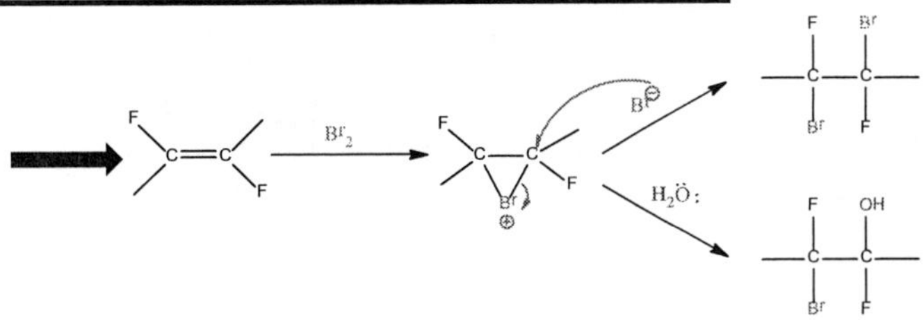

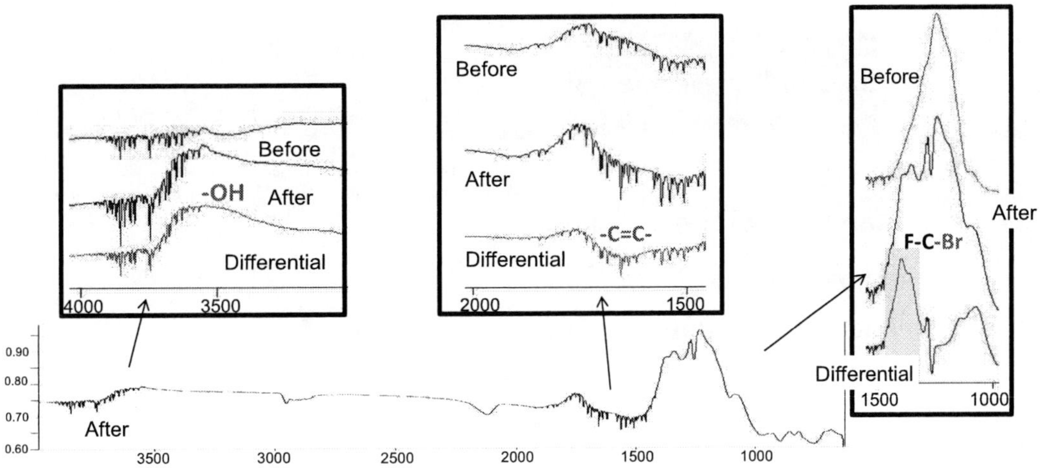

Mukherjee et. al, *ECS Solid State Lett.*, **2**, N11 (2013)

Identify Fluoropolymer Backbone $-[CF_2-CF_2]_n-$ By Reductive Defluorination on C-F bond

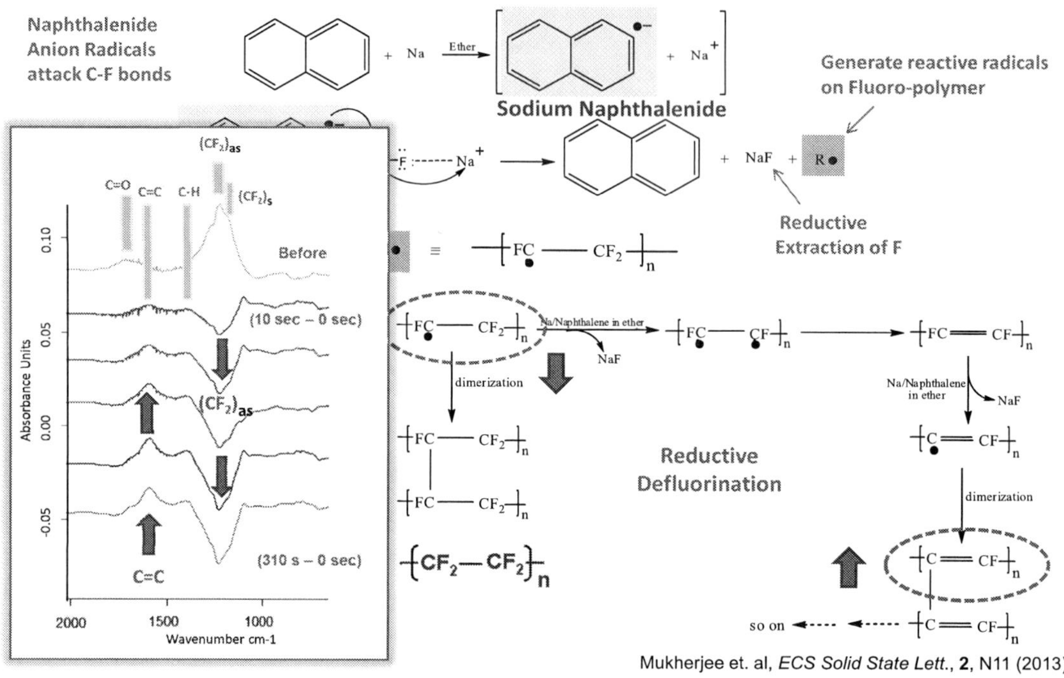

Mukherjee et. al, *ECS Solid State Lett.*, **2**, N11 (2013)

MIR-IR : UV-assisted Cleaning on Post-etch Residues

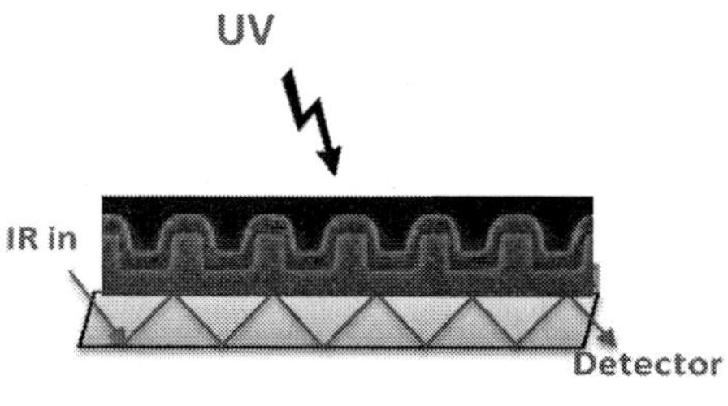

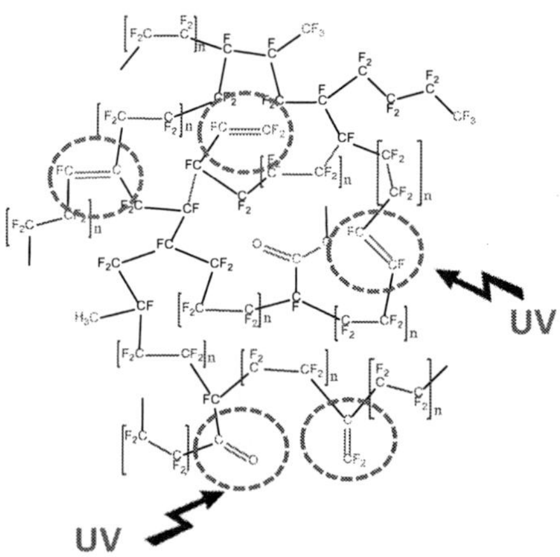

Kanwaljit Singh	***Intel***
Ian Brown	***TEL***
Jacques Faguet	***TEL***

Interfacial Electrochemistry & Material Research Lab UNT

SEM Images of UV & UV+Wet Clean Samples

Kanwal Jit Singh

SEM Images of UV & UV+Wet Clean Samples

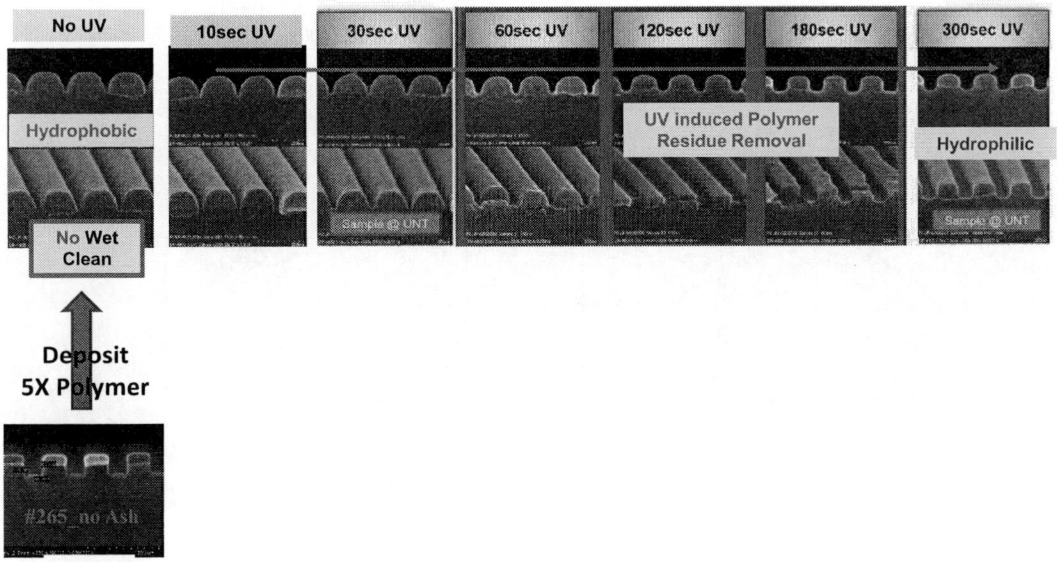

Kanwal Jit Singh

SEM Images of UV & UV+Wet Clean Samples

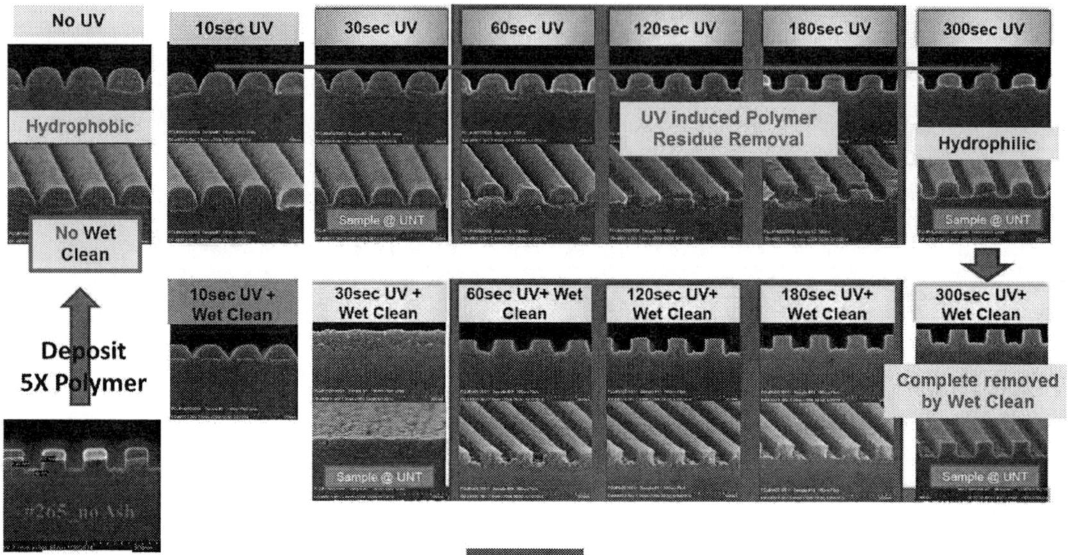

Kanwal Jit Singh

Ian Brown

C=O Chromophore Required for UV-assisted Clean

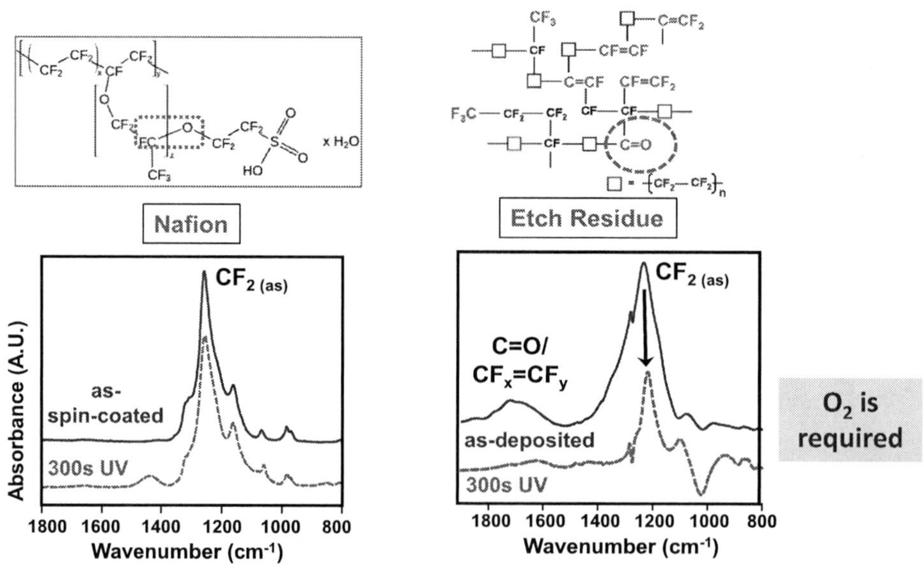

With -C–O, **no** UV clean

Need -C=O, for UV clean

Chemical Bonding Transformation Mapping - UV Radiation

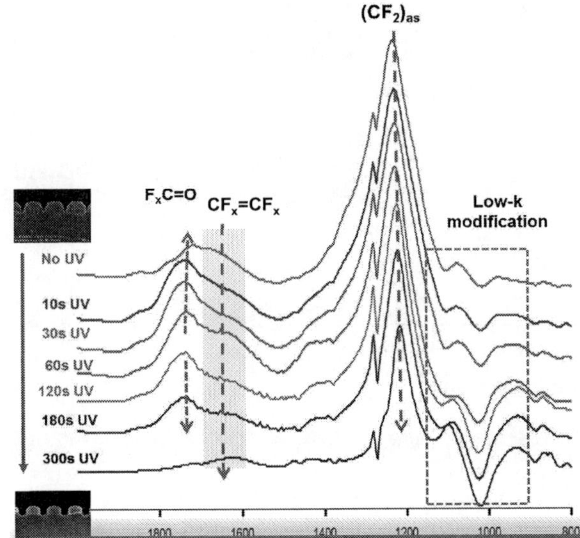

- Fluorocarbon polymer backbone $-(CF_2)_n-$ steadily decrease as UV treatment time increases
- Gradual reduction of $CF_x=CF_x$ band
- Fluoro-carbonyl $F_xC=O$ forms by photo-cleavage of $CF_x=CF_x$ bonds

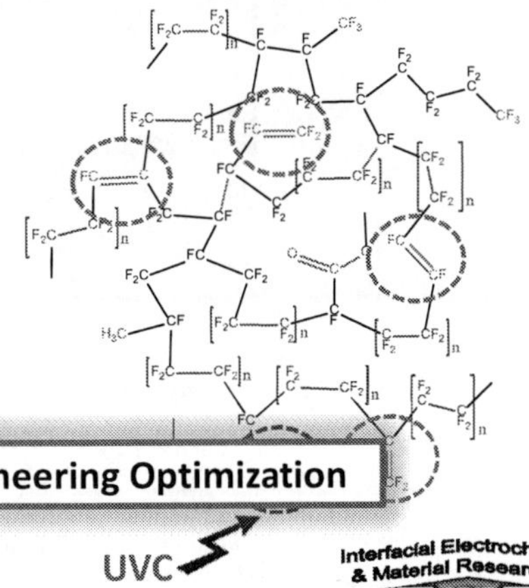

Provide Critical Process Evolution Info for Engineering Optimization

Interfacial Electrochemistry & Material Research Lab
UNT

Summary of Chemical Bonding Transformation with UV Radiation

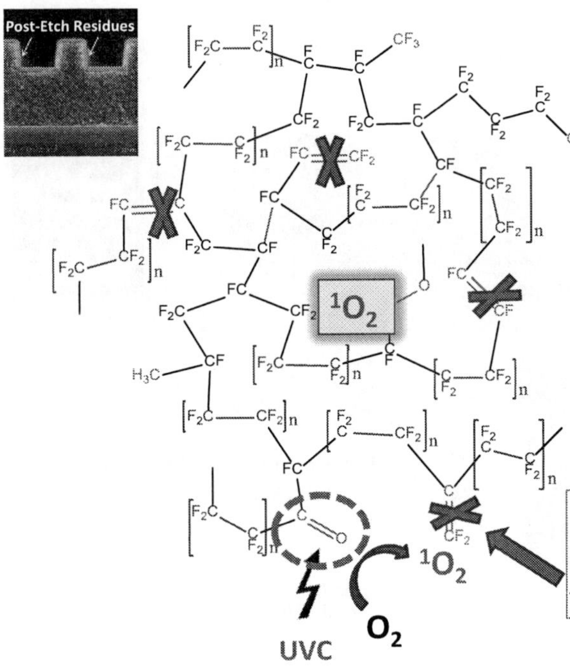

Volatile Fragments	BP
Carbonyl Fluoride (CF_2O)	-84.57
Trifluoroacetyl Fluoride (C_2F_4O)	-59
Hexafluoroacetone (C_3F_6O)	-28
Octafluorobutan-1-one (C_4F_8O)	7-9
Nonafluoropentanoyl fluoride ($C_5F_{10}O$)	52.2
Perfluoro (2-methyl-3-pentanone) ($C_6F_{12}O$)	49

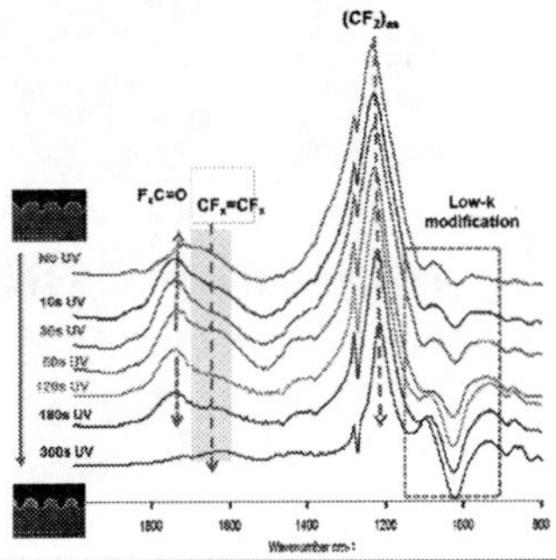

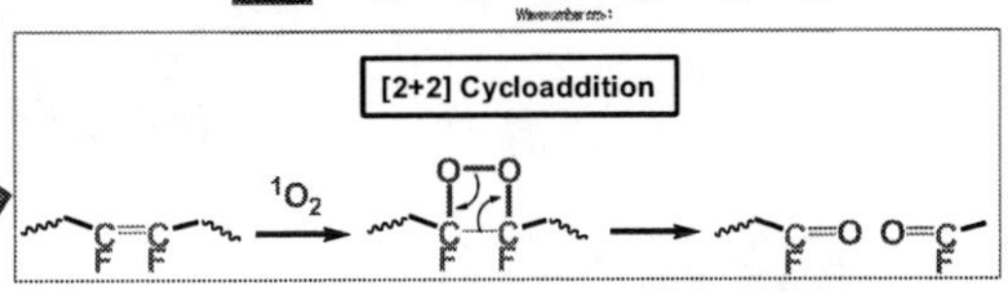

Interfacial Electrochemistry & Material Research Lab
UNT

IR Spectroscopic Evidence for Hydroxyl Formation during UV/Air Treatment

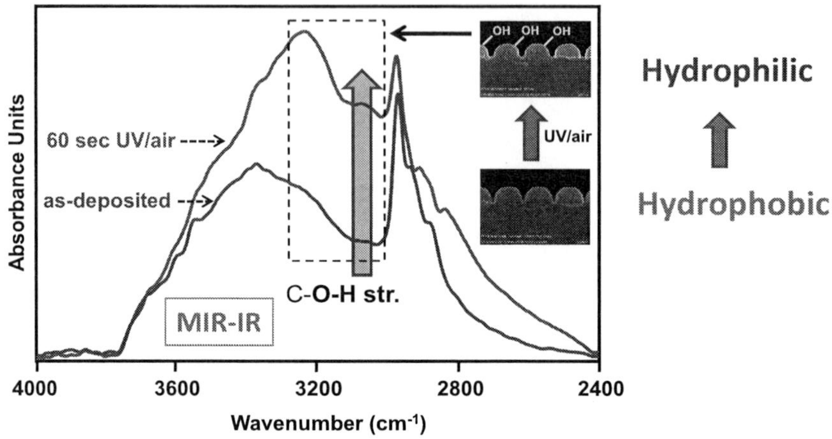

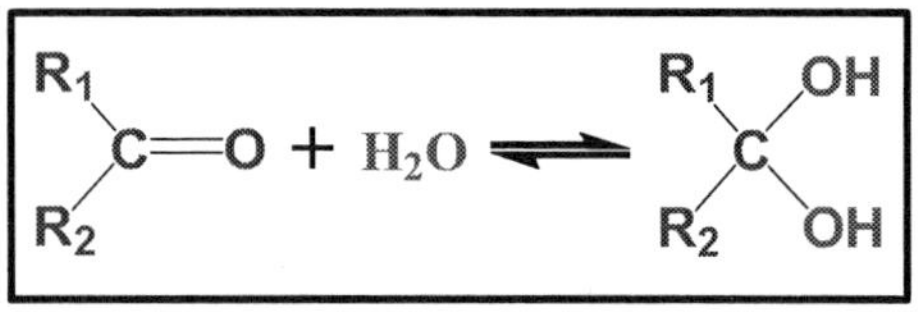

R_1	R_2	K_{eq} = [Hydrate]/[Carbonyl]
CH_3	CH_3	1.4 x 10^{-3}
CF_3	CH_3	35
CF_3	CF_3	1.2 x 10^{6}

~10^3 times faster; ~10^9 times faster

Ref: Guthrie, J. P. Carbonyl Addition Reactions: Factors Affecting the Hydrate-Hemiacetal and Hemiacetal-Acetal Equilibrium Constants. *Can. J. Chem.* **1975**, *53*, 898-906

Chemical Bonding Transformation Insights – UV Clean Optimization More Efficient

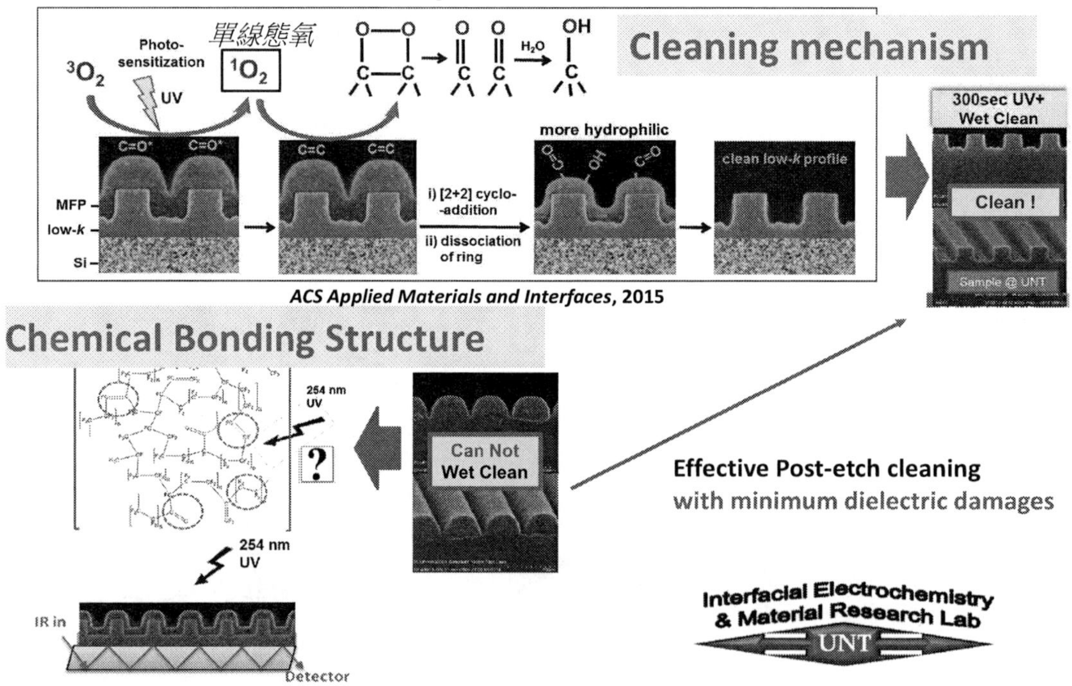

BEOL Etch and Cleans Process Integration

ULK Materials Engineering
- Low K_{eff}
- UV cured PECVD
- HF resistant

Reactive Ion Etching
- CD control
- Clean friendly
- Damage control

Post Etch Cleans
- Less damage
- Fast clean
- Keep CD, K_{eff}

Chemical Bond Formation & Bond Breaking - **Highly Selective in time & space !**

BEOL Integration Challenges

Metrology Tools

Physical	Not complete
Electrical	Costly
Chemical	

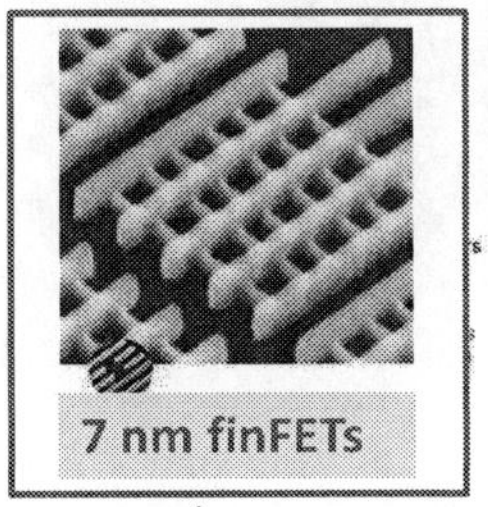

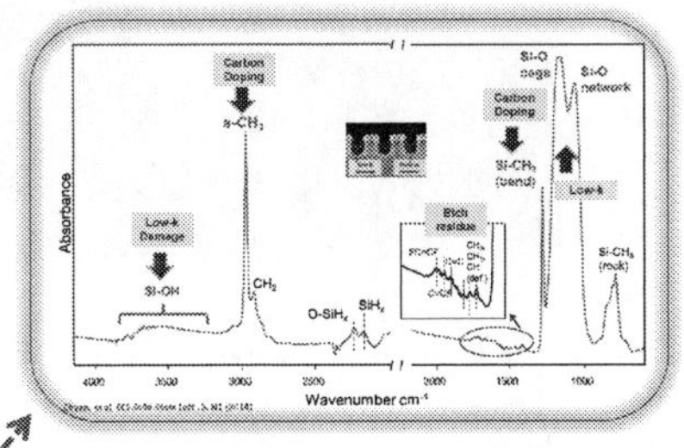

Chemical bonding transformation map

Patterned ILD

PI: Oliver Chyan*, *Chyan@unt.edu*

Interfacial Electrochemistry & Material Research Lab
UNT

MIR-IR Applications in Advanced IC Fabrication

- **FEOL (Ge, SiGe etc) , BEOL etching and cleaning, formulation**
- **Monitor Post-etch residue removal & minimize Low-k damages**
- **Optimize Plasma Etch/Strip/Clean Process integration**
- **Monitor Low-k damages and optimize restoration process**
- **UV curing on porous Low-k dielectrics materials**
- **Evaluate TiN hard mask for Low-k pattern fabrication**
- **Flowable low-k dielectrics for gap filling in nanostructure**
- **Atomic Layer Deposition/Etching : provide critical interfacial chemical bonding info for better atomic layer control.**

PI: Oliver Chyan*, *Chyan@unt.edu*

Interfacial Electrochemistry & Material Research Lab
UNT

Goal: Achieve better understanding of fundamental materials properties at the critical **interfaces** of practical applications.

Interfacial Electrochemistry & Material Research Lab

UNT

Group members:

Alex Lambert, Issac Goutham, Muthappan Asokan, Seare Berhe

Alumni :

Dr. Sirish Rimal (*JSR-Micro*)

Dr. Tamal Mukherjee (*LAM)*

Dr. Nick Ross (*Intel*)

Dr. Kyle Yu (*TEL*)

Dr. Karthik Pillai (*TEL*)

Dr. Simon Koskey (*Intel*)

intel ATMI NOVELLUS TEL Lam RESEARCH

43

END

44

imec

WET CLEAN CHALLENGES IN 22 NM ½ PITCH AND 16 NM ½ PITCH STRUCTURES

E. KESTERS*, Q.T. LE, C. LORANT, X.XU AND F. HOLSTEYNS

Email address: kesterse@imec.be
Address: Kapeldreef 75
3001 Heverlee
Phone: +32 16 288609

SPCC 2017

PUBLIC

OUTLINE

- Introduction
- PERR clean for 22 nm ½ pitch structures:
  - Successful removal of fluorinated residues together with TiN HM
- Impact of dissolved oxygen in dilute HF on metal loss
  - Effect of fluid dynamics, chamber atmosphere and dissolved oxygen concentration in HF on Cu etch
- Prevention of pattern collapse by using hot IPA and SFC (surface functionalizing chemistry)
  - Parameters affecting pattern stability
  - Approaches
  - Method for surface functionalization: typical reaction
  - Impact of SFC on blanket OSG2.55, thermal oxide and 90 nm pitch high AR BEOL trench structures
  - Prevention of pattern collapse on 16 nm ½ pitch wafers after V1M2 etch using hot IPA and SFC: morphological study
- Summary

imec

2

INTRODUCTION

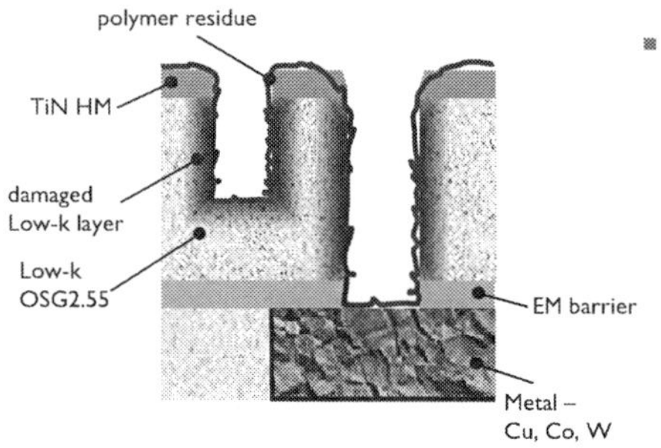

- What is challenging regarding the PERR step?
  - Remove/ pullback or preserve TiN HM
  - Remove fluorinated polymer residues
  - Compatibility requirements:
    - Cu, Co, W, liner and barrier ➔ not to induce corrosion
      1. HF is one of the commonly used chemistries for DD clean
      2. HF based low dissolved oxygen (<20 ppb for the liquid & 500ppm (air))
    - Advanced OSG LK (lower k-value and higher porosity), including the LK damaged layer
  - Prevention of pattern collapse
    - Transfer from 22 nm ½ pitch towards 16 nm ½ pitch, makes the structures more prone to pattern collapse

PART 1

PART 2

imec

PERR CLEAN FOR 22 NM ½ PITCH STRUCTURES: SUCCESSFUL REMOVAL OF FLUORINATED RESIDUES TOGETHER WITH TIN HM

PERR CLEAN FOR 22 NM1/2 PITCH STRUCTURE

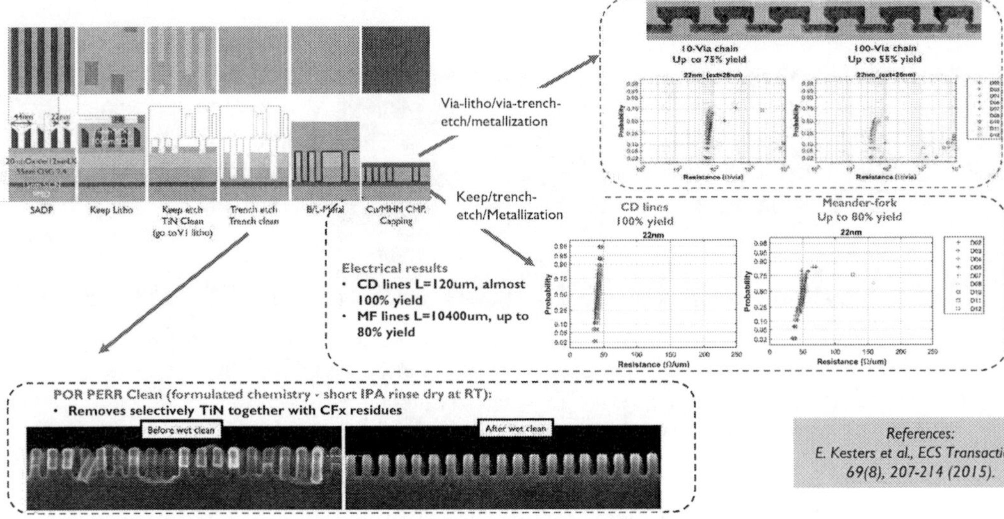

References:
E. Kesters et al., ECS Transactions, 69(8), 207-214 (2015).

imec 5

IMPACT OF DISSOLVED OXYGEN IN DILUTE HF ON METAL LOSS

INTRODUCTION

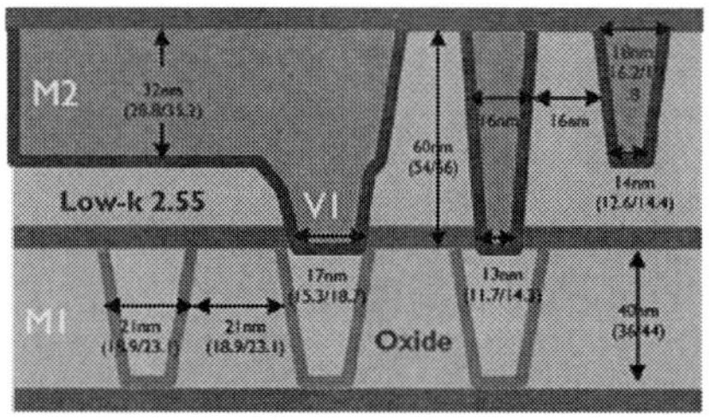

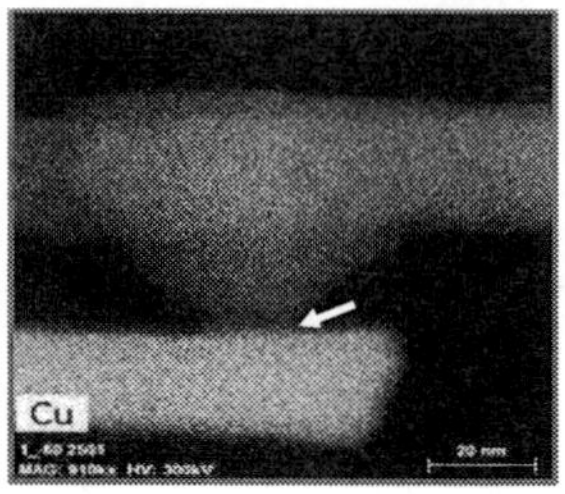

- Cross-sectional representation of the two level metal structure (16 nm ½ pitch structure)
- 16nm M2 dual-damascene structures metallized with Copper

Briggs et al., to be presented at IITC, 2017.

- Clean concept transferred from 22 nm ½ pitch to 16 nm ½ pitch structures:
  - Further reduction Cu loss during cleaning sequence required
    - HF based low dissolved oxygen cleans are reported to be crucial for DD cleans

E. Kesters et al., Solid State Phenomena, 1012-0394, Vol. 255,pp. 251-254.

L. Broussous et al., Solid State Phenomena, 1012-0394, Vol. 255,pp. 260-264.

    - Rinse optimization using dNH_4OH vs. dCO_2 (not discussed)

imec

MATERIALS AND METHODS

- SU-3200 platform, SCREEN Single wafer cleaning tool

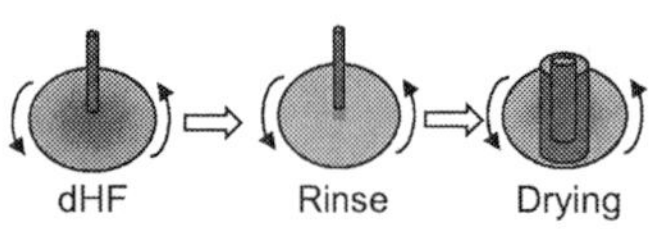

- [HF]: 0.05 – 0.1%
- DO: 70 - 3000 ppb
- Process time : 20, 60 and 120 s
- Other variables:
  - ambient oxygen (controlled vs. non-controlled)
  - fluid dynamics (reference vs. improved)

- Materials:
  - 500 nm blanket ECD Cu

- Characterization:
  - 4-point probe measurement (sheet resistance)

imec

EFFECT OF FLUID DYNAMICS ON CU ETCH

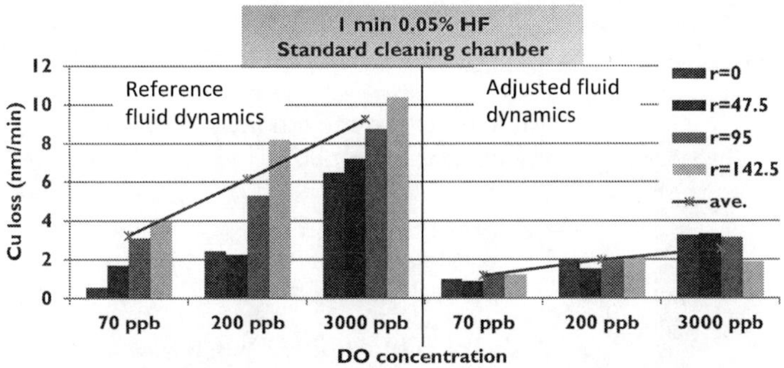

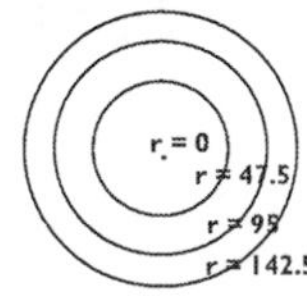

- Copper loss increases with increased DO concentration in dHF solution (70 to 3000 ppb DO)
- Improved fluid dynamics reduces the amount of Cu loss
  - → prevents copper losses towards wafer edge are suppressed, even at increased DO (= 3000 ppb)

If you are not able to control the ambient atmosphere, fluid dynamics can improve the metal compatibility when using dHF

imec

EFFECT OF CHAMBER ATMOSPHERE ON CU ETCH

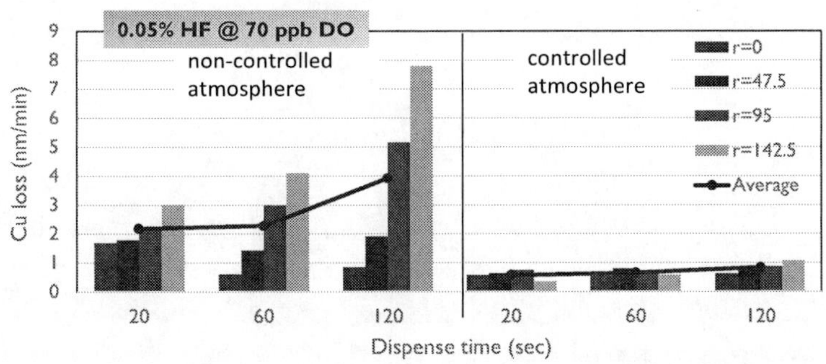

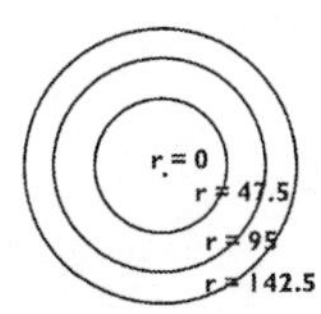

- Non-controlled ambient combined with low DO 0.05% HF process:
  - It was observed that the Cu etch was higher toward the outer peripheral side of the wafer compared to the controlled ambient.
  - Cu loss increased with dispense time.
- Controlled ambient (low oxygen ambient) combined with low DO 0.05% HF:
  - did not attack bulk Cu, even after 2 min dispense

imec

10

EFFECT OF DISSOLVED OXYGEN CONCENTRATION IN DHF SOLUTION ON CU ETCH

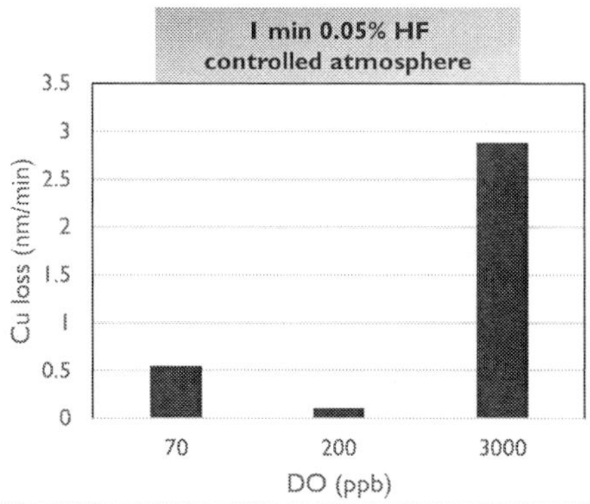

- Cu loss increased with increasing DO concentration in dHF
- Cu loss with 0.05% HF:
  - ≤ 200 ppb DO: less than 1 nm Cu loss was observed
  - 3000 ppb DO: Cu loss > 1 nm
- A similar trend in etching behavior was observed using 0.1 and 0.2% HF (not shown)

The etching behavior of Cu strongly depended on:

- DO concentration and was not affected by the HF concentration (0.1 – 0.2% range).
- If HF is used for PERR (in combination with formulated chemistries) , be aware that DO concentration is low enough

PREVENTION OF PATTERN COLLAPSE BY USING HOT IPA AND SFC (SURFACE FUNCTIONALIZING CHEMISTRY)

PARAMETERS AFFECTING PATTERN STABILITY

$$F \propto \frac{EI}{h^3}\delta_x \sim \frac{E\,w\,s}{AR^3}$$

- E: Young's modulus
- AR: aspect ratio
- w: width in transvers direction
- s: space between nearby structures
- Surface chemistry of the dielectric sidewall
- Wet clean chemistry and rinse liquid

Non-rigid materials, collapse at low AR | Rigid materials, collapse at high AR | Rigid materials, moderate AR

Resist: 150 nm L/S, AR~ 6, *Tanaka, 1993, Jpn. J. Appl. Phys,* | EUV resist: 28/32 nm L/S, AR ~ 2.8, *Yoshimoto, 2011, SPIE* | Low k: 45 nm L/S, AR~ 5 | MEMS AR~20-100 | DRAM capacitors | Si pillars AR~ 20 | Si fins AR<10

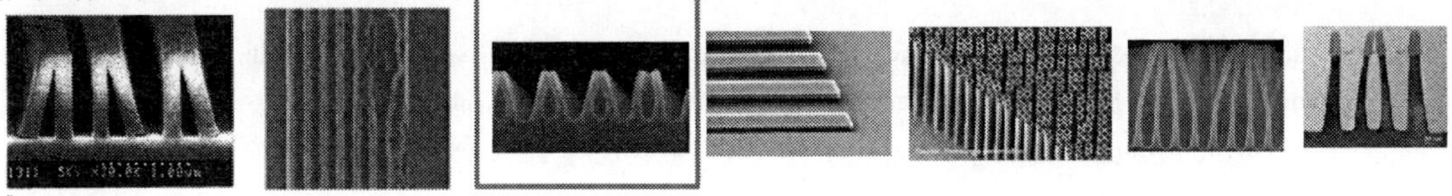

imec

13

PREVENTION OF PATTERN COLLAPSE: APPROACHES

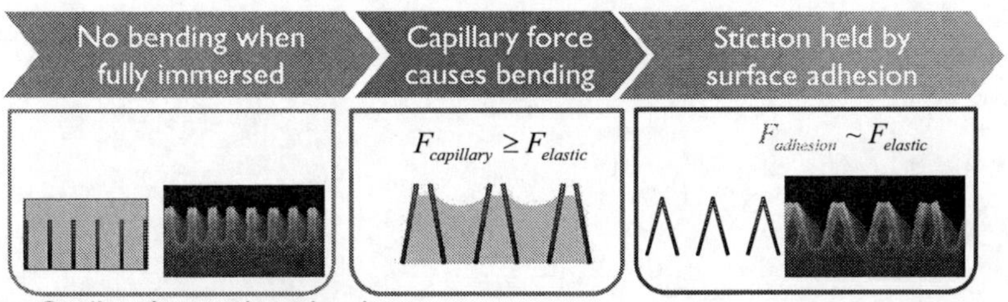

X. Xu et al., ACS Nano 8(1), 885-893 (2014)

M. Sankarapandian et al., Solid State Phenom. 195, 107 (2013)

- Capillary force induces bending
  → reduce capillary force such that

1. During spin-rinse drying by an IPA final rinse at elevated temperatures
   → Low surface tension liquid: e.g. IPA drying, max 3x reduction in force
   - Water: γ= 0.072 N/m
   - IPA: γ= 0.021 N/m
2. Capillary force can be reduced further by changing the surface energy of low-k lines.
   - This can be done by modification of the structures surface wetting properties by deposition of an organic monolayer providing a contact angle of 90 deg or above

- Reduce collapse force: Modify exposed surface without damaging the low-k

imec

14

METHOD FOR SURFACE FUNCTIONALIZATION: TYPICAL REACTION

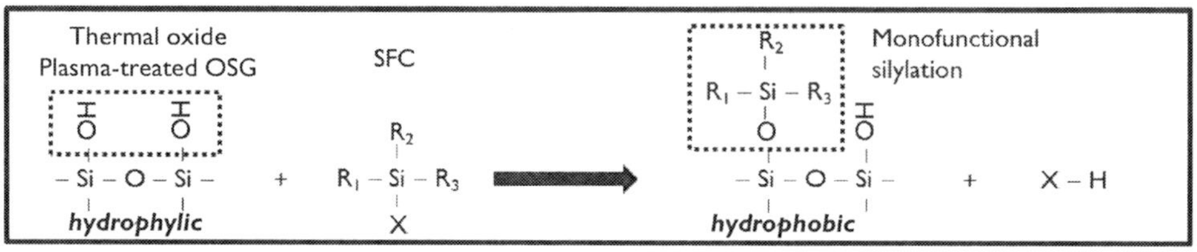

- Restoration of damaged layer is performed by a silane-coupling reaction using organic solvent
- Silylation process
  - Success criteria: reactivity of surface (sidewall and bulk - pore wall), solubility of silylating agents in solvent
  - Limited at the surface of the damaged low-k if silylating molecule size > pore diameter
  - Incorporation of silylating molecules in low-k bulk if silylating molecule size < pore diameter
- *Use of surface functionalizing agents as part of the rinsing sequence*

imec 15

IMPACT OF SFC ON BLANKET OSG2.55, THERMAL OXIDE AND 90 NM PITCH HIGH AR BEOL TRENCH STRUCTURES

SURFACE FUNCTIONALIZATION: EXPERIMENTAL

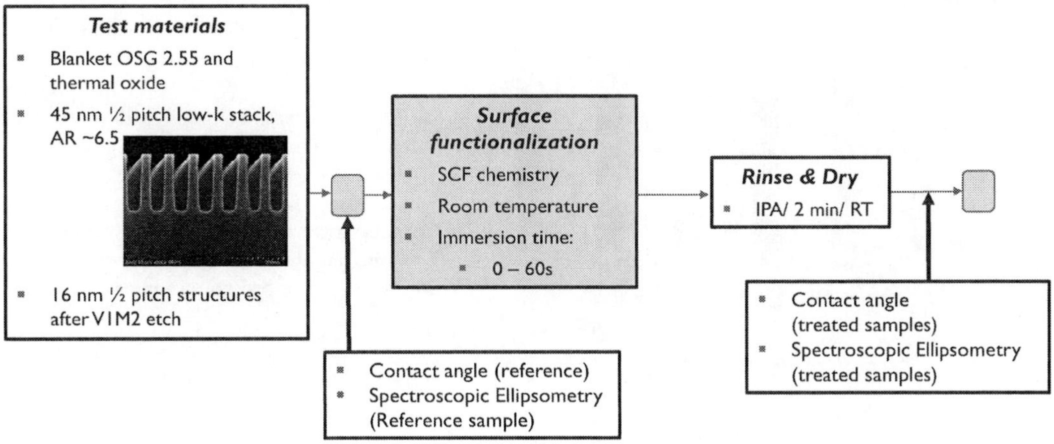

imec

RESULTS

PLASMA-TREATED OSG 2.55

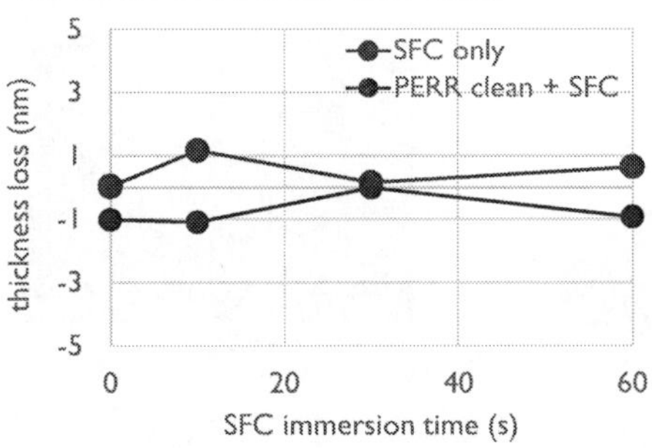

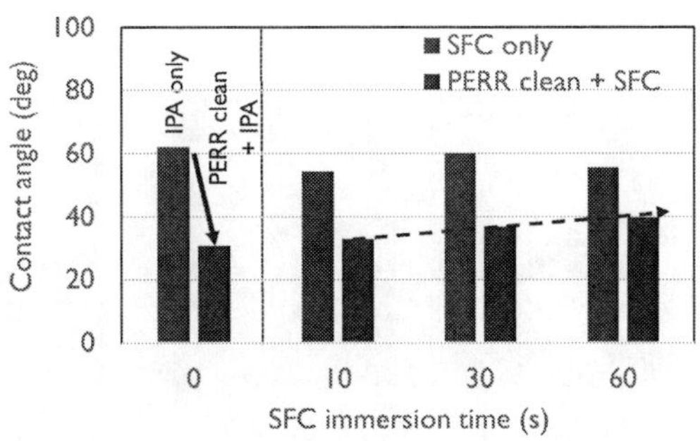

- Negligible change in thickness because of PERR clean
- PERR clean induces a decrease in contact angle of ~ 30 deg
- Thickness remains similar after SFC immersion, while contact angle is slightly increasing with SFC immersion time

SFC shows a limited reaction with plasma-treated OSG 2.55 surface

imec

RESULTS
THERMAL OXIDE

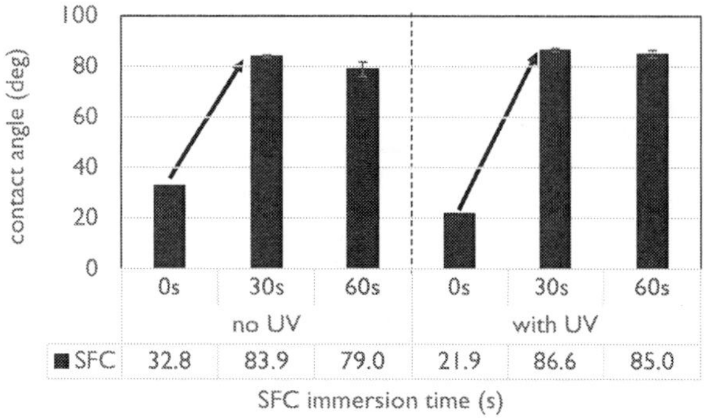

- Thermal oxide surface is hydrophilic before SFC treatment
- UV/O_2 pre-treatment in order to condition the surface (to increase [OH] on surface)
- Contact angle substantially increase to ~ 90 deg independent from immersion time and surface pre-conditioning
- Without UV pre-conditioning: good surface to start reaction of SFC

SFC: Reactivity thermal oxide >>plasma-treated OSG 2.55

imec

19

CONTACT ANGLE HIGH A/R 90 NM PITCH BEOL TRENCH STRUCTURE

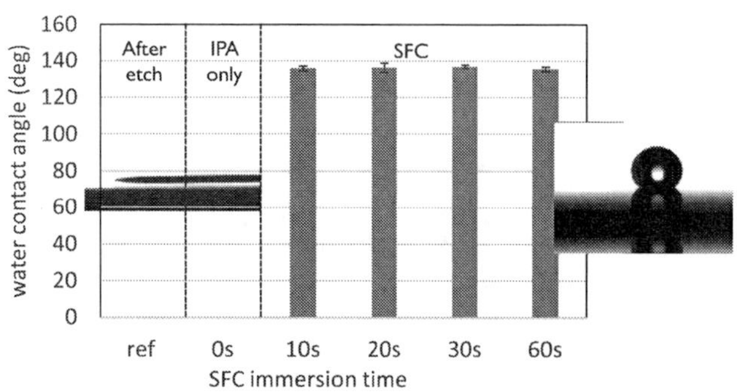

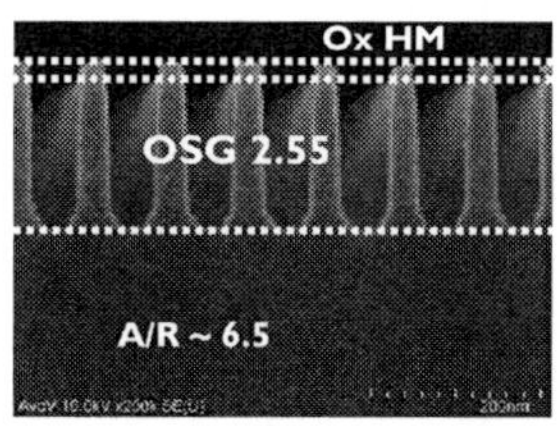

- Large change in contact angle (> 100deg) after treatment for 10 s in SFC
- SFC present at the surface
- Functionalizing oxide HM is successful

Functionalization of the top surface plays a key role to prevent lines from pattern collapse

imec

20

PREVENTION OF PATTERN COLLAPSE ON 16 NM ½ PITCH WAFERS AFTER V1M2 ETCH USING HOT IPA AND SFC: MORPHOLOGICAL STUDY

TRANSFER FROM 22 NM ½ PITCH TOWARDS 16 NM ½ PITCH STRUCTURES

22 nm ½ pitch structure	16 nm ½ pitch structure
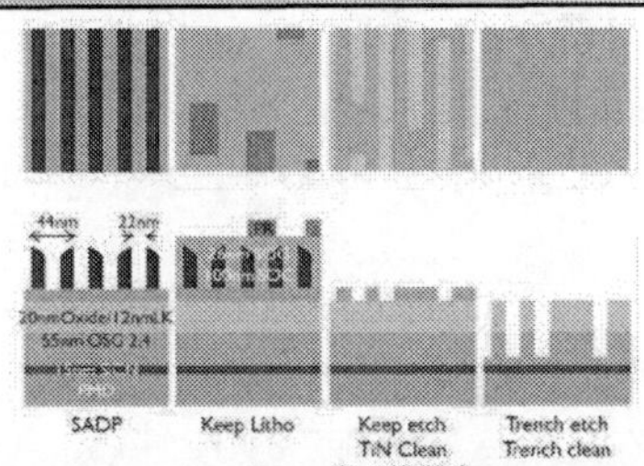	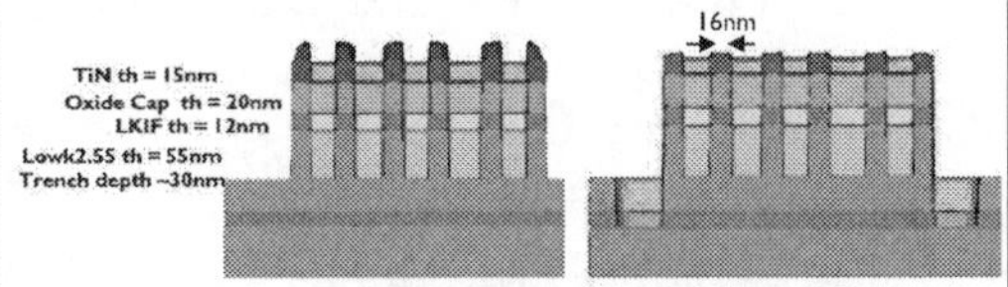
	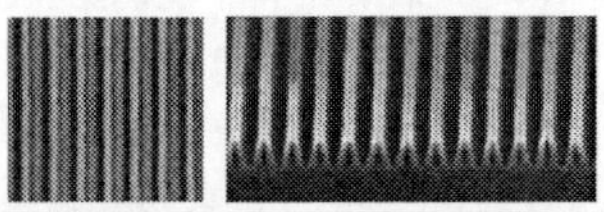

Transfer from 22 nm ½ pitch towards 16 nm ½ pitch:
→ structures are more prone to collapse after POR PERR clean (formulated chemistry + short IPA rinse at RT)

imec

22

EFFECT OF HF TIME

[HF] = 0.05% IPA = 120s

HF	PERR	RINSE - DRY	CD (nm)	(0,0) / (0,3) / (0,5)	remark
0s	120s	IPA (65C)	19.8	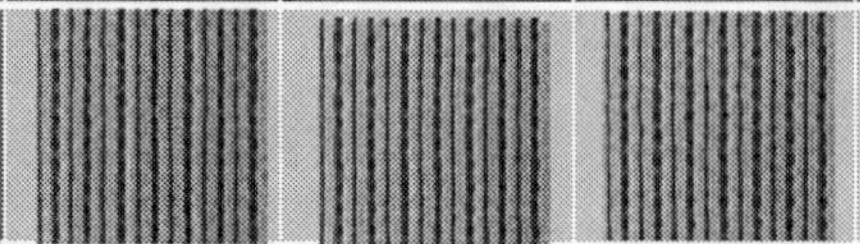	No collapse seen without HF pre treatment
10s	120s	IPA (65C)	19.1	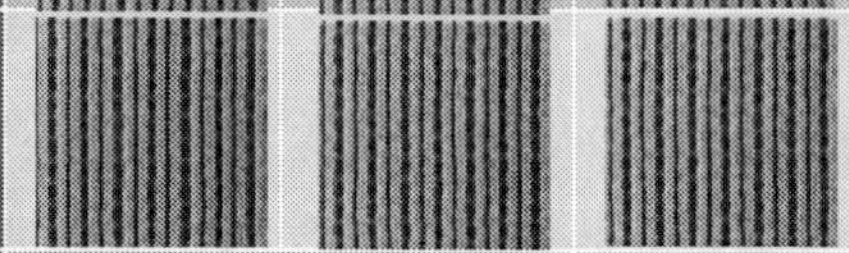	Reducing HF time from 20s to 10s helps to prevent lines to collapse
20s	120s	IPA (65C)	18.5	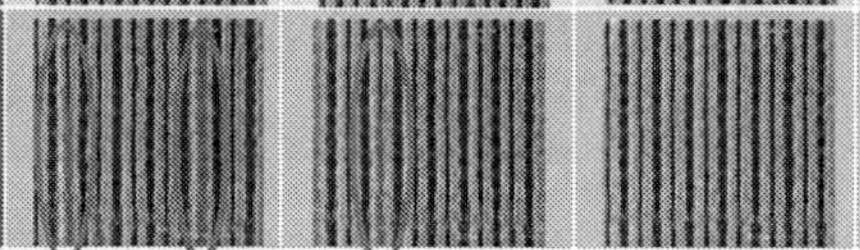	Collapse observed Difference in line CD of ~1.3nm between the 0s and the 20s dHF
20s	120s	SFC + IPA (RT) SFC + IPA (65C)	18.5	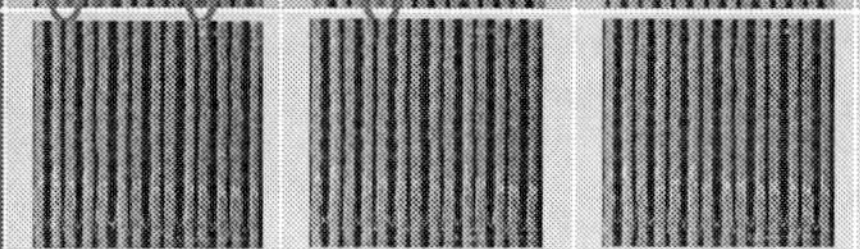	Adding SFC prevents lines to collapse even with 20s dHF Even rinsing using IPA at RT, no collapse was observed

A threshold w.r.t. CD has been observed and SFC has a positive impact on pattern collapse reduction

SUMMARY

1. Impact of dissolved oxygen in dilute HF on metal etch
   1. Tuned fluid dynamics reduces copper losses towards wafer edge, even at increased DO (= 3000 ppb) concentrations
   2. The Cu loss was also strongly dependent on the chamber atmosphere condition.
   3. The etching behavior of Cu strongly depended on the DO concentration and was not affected by the HF concentration (within 0.1 – 0.2 % range).

1. Pattern collapse prevention
   1. Functionalization of the top surface plays a key role to prevent lines from pattern collapse
   2. How to prevent line collapse for 16 nm ½ pitch structures?
      1. Increase IPA rinse from RT to 65C
      2. CD control: A threshold w.r.t. CD is observed for pattern collapse
      3. Make use of SFC in rinse-dry sequence:
         1. Adding SFC step prevents the lines to collapse, even with 20s 0.05%HF pre-treatment and IPA RT

imec 24

ACKNOWLEDGEMENT

SCREEN

Akihisa Iwasaki

imec 25

imec

embracing a better life

imec PUBLIC

High Temperature Water Clean and Etch Reactions with Low-k and SiO_2 Films: Experiments and Simulations

Joshua Barclay, Lu Deng, Oseoghaghare Okobiah,
Tina Sengphanlaya, Jincheng Du, Rick Reidy
University of North Texas

UNT UNIVERSITY OF NORTH TEXAS

Rationale

- High temperature water (HTW) has demonstrated some interesting capabilities etching SiN (SPCC 2015)
- Given HTW's enhanced reactivity, is HTW selective for hydrophilic (SiO_2) or hydrophobic (low-k) films?

UNT UNIVERSITY OF NORTH TEXAS

Outline

- Experimental Apparatus and Conditions
- Low k (k=2.3 porosity ~35%)
- Silica (thermal oxide)
- Low-k and SiO_2 characterization following exposure to HTW
  - Ellipsometry—thickness changes
  - FTIR—changes in the chemistry and thickness
  - Profilometry—define boundaries
- Simulations of HTW reactions with porous low-k films

UNT UNIVERSITY OF NORTH TEXAS

Experimental Conditions

- samples placed in the reactor with 98°C DI water
- heated to 120, 140, 160, and 180°C for 5, 10, and 20 minutes (heat up takes ~11-38 minutes)
- Reactor removed from heat and quenched in water for 2 min
- Samples rinsed with DI water and allowed to dry.

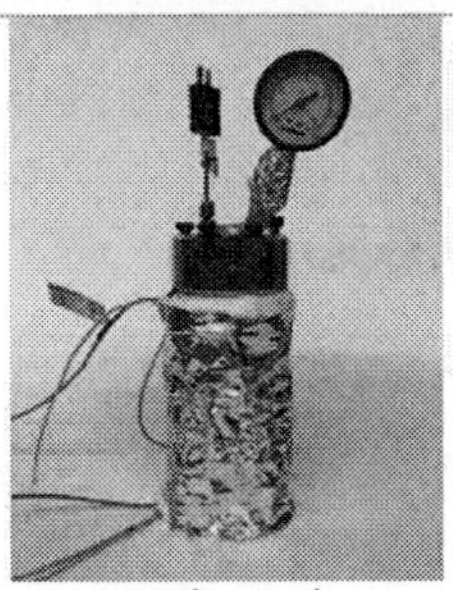

Heating mantle and reactor (600ml)

Reactor can be easily removed from mantle

Sample holder

UNT UNIVERSITY OF NORTH TEXAS

Previous work on SiO_2 in HTW

- Solubility increases **~6x** at 160°C (vs RT)
- solubility at 160°C sufficient to remove >1000 nm of oxide (much greater than current sample)

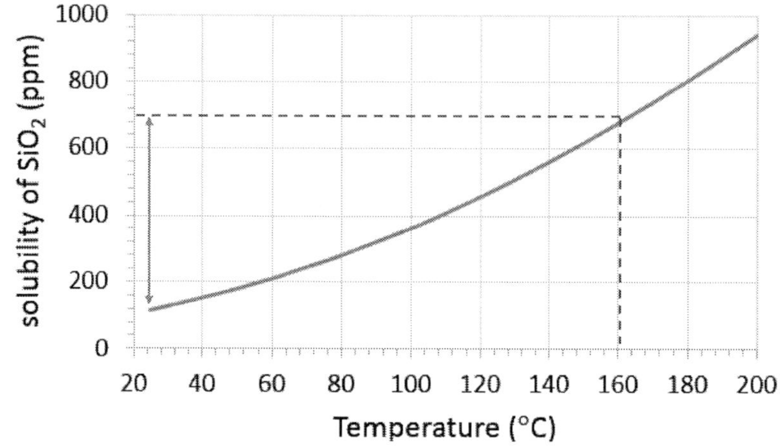

RO Fournier and JJ Rowe, Amer. Miner, 62 1052-1056 1977

UNT UNIVERSITY OF NORTH TEXAS

Ionization of Silica in HTW

- Ionization of silica-- ~10x than room temperature
- First ionization (formation of silicic acid) maxes at ~160°C
- Second ionization (formation of silicic acid) maxes at ~120°C

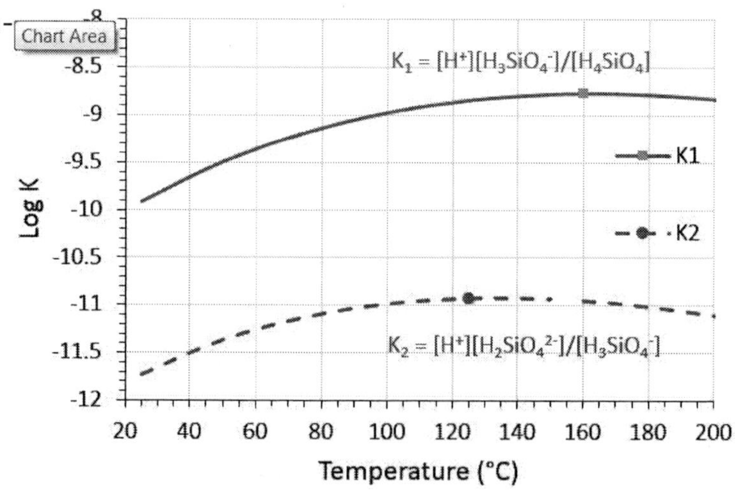

UNT UNIVERSITY OF NORTH TEXAS

Ellipsometry Results and Etch Rates

Type	Unetched Thickness (Å)	0 Minutes Thickness (Å)	5 Minutes Thickness (Å)	10 Minutes Thickness (Å)	20 Minutes Thickness (Å)	Etch Rate (Å/min)
Low K 160°C	4123	3969	3955	3977	3941	Negligible
Low K 180°C	4123	3979	3956	3939	3856	Negligible
SiO_2 120°C	322	300		202	58	12
SiO_2 140°C	322	149	59	0		18
SiO_2 160°C	322	0			0	
SiO_2 160°C	3098	2828		2408	1864	57
SiO_2 180°C	3098	1817	938	355	0	162

UNT UNIVERSITY OF NORTH TEXAS

Etch Rates: low-k vs SiO_2

- SiO_2 etch rate quickly increases around 160°C
- low k showed a negligible etch rate at temperatures where SiO_2 was significantly etched

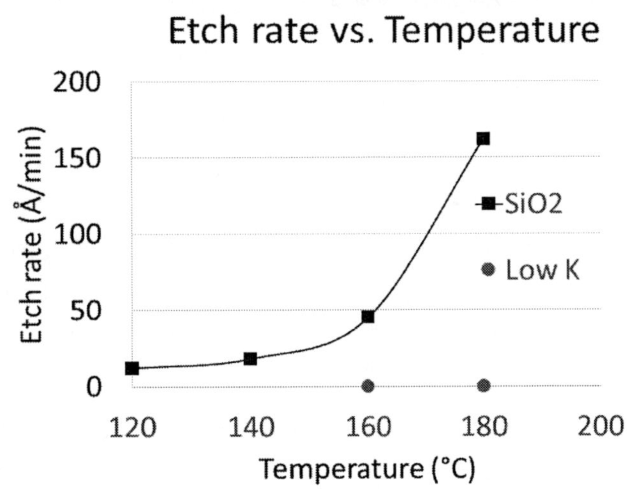

UNT UNIVERSITY OF NORTH TEXAS

Profilometry Profile SiO_2: from Protected to Exposed to 160°C Water

- Region exposed to HTW showed etching
- Etching increased with time

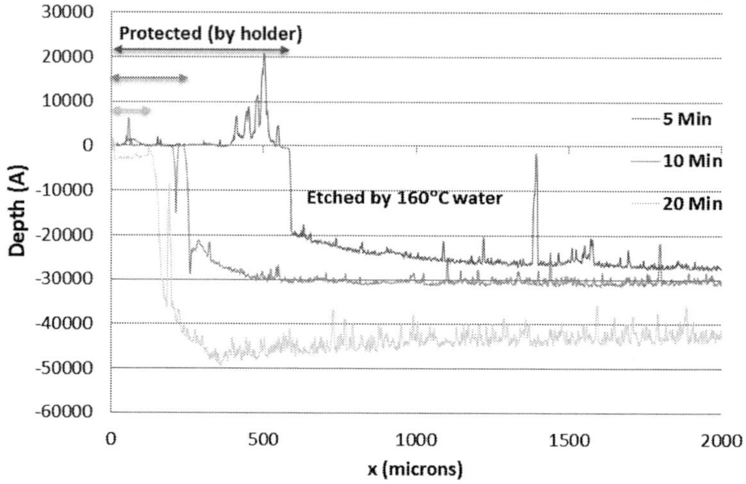

UNT UNIVERSITY OF NORTH TEXAS

GATR FTIR of SiO_2

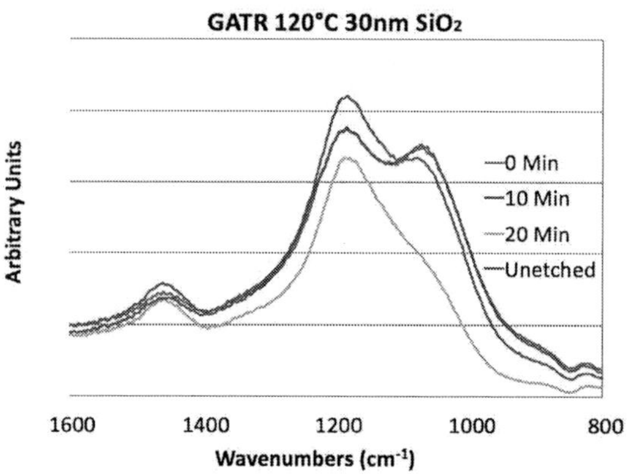

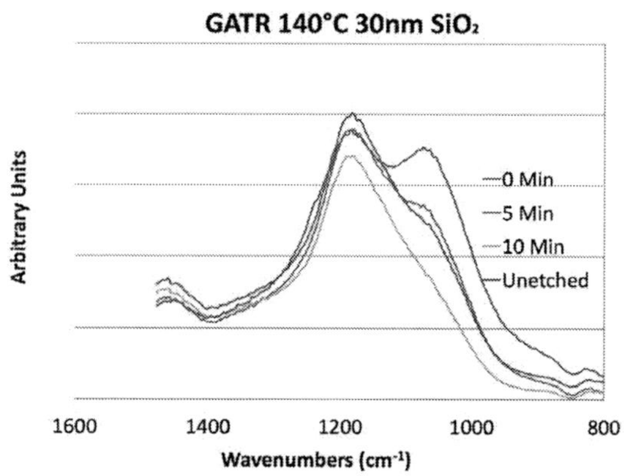

- Shows SiO_2 thickness decreasing with time

UNT UNIVERSITY OF NORTH TEXAS

GATR FTIR of low-k

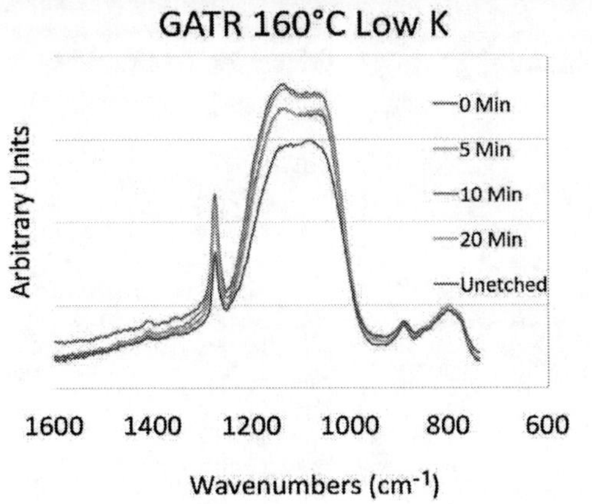

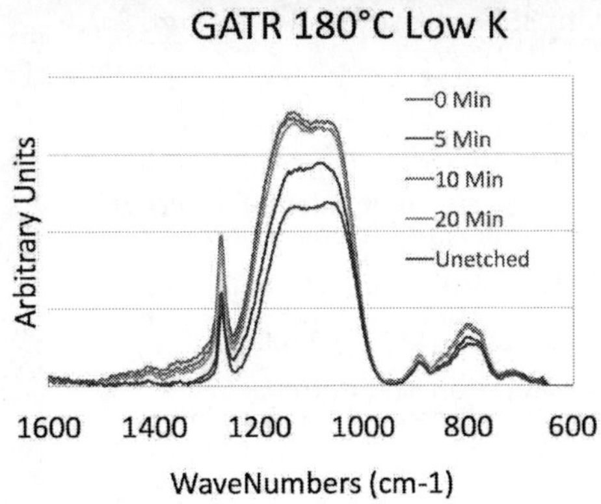

- This is interesting because?

UNT UNIVERSITY OF NORTH TEXAS

Contact Angle: low-k

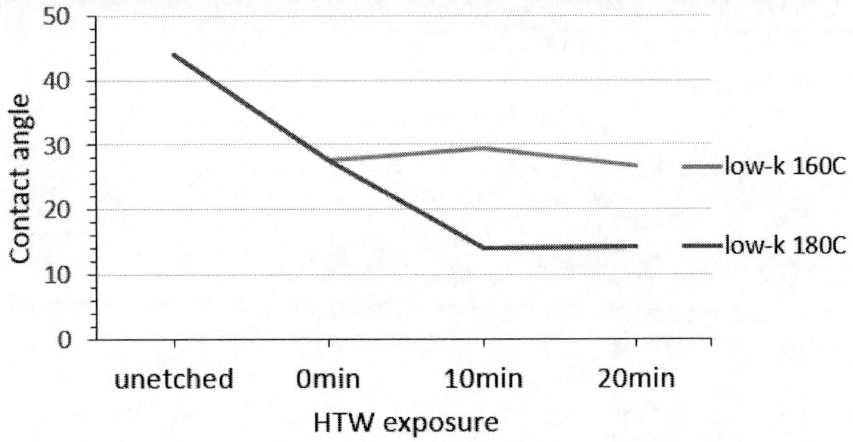

- Contact angle decreases with time at elevated temperatures

UNT UNIVERSITY OF NORTH TEXAS

Simulations of nanoporous low-k and water system using Reactive Force Field (ReaxFF) based MD

- Nanoporous silica was generated with 35% porosity (*J. Am. Ceram. Soc.* 97(2015)2772).
- Organic component introduced to nanoporous silica to form OSG (*J. Non-Cryst. Solids*, 431(2016)103).
- Surface is formed and water added to the surface and forms an interface
- Water/silica system studied using Reactive Force Field (ReaxFF) (*J. Phys. Chem. C* 120 (2016) 24803)
- Equilibration at different temperatures for 1 nano-seconds ($4x10^6$ steps)
- Monitored reactions such as Si-O-Si dissociation via hydrolysis and $Si-CH_3$ breakage

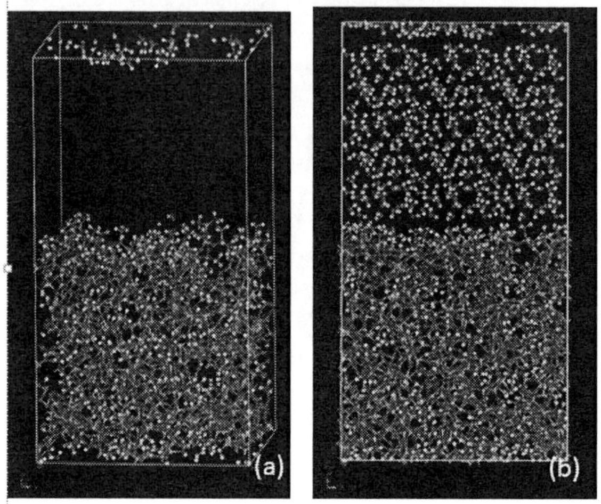

UNT UNIVERSITY OF NORTH TEXAS

Reaction steps in Si-O-Si breakage

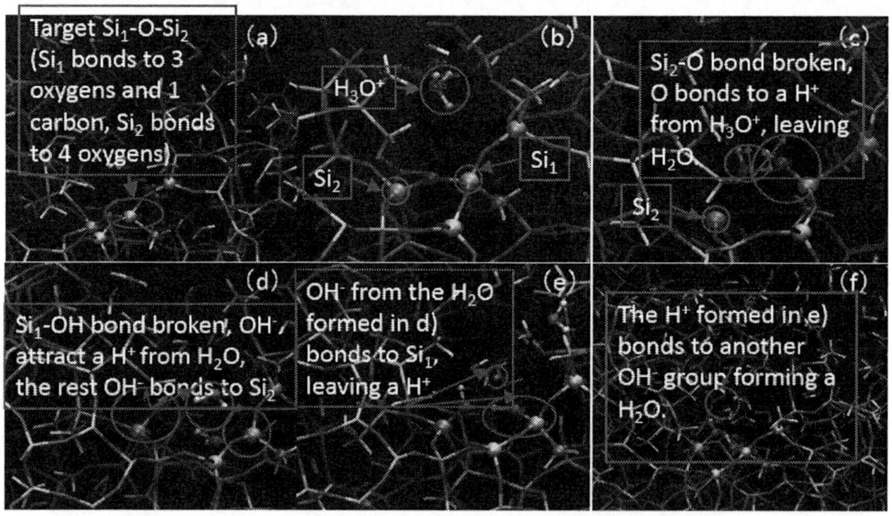

H_2O+Si-O-Si - → Si-OH +HO-Si Via a hydronium ion.

UNT UNIVERSITY OF NORTH TEXAS

Summary (Experimental)

- HTW SiO_2 etch rate increases dramatically around 160°C
- Low K samples were not measurably etched
- Decreasing hydrophobicity of Low K with treatment time→ some changes in film surface
- Future work
  - Confirm etching mechanism
  - Etching of other films
  - Use HTW to remove particles from low-k films

UNT UNIVERSITY OF NORTH TEXAS

Summary (Simulations)

- Reactive force field based MD simulations generated nanoporous organosilicate glass structures with 35% porosity and 10% carbon.
- Water/OSG interface formed and reaction studied at different temperatures.
- Si-O-Si hydrolysis reactions due to hydronium ions. HO-SiO_3 and SiO_4 are more vulnerable than CH_3-SiO_3 (concurs with experiments)
- Source of hydronium ions: enhanced dissociation of water and subsequent silicic acid formation.
- Future work:
  - to analyze reactions as a function of temperature and time to obtain statistical data.
  - Compare reactivity of OSG and nanoporous silica with similar porosity.

UNT UNIVERSITY OF NORTH TEXAS

Acknowledgements

- Sematech (wafers)
- Center of Advanced Research and Technology (CART)

UNT UNIVERSITY OF NORTH TEXAS

Contact cleaning opportunities on single wafer tool

[1]L. Broussous[a], [1]S.Zoll, [2]H.Ishikawa, [2]F.Buisine, [2]A.Lamaury,

[1] STMicroelectronics, 850 rue Jean Monnet, 38926 Crolles, France
[2] SCREEN SPE Germany GmbH, Fraunhoferstr. 7, 85737 Ismaning, Germany
a _ lucile.broussous@st.com

SPCC2017_ Austin, March 2017

Content

Introduction : Contact Cleaning Challenges

Experimental : Technologies, Materials, Chemistries

Results : Cleaning and Drying performances

Conclusion : New opportunities for contact cleaning & Drying

SPCC2017_ Austin, March 2017

Contact Cleaning overview

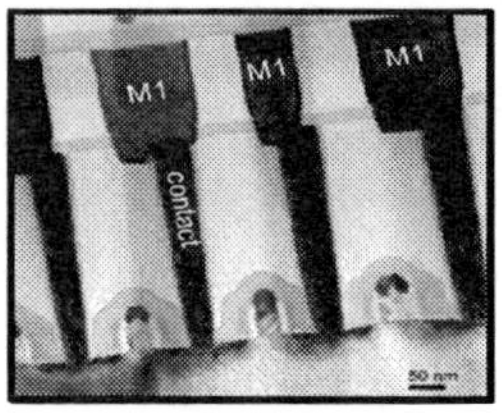

500x500 nm²

What was contact cleaning ?

- A « standard » post-etch clean :
  - ❖ Batch sprays and Wet-benches
    - Cleaning process « BEOL like » with EKC solvents
    - Cleaning process « FEOL » with SPM_SC1_SC2
  - ❖ Transfer to Single wafer for defectivity reduction
    - « usual » recipes : SPM_SC1_SC2_Nanosprays
    - New recipes : HF/O3, …
- Almost same clean what ever the silicide CoSi / NiSi

However, a lot of recent evolutions were due to new materials introduction, patterning scheme evolutions, Contact A/R increase

Contact Cleaning Challenges

Cleaning Challenges due to new materials introduction :

- Metal Gate technologies → 28nm FDSOI & Silicide First Architecture
  → Gate metal etch/corrosion → SPM step removed
  → **SC1 + HCl only is one alternative**
- Mix of std contact landing on silicide with contact on sensitive materials :
  → W in stacked CT or W trenches for 14nm technologies
  → SPM & SC1 steps not compatible with W
  → **Need to evaluate other solutions**

Drying Challenges :

- Contact CD ↘ & Aspect Ratio ↗
- Q-Time management after cleaning
  → **IPA Drying evaluation**

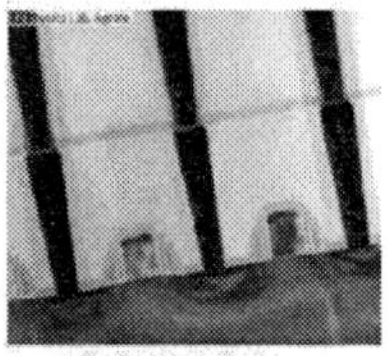

stacked contact

This work aims to :

- **compare cleaning efficiency, by using a High Yielding technology**
- **Investigate IPA Drying for high aspect ratio contacts**

Experimental _ Tool & Chemistries
5

- Chemistries & recipes
  - Reference clean : SPM _ SC1 _ SC1NS*, std spin Dry
  - "SC1" : SC1_SC1NS_HCl : std spin Dry & IPA Dry
  - "HF02" : dHF (0,2%wt.), DIW rinse with NS*, std spin Dry & IPA Dry
  - "GLYHF" : Mix HF0,025% + Glycolic acid 1% @ 60° C : std spin Dry
- Wet tool & options
  - Single wafer cleaning SU-3100
  - Central dispense + Nanospray for SC1
  - [O2] control over the wafer with Shieldplate @ low position for dHF
  - IPA Dry after SC1 or dHF

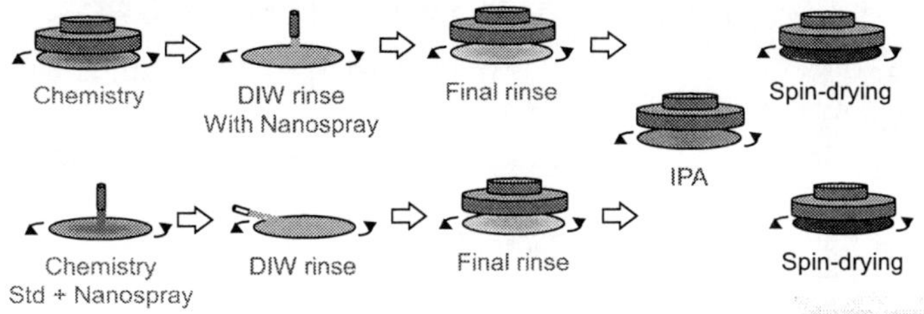

life.augmented SPCC2017_ Austin, March 2017 SCREEN

Experimental _ Materials and Technologies
6

- Wet etch-rate
  - NiSi thin film 25nm thick, measured by ellipsometry
  - SiO2 thin film (PMD stack), measured by ellipsometry (not shown here)
  - TiN, W thin films, measured by 4 point probe resistivity (not shown here)
  - Contact CD variation, measured by SEM_CD
- Technologies & Tests wafers

Techno type	Silicide type	Contact CD & A/R	Cleaning Process vs SPM_SC1 reference
C014	NiSi (Pt10%)	40nm	HF0,2%
C028	NiSi (Pt10%)	40nm	HF0,2% / GLYHF / SC1_HCL
C040	NiSi (Pt10%)	55nm max A/R = 7	HF0,2% / GLYHF / SC1_HCL

life.augmented SPCC2017_ Austin, March 2017 SCREEN

Results _ NiSi thickness & CD Variations

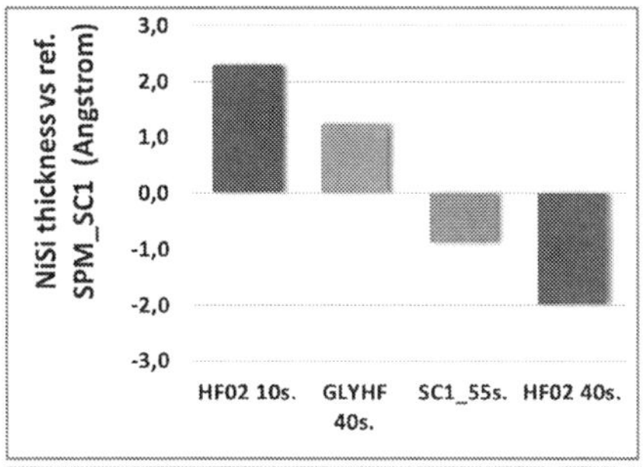

Silicide thickness variation vs Ref. SPM_SC1 (NiSi)

- ✓ Range -2 to +2 Å
- ✓ **Very small variations vs PoR**

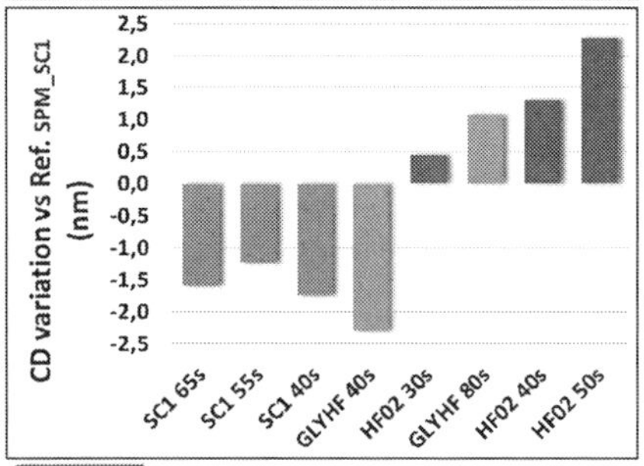

Contact CD Variations vs Ref. SPM_SC1 :

- ✓ **Range -2 to +2 nm**
- ✓ SC1 & low HF budget : CD < Ref.
- ✓ HF budget ↗ : CD > Ref. & CD ↗

CD variation sensitivity depends on technology node and possible adjustment to be done in Litho & Etch

life.augmented SPCC2017_ Austin, March 2017 SCREEN

Preliminary Results with HF0,2%

TEM view of 14nm contact landing on W trench, clean HF02, 50s :

Contact filling for TEM lamella preparation

TiN barrier

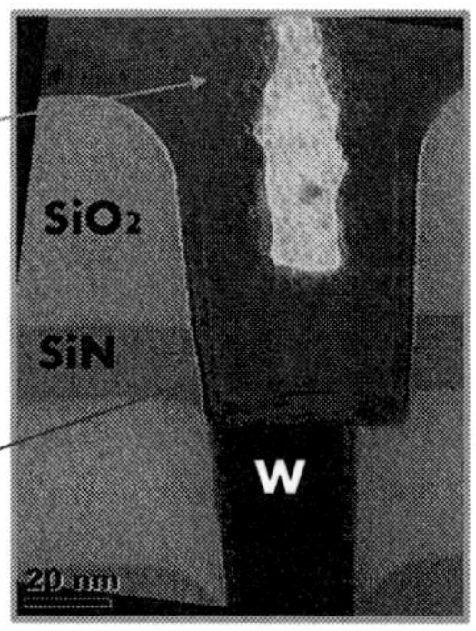

- ✓ No Oxide over - etch vs SiN
- ✓ No degradation of contact profile
- ✓ No over-etch on W trench

life.augmented SPCC2017_ Austin, March 2017 SCREEN

Preliminary Results _ cleaning wo SPM

9

Example of SRAM Yield mappings for 28nm technology (metal gate)

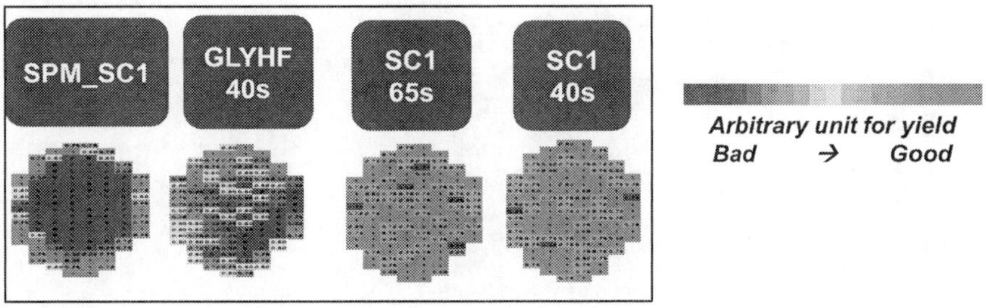

PoR strongly impacted by voids in Metal gate (metal etch by SPM)

- ✓ SC1 : Best Split
- ✓ GLYHF 40s. Poor cleaning → Yield loss

life.augmented *SPCC2017_ Austin, March 2017* SCREEN

Results : Cleaning efficiency comparison on High Yielding Technology (40nm node)

10

Example of SRAM Yield mappings for 40nm technology

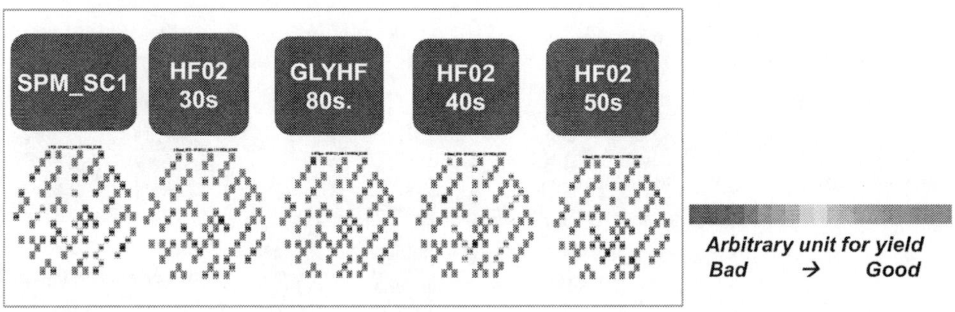

No materials compatibility concerns → Ok for cleaning comparison

- ✓ HF0,2% : good yields (30 – 50sec. Process)
- ✓ GLYHF 80s. = better than 40s.

life.augmented *SPCC2017_ Austin, March 2017* SCREEN

Results : Cleaning efficiency comparison on High Yielding Technology (40nm node)

11

40nm Contact resistance variation vs. Ref. (Contact on P+ active)

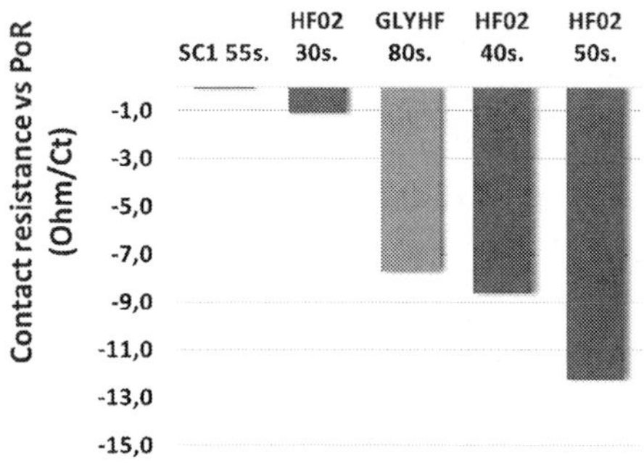

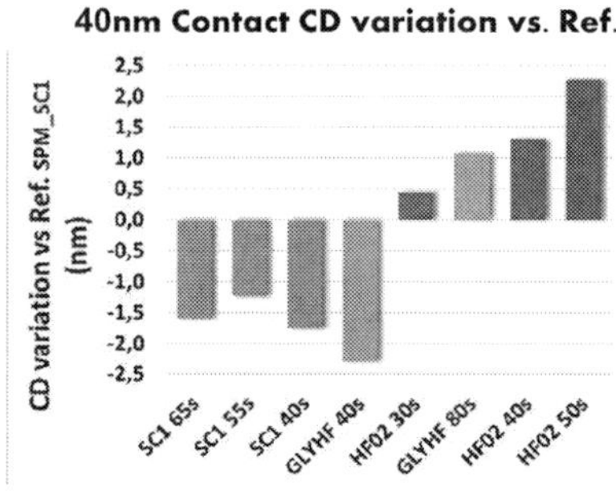

- ✓ **No Resistance increase vs PoR → confirm efficient clean**
- ✓ SC1 : No CD Loss & Good efficiency (R = same as PoR)
- ✓ GLYHF & HF02 : Resistance ↘ well correlated to CD ↗

life.augmented *SPCC2017_ Austin, March 2017* SCREEN

Results : IPA Drying efficiency

12

40nm Contact resistance variation vs. PoR. Contact std & High A/R contact

NEW Parameters :
- ✓ IPA DRY
- ✓ Contact A/R

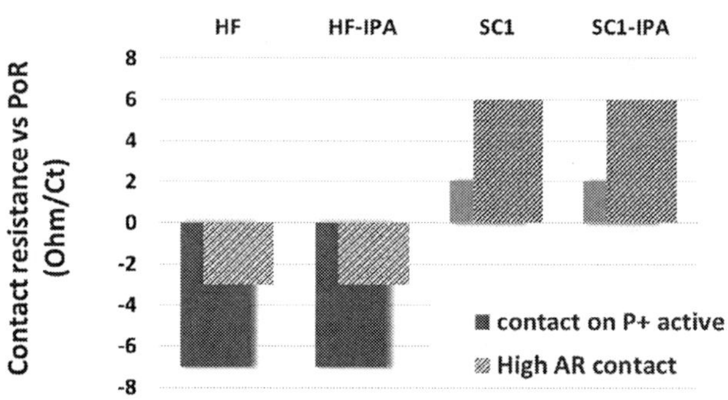

- ✓ No Resistance increase with IPA vs No IPA → No issue
- ✓ IPA dry implementation possible without process degradation

life.augmented *SPCC2017_ Austin, March 2017* SCREEN

Results : Qtime effect

13

NEW Parameter :
Qtime WET – TiN barrier dep

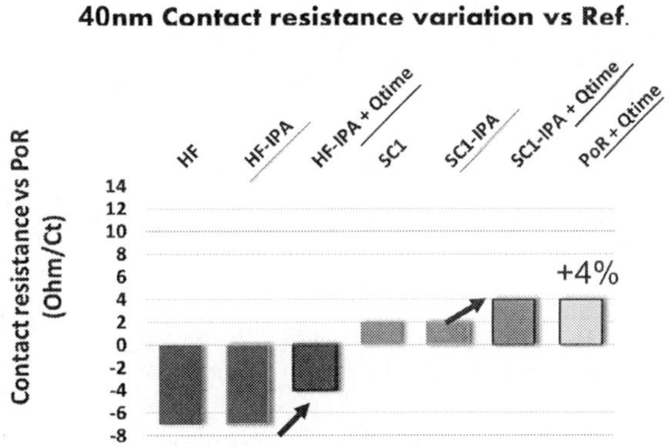

- ✓ **No drastic PoR degradation with Qtime (4%) → No clear conclusion**
  (maybe due to physical split effect with few wafers in the Foups)
- ✓ **Slight Resistance increase with Qtime (all wet process), same on high A/R contact**
- ✓ **No specific effect of IPA dry vs Qtime management on this lot**

Conclusion

14

- Single wafer tool improved cycles of learning for developing new cleaning recipes sequences
- New materials / new integration schemes requires changes in traditional wet cleans
  - SPM is no longer compatible
  - Diluted HF is an effective replacement but consideration is required for CD changes
- IPA drying is compatible although no benefits shown in this work
  - Good candidate for future technology nodes if N2 drying issues arise.

life.augmented *SPCC2017_ Austin, March 2017* SCREEN

Thank you for your attention.

life.augmented

SPCC2017_ Austin, March 2017

SCREEN

BS Poly Removal Defect Reduction

Hong Zhai, Colin Weidner, Dhiman Bhattacharyya, Norberto DeOliveria, and Vincent Sih

GLOBALFOUNDRIES
Malta, NY

Removal of edge cluster defects by improving recipe and hardware for backside polysilicon wet etch process

TOC

Table of contents:

1) Mission: repurposing an existing tool
2) Challenge: gap and performance
3) Approach and methodology
4) Current Result & status
5) Concluding remarks

2

PC: Particle Count

ER: Etch Rate

3

PC performance GOTO vs POR

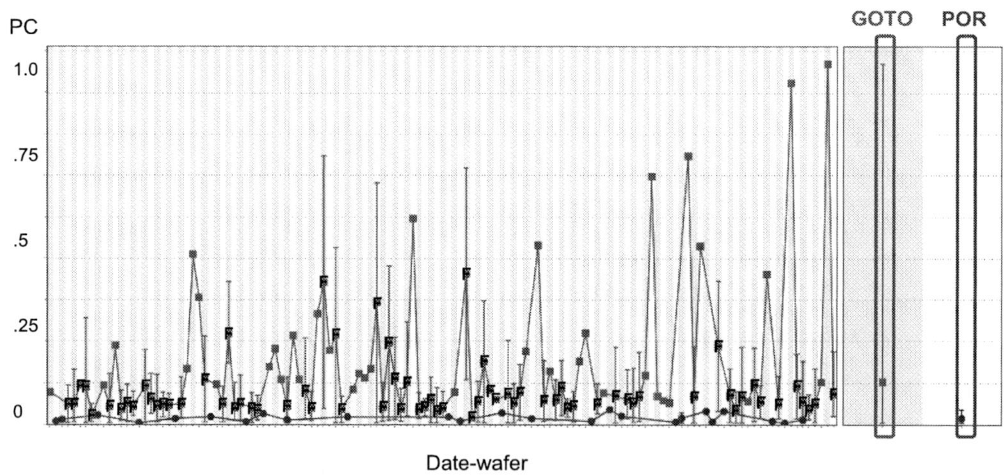

4

Half Year Later …

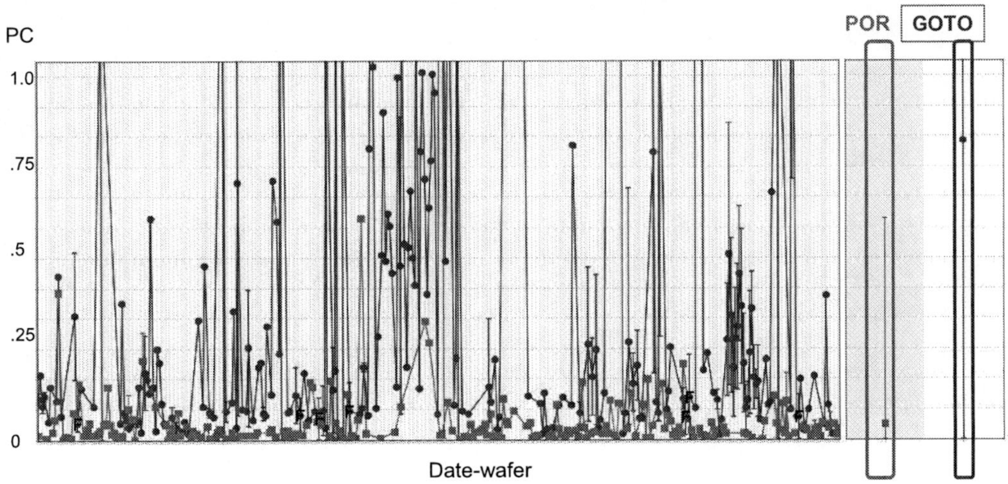

5

GOTO PC Qual

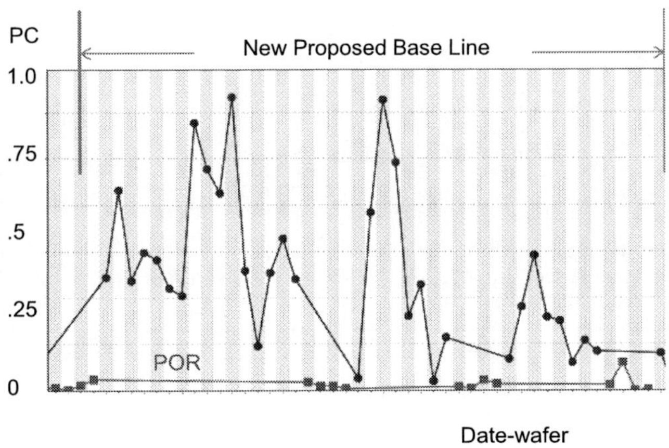

Stage of starting manufacture Qualification

7

Deep dive

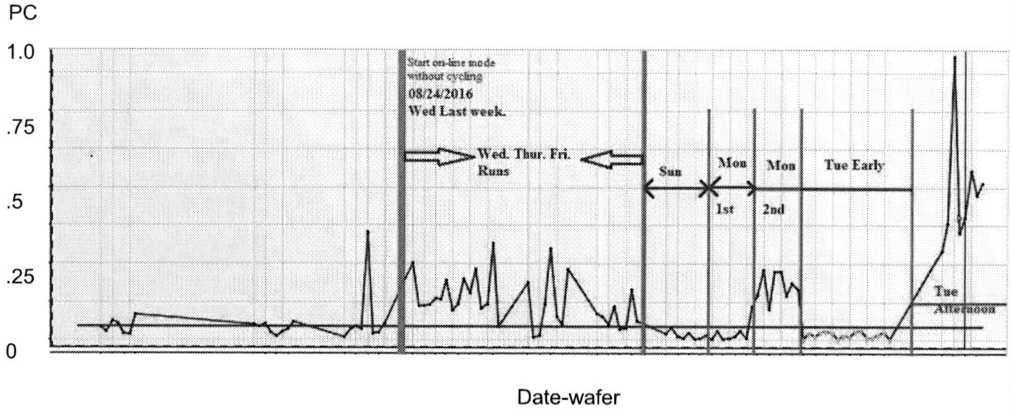

Re-Categorize the previous busy chart:

(1) All wafer check one by one for history: 88 past runs wfrs.

(2) Red boxed: all new Bare Si wfr.

(3) Black box from one tool family.

(4) Non-boxed from any other tools in chart.

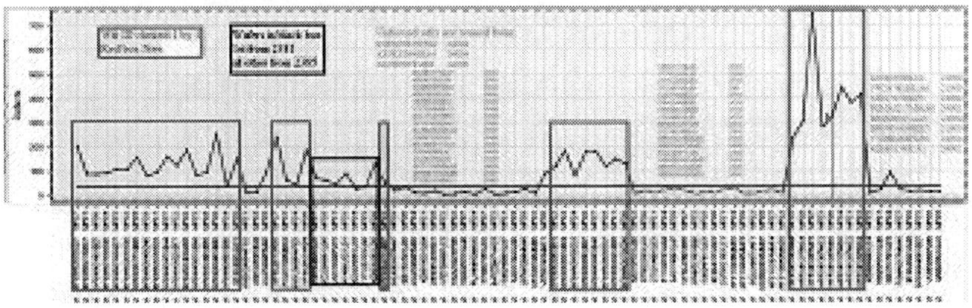

A few runs with controlled parameters

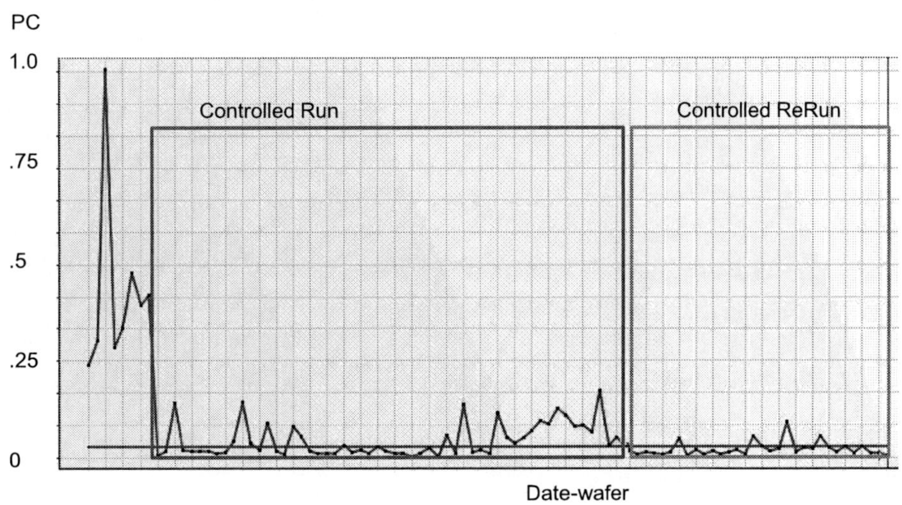

10

New Step

(1)Addition Pre Step.
(2)Pre Defect measurement.
(3)Process Step.
(4)POST Defect measurement.
(5)Additional Steps, like SEM etc

11

Two weeks watch after new plan implementation

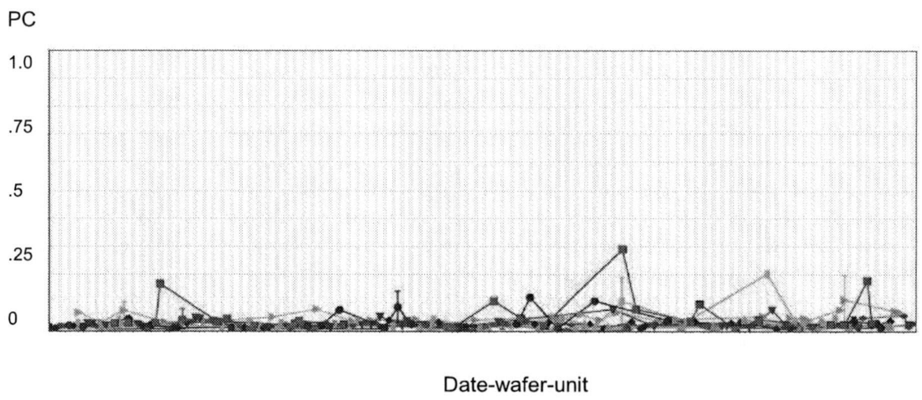

12

SUMMARY REVIEW

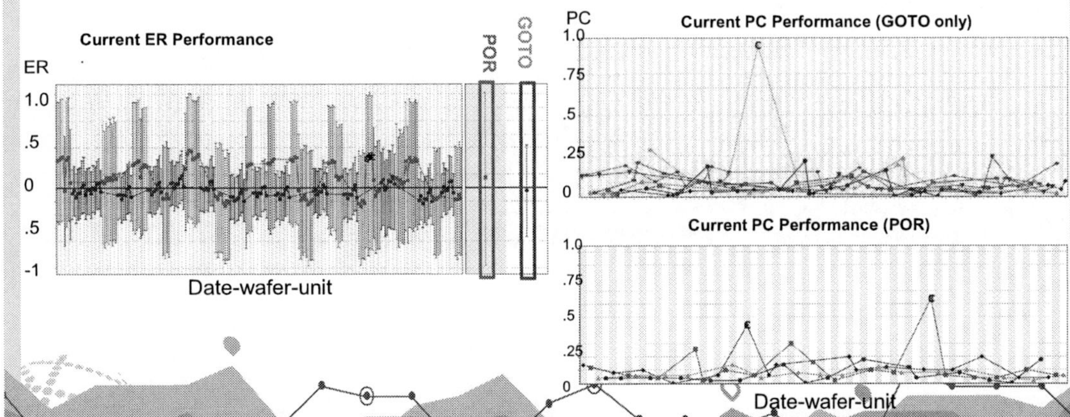

Check out this slide again!
Two weeks watch after new plan implementation
Edge cluster as Pin signature observed!

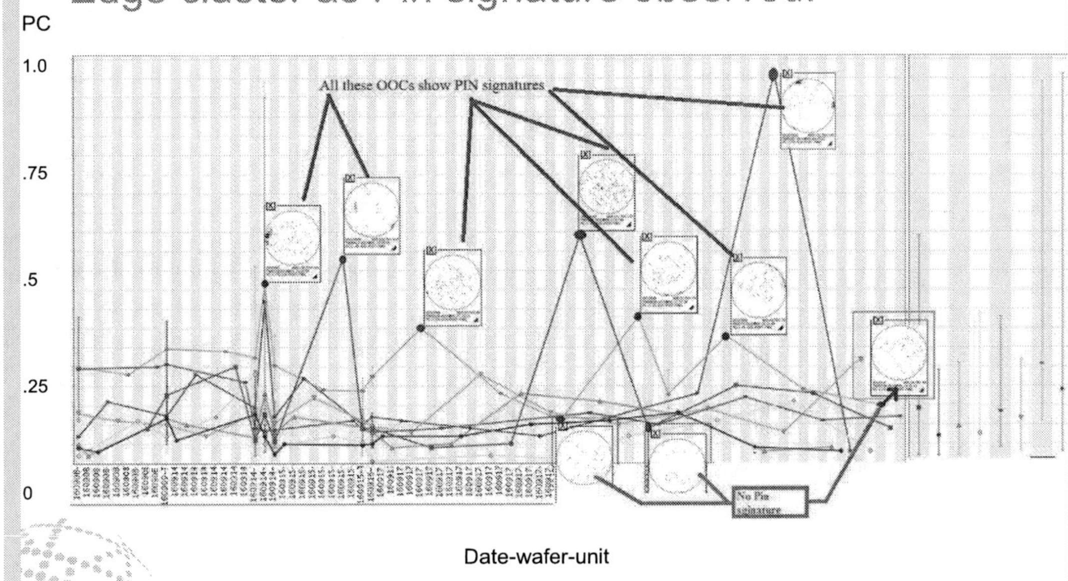

14

An typical example

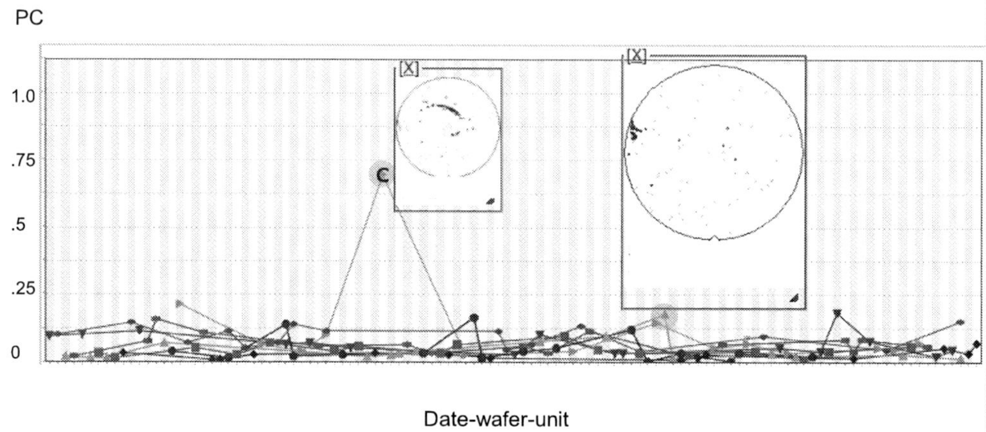

15

Actions taken and still continue…

Process

- Spin speed.
- Optimizing opening/closing the A and B chunk pins.
  1) Exploring possible water droplets
  2) Pins rubbing the wafer?
- A few different things:
- - Tool family sequence compare.
- - Adjust open/close sequences.
- - Perform rinse segmentation.

Hardware

- Handling and pocket size.
  To try:
  - Adjusting pocket size.
  - Perform Chuck A/B sequence

Recipe

- Current (Rev 12) vs Rev 14 (150% increase spin speed) vs Rev 15 (A-B Dry) options.

Pocket size:

- Water is pooling?
- 150% increased spin speed test.

Chuck Pin Open/Close

- Chuck Pin Open/Close vs A/B switching times.
- Two chambers showed promising data (reduced pin signature) with the pocket size at optimizing counts.
- Change another set of MPCs to different pocket size.
- Data that we gathered from these two pocket size.

16

Path forward:

(1)Pin Signature. GAP to close.
(2)Correlation Pin signature with Prod.
(3)Yield gain?

Concluding Remarks:

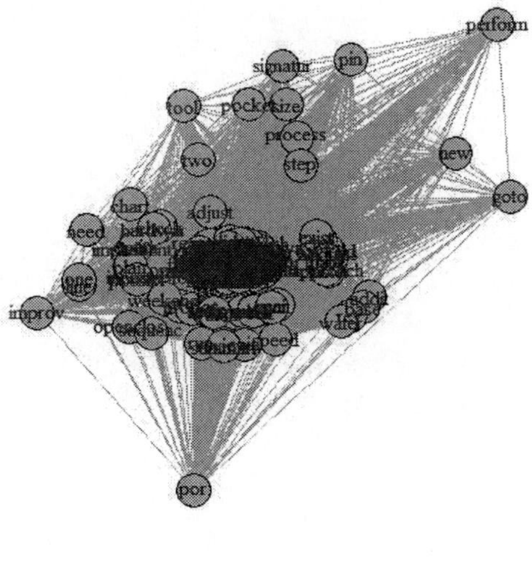

Thank you!

The Future of Micro-Contamination Control in Chemical Delivery Systems for Advanced Lithography, Wet Etch & Clean Semiconductor Processes

SPCC 2017

Dr. Archita Sengupta

Senior Technologist & Rel. MCSC Program Manager (Intel TMG_GSM)

Co-Authors: Thomas Phely-Bobin*, Paul Trio*, Robert McIntosh#

- Thomas Phely-Bobin*, Director, Applications Development and Metrology Group, Microcontamination Control Division, Entegris Inc.
- Paul Trio*, Senior Manager, Strategic Initiatives, SEMI
- # Robert McIntosh, Technical Marketing Manager Georg Fischer Piping Systems
- *: Members of the Semiconductor Components, Instruments and Systems (SCIS) special interest group, part of SEMI® organization, focusing on defectivity for Liquid Delivery Systems, SECC and SEMI Standard groups. *: Member of SEMI Organization

Acknowledgement:

Intel: Don Hadder, Kevin Pate,, James Staffa, Jeff Hemphill, Jeremy Moberg, Kevin Koga, John O'Sullivan, Thomas Quinn, Janice Golda, John Simmons, Sophia Wang, Jun Wang, Ted Jeffries.

Outside Intel: Tim Hendry, Abbas Rastegar, Intel POR Suppliers, SCIS team, Rao Varanasi, Debra Rafferty, Paul Trio, Thomas Phily-Bobin, Slava Libman, Rocky Gipson, Bob Halliday & Additional references are on the slides.

Overview

- Beyond 14nm: New Challenges with Supply line contamination and Nano-defectivity
- IDM Perspective of Problem Statement and End Goal
- Examples of Wafer Contamination Sources and Importance
- Addressing the Gaps:
  - Criticality of Metrology Requirement: Defect Identification/Characterization
  - Supply Chain Responsibility & Quality Control
- Need for Industry Collaboration:
  - Joint Responsibility
  - Expectation from POR Suppliers
- End remarks

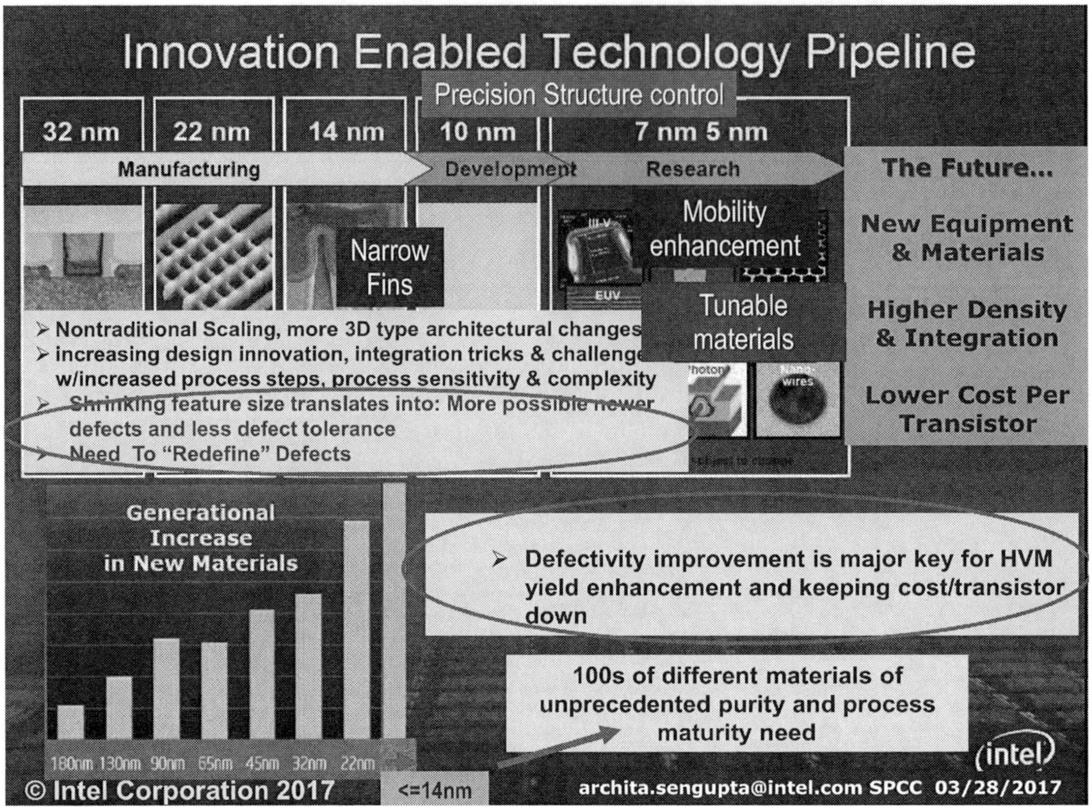
Innovation Enabled Technology Pipeline
Precision Structure control
32 nm
22 nm
14 nm
10 nm
7 nm 5 nm
Manufacturing
Development
Research
The Future...
New Equipment & Materials
Higher Density & Integration
Lower Cost Per Transistor
Narrow Fins
Mobility enhancement
Tunable materials
EUV
➢ Nontraditional Scaling, more 3D type architectural changes
➢ increasing design innovation, integration tricks & challenge w/increased process steps, process sensitivity & complexity
➢ Shrinking feature size translates into: More possible newer defects and less defect tolerance
➢ Need To "Redefine" Defects
Generational Increase in New Materials
180nm 130nm 90nm 65nm 45nm 32nm 22nm
<=14nm
➢ Defectivity improvement is major key for HVM yield enhancement and keeping cost/transistor down
100s of different materials of unprecedented purity and process maturity need
© Intel Corporation 2017
archita.sengupta@intel.com SPCC 03/28/2017

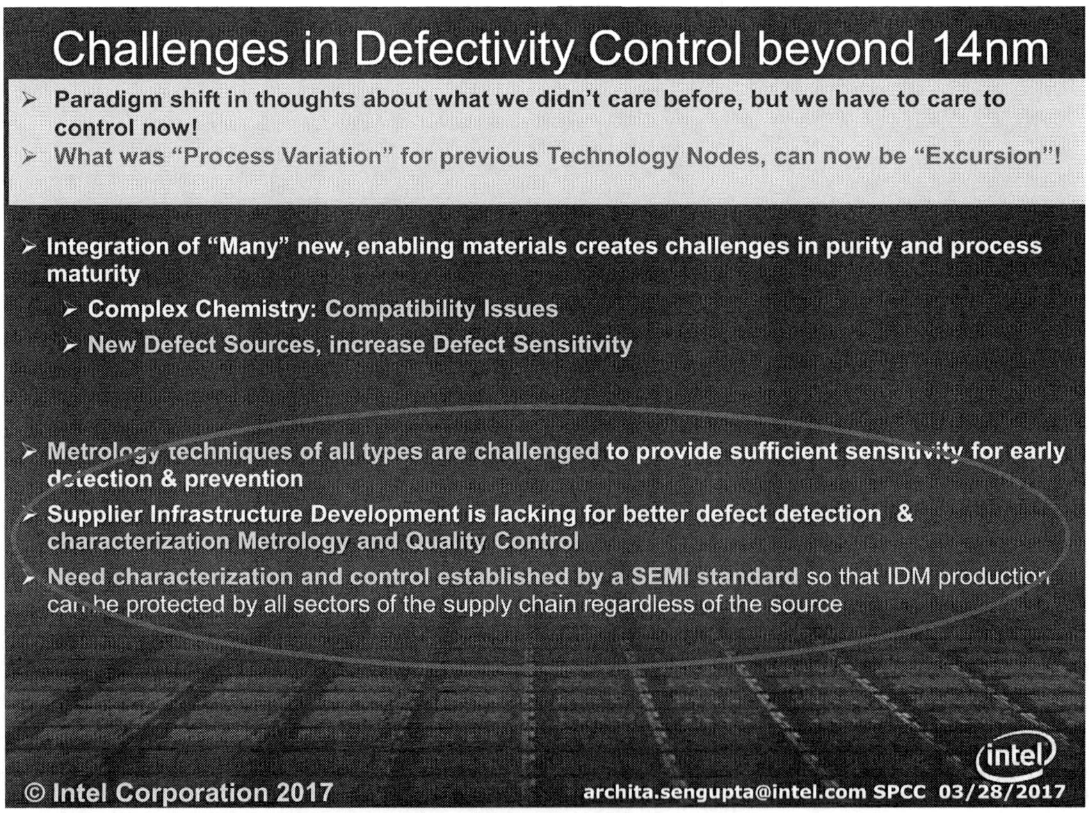
Challenges in Defectivity Control beyond 14nm
➢ Paradigm shift in thoughts about what we didn't care before, but we have to care to control now!
➢ What was "Process Variation" for previous Technology Nodes, can now be "Excursion"!
➢ Integration of "Many" new, enabling materials creates challenges in purity and process maturity
➢ Complex Chemistry: Compatibility Issues
➢ New Defect Sources, increase Defect Sensitivity
➢ Metrology techniques of all types are challenged to provide sufficient sensitivity for early detection & prevention
➢ Supplier Infrastructure Development is lacking for better defect detection & characterization Metrology and Quality Control
➢ Need characterization and control established by a SEMI standard so that IDM production can be protected by all sectors of the supply chain regardless of the source
© Intel Corporation 2017
archita.sengupta@intel.com SPCC 03/28/2017

HVM Requirements

- **Consistency/Predictability of and Less Start Up Time:**
  - Less Flushing for initial particles and metal ions extraction
  - Always Up and Running tools/Chemical Delivery Systems
  - **Need a very tight distribution of performance to bring up new tools**
- **Less Process Down Time:** Productivity Improvement
  - **Reduce excursion** of particles/metal ions
  - **Environmental Control** (No adders from components)
- **Meet "On wafer" Defectivity Baseline** (Process/Materials/Equipment)
  - Smaller particles
  - Metal ions
  - Organics/NVR
- **Consistency and Quality Control of "All" the products for HVM use**
  - **Across Supply Chain**

 SPCC 03/28/2017

Global Control Needed for Defectivity

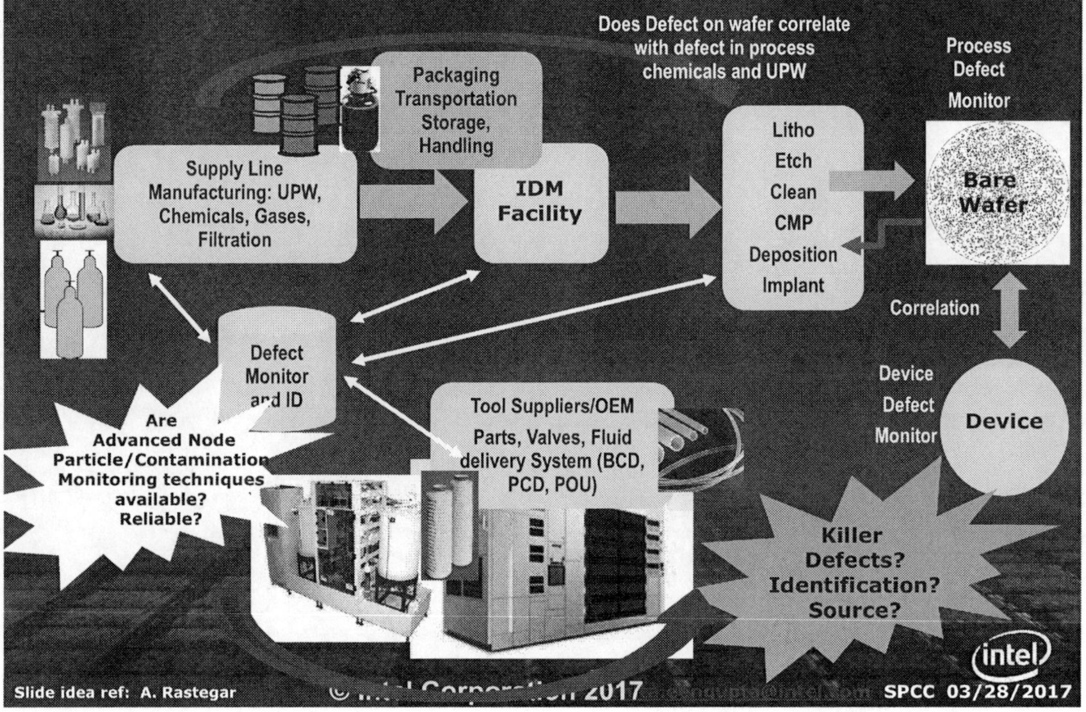

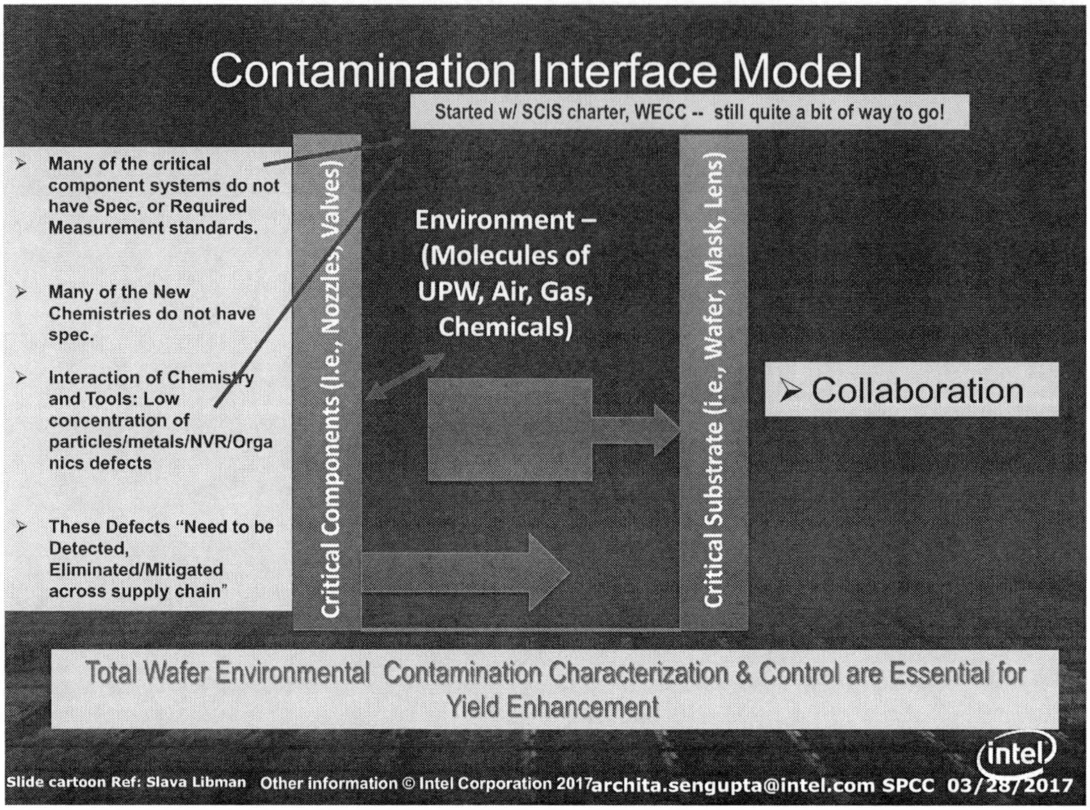
Contamination Interface Model
Started w/ SCIS charter, WECC -- still quite a bit of way to go!
Many of the critical component systems do not have Spec, or Required Measurement standards.
Many of the New Chemistries do not have spec.
Interaction of Chemistry and Tools: Low concentration of particles/metals/NVR/Organics defects
These Defects "Need to be Detected, Eliminated/Mitigated across supply chain"
Critical Components (I.e., Nozzles, Valves)
Environment – (Molecules of UPW, Air, Gas, Chemicals)
Critical Substrate (i.e., Wafer, Mask, Lens)
Collaboration
Total Wafer Environmental Contamination Characterization & Control are Essential for Yield Enhancement
intel
Slide cartoon Ref: Slava Libman Other information © Intel Corporation 2017 archita.sengupta@intel.com SPCC 03/28/2017

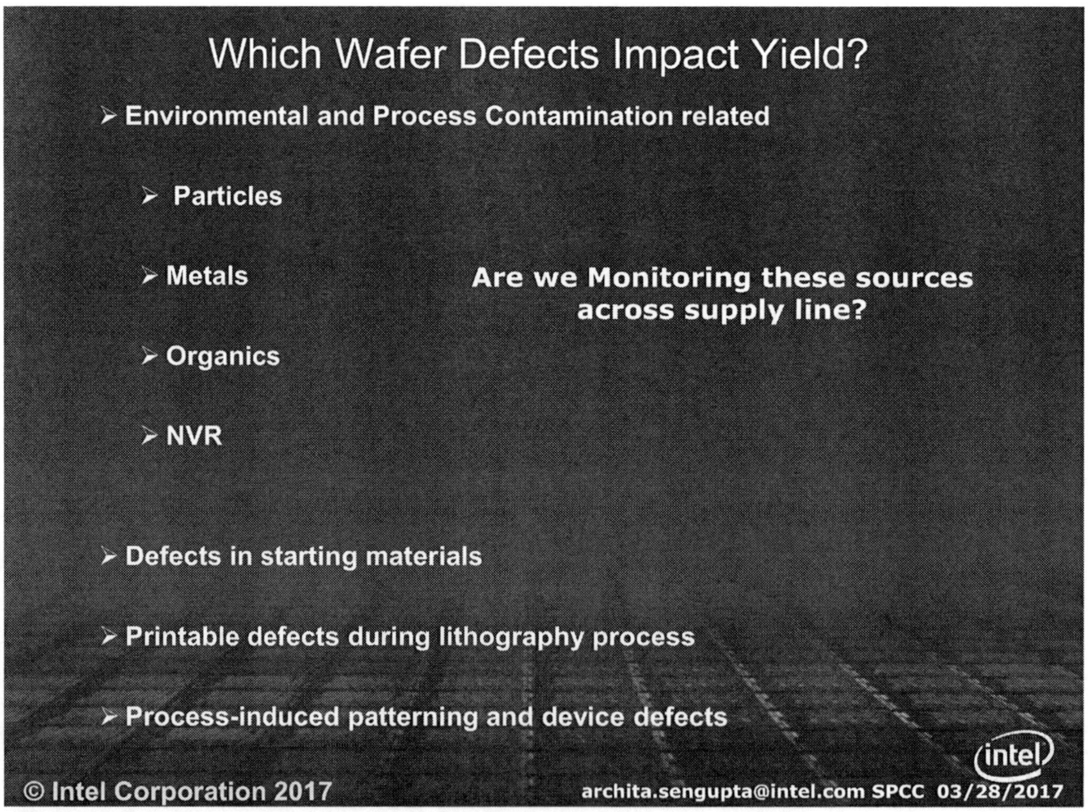
Which Wafer Defects Impact Yield?
Environmental and Process Contamination related
Particles
Metals
Organics
NVR
Are we Monitoring these sources across supply line?
Defects in starting materials
Printable defects during lithography process
Process-induced patterning and device defects
intel
© Intel Corporation 2017
archita.sengupta@intel.com SPCC 03/28/2017

Example of Metrology Challenges: Chems

Material	Major Challenge	Gap in Metrology for Control	Gap in Metrology for Characterization
Wet chems and formulated chems, Solvents, Litho chem	• Residues (Trace Metals/organics/Particles) • Interaction w/critical components	• Main component assay is insufficient for custom blend chem • Filtration may not be sufficient? • LPC: HVM monitor needs <30nm bin size.	• Specific detectors to enhance detectability • High resolution to help ID the unknown • ICPMS DL, MS techniques to provide better ID capability.
CMP	• MPS trending <<100nm, • Complex interactions between chemicals/consumables: Filter+Nano-particle interaction	• Performance-based monitors may be required • Normal impurity profile variation of commonly used Industrial-grade additives no longer acceptable	• Improved analytical metrology to characterize abrasive colloids (<=20nm sized particle, a minute mass fraction of the hetero-dispersed abrasive PSD) • Traditional CofA parameters not sufficient to predict performance

 archita.sengupta@intel.com SPCC 03/28/2017

Defects caused by Chemical Handling and Delivery System (Chemical Pass)

- **As chemical travels within any dispense system, contamination is added (leaching and/or reaction with components)**
- PFA Tubing → May lead to "Metal Leaching"

Changing the Industry: New SEMI C90-1015 standard has been published: for limiting the amount of Fe in PFA materials (tube, valves, fittings, resin) used in the semiconductor industry for liquid chemical distribution.(Oct 2015)

Over time:
- **Metal Leaching**
- **Deposition of particles/organics (from hydrodynamic boundary layer)**
- **Dislodgement of particles**

Chemical Entering (Post filtration)

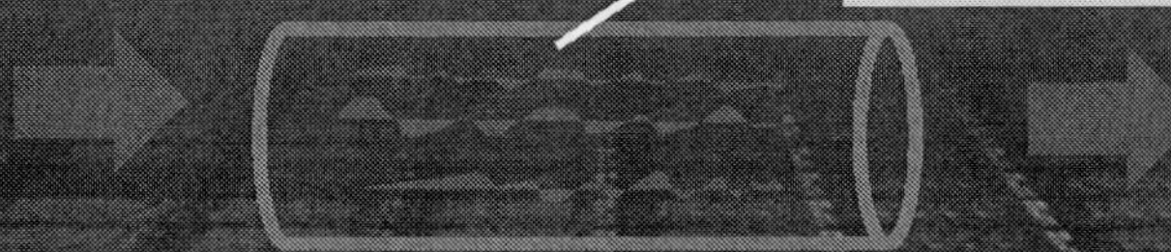

Chemical Exit (PM, Dispense rate, idle time etc. determine the residence time)

 archita.sengupta@intel.com SPCC 03/28/2017

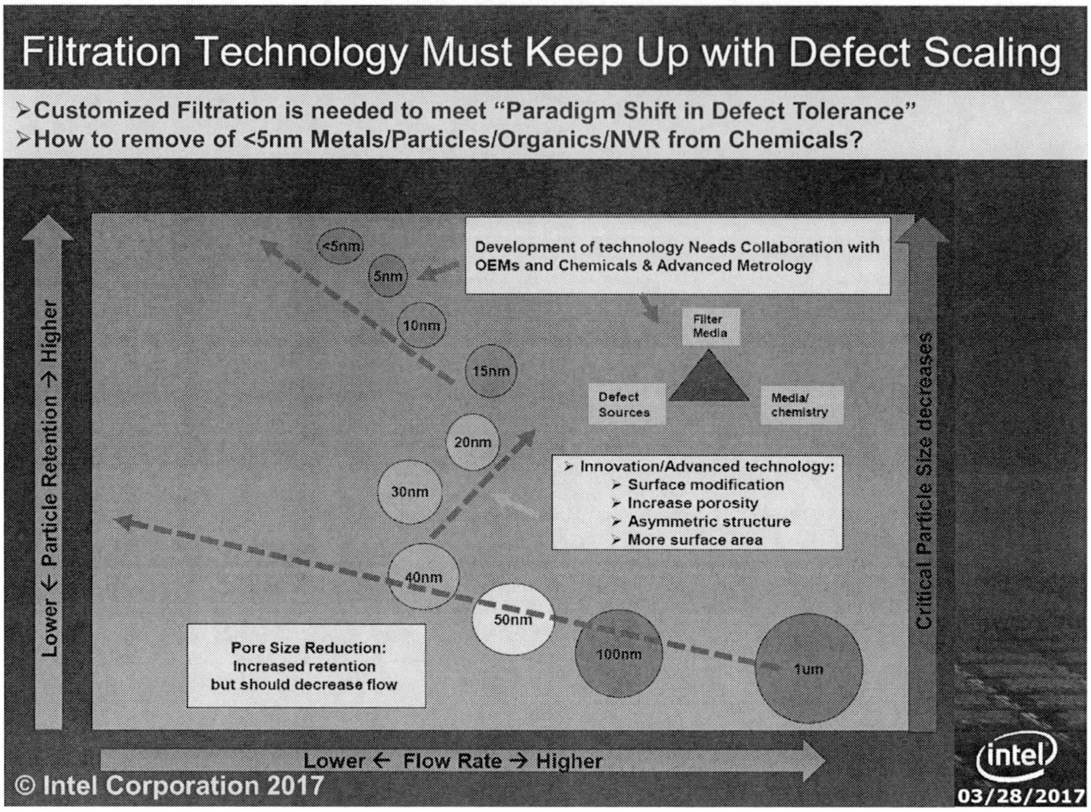
Filtration Technology Must Keep Up with Defect Scaling
➢Customized Filtration is needed to meet "Paradigm Shift in Defect Tolerance"
➢How to remove of <5nm Metals/Particles/Organics/NVR from Chemicals?
Development of technology Needs Collaboration with OEMs and Chemicals & Advanced Metrology
Filter Media
Defect Sources
Media/ chemistry
➢ Innovation/Advanced technology:
➢ Surface modification
➢ Increase porosity
➢ Asymmetric structure
➢ More surface area
<5nm
5nm
10nm
15nm
20nm
30nm
40nm
50nm
100nm
1um
Pore Size Reduction: Increased retention but should decrease flow
Lower ← Particle Retention → Higher
Critical Particle Size decreases
Lower ← Flow Rate → Higher
© Intel Corporation 2017
intel
03/28/2017

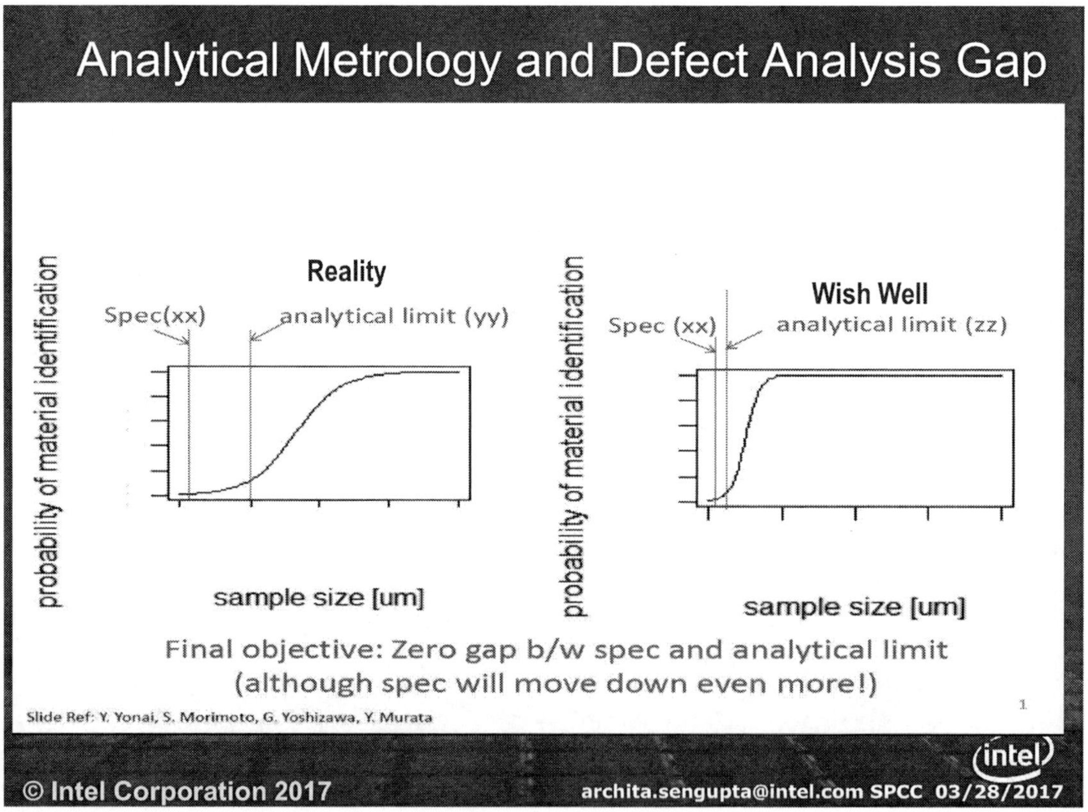
Analytical Metrology and Defect Analysis Gap
Reality
Spec(xx)
analytical limit (yy)
probability of material identification
sample size [um]
Wish Well
Spec (xx)
analytical limit (zz)
probability of material identification
sample size [um]
Final objective: Zero gap b/w spec and analytical limit
(although spec will move down even more!)
Slide Ref: Y. Yonai, S. Morimoto, G. Yoshizawa, Y. Murata
1
© Intel Corporation 2017
intel
archita.sengupta@intel.com SPCC 03/28/2017

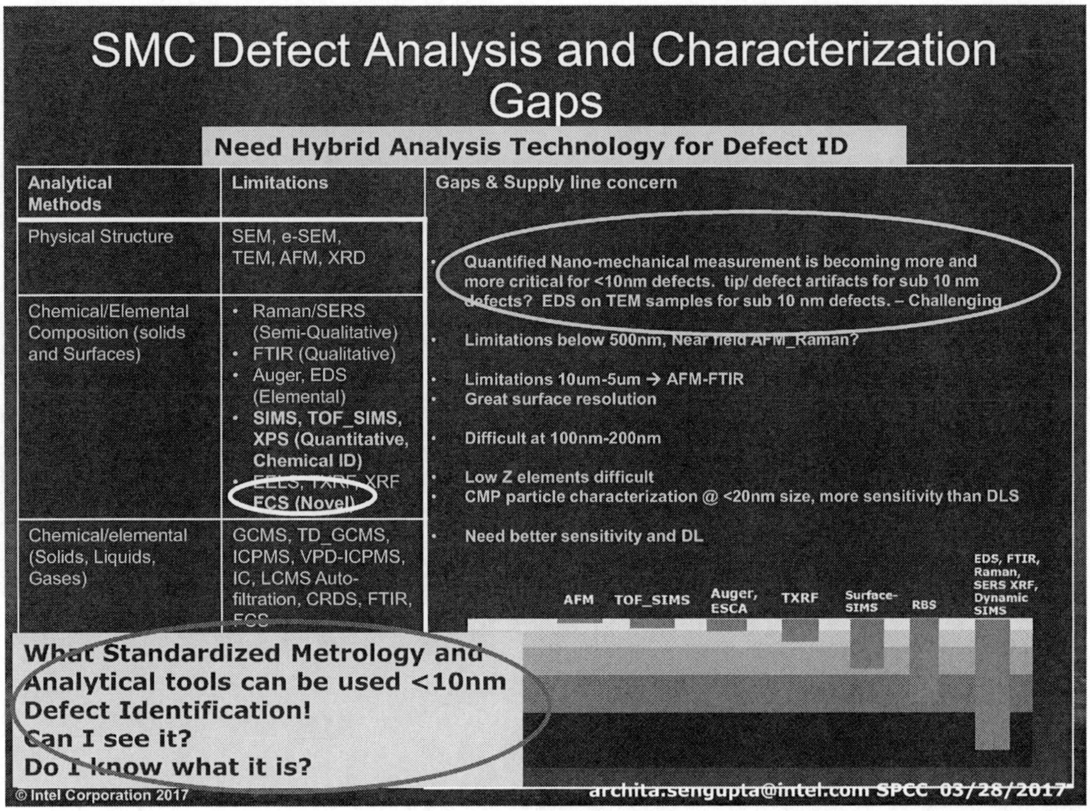
SMC Defect Analysis and Characterization Gaps
Need Hybrid Analysis Technology for Defect ID
Analytical Methods
Limitations
Gaps & Supply line concern
Physical Structure
SEM, e-SEM, TEM, AFM, XRD
Quantified Nano-mechanical measurement is becoming more and more critical for <10nm defects. tip/ defect artifacts for sub 10 nm defects? EDS on TEM samples for sub 10 nm defects. – Challenging
Chemical/Elemental Composition (solids and Surfaces)
Raman/SERS (Semi-Qualitative)
FTIR (Qualitative)
Auger, EDS (Elemental)
SIMS, TOF_SIMS, XPS (Quantitative, Chemical ID)
EELS, TXRF, XRF
FCS (Novel)
Limitations below 500nm, Near field AFM_Raman?
Limitations 10um-5um → AFM-FTIR
Great surface resolution
Difficult at 100nm-200nm
Low Z elements difficult
CMP particle characterization @ <20nm size, more sensitivity than DLS
Chemical/elemental (Solids, Liquids, Gases)
GCMS, TD_GCMS, ICPMS, VPD-ICPMS, IC, LCMS Auto-filtration, CRDS, FTIR, FCS
Need better sensitivity and DL
AFM
TOF_SIMS
Auger, ESCA
TXRF
Surface-SIMS
RBS
EDS, FTIR, Raman, SERS XRF, Dynamic SIMS
What Standardized Metrology and Analytical tools can be used <10nm Defect Identification!
Can I see it?
Do I know what it is?
© Intel Corporation 2017
archita.sengupta@intel.com SPCC 03/28/2017

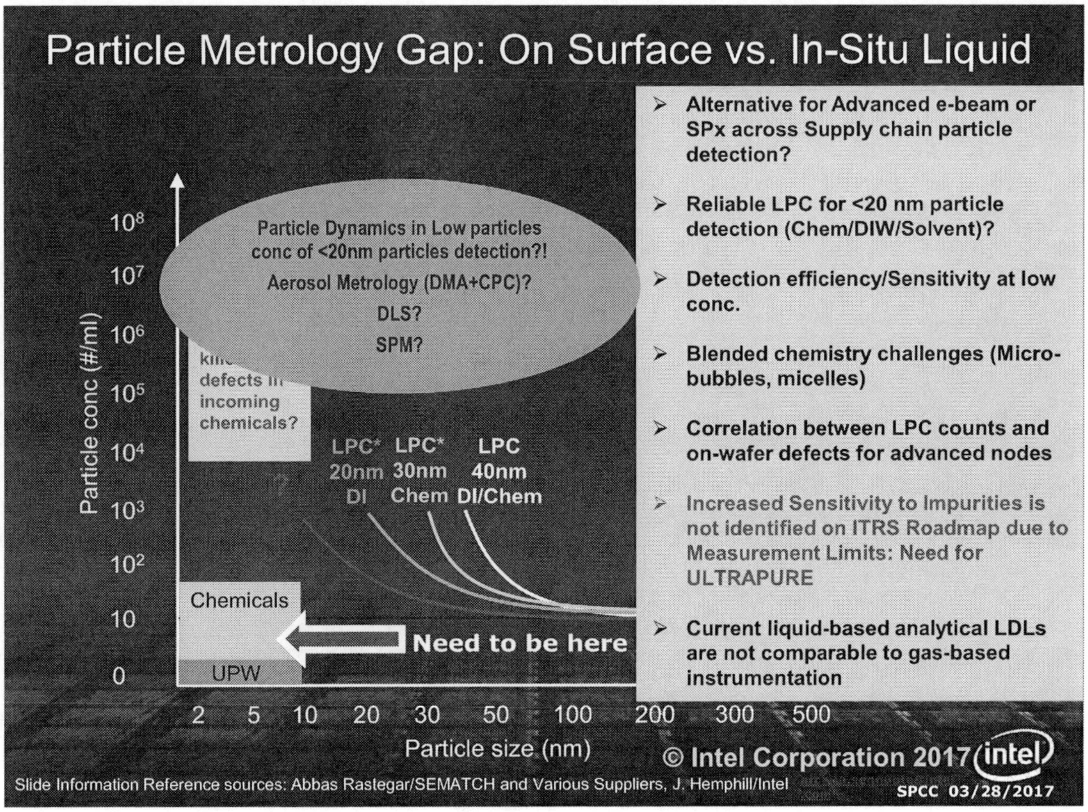
Particle Metrology Gap: On Surface vs. In-Situ Liquid
Particle Dynamics in Low particles conc of <20nm particles detection?!
Aerosol Metrology (DMA+CPC)?
DLS?
SPM?
defects in incoming chemicals?
LPC* 20nm DI
LPC* 30nm Chem
LPC 40nm DI/Chem
Chemicals
UPW
Need to be here
Particle conc (#/ml)
10^8
10^7
10^6
10^5
10^4
10^3
10^2
10
0
2 5 10 20 30 50 100 200 300 500
Particle size (nm)
Alternative for Advanced e-beam or SPx across Supply chain particle detection?
Reliable LPC for <20 nm particle detection (Chem/DIW/Solvent)?
Detection efficiency/Sensitivity at low conc.
Blended chemistry challenges (Micro-bubbles, micelles)
Correlation between LPC counts and on-wafer defects for advanced nodes
Increased Sensitivity to Impurities is not identified on ITRS Roadmap due to Measurement Limits: Need for ULTRAPURE
Current liquid-based analytical LDLs are not comparable to gas-based instrumentation
© Intel Corporation 2017
intel
Slide Information Reference sources: Abbas Rastegar/SEMATCH and Various Suppliers, J. Hemphill/Intel
SPCC 03/28/2017

How do we Control/Prevent the Contaminants?

1st Request: Three major control of the contaminants before ship the "Best" products to the factory:

- The Materials
- The Equipment
- The Components

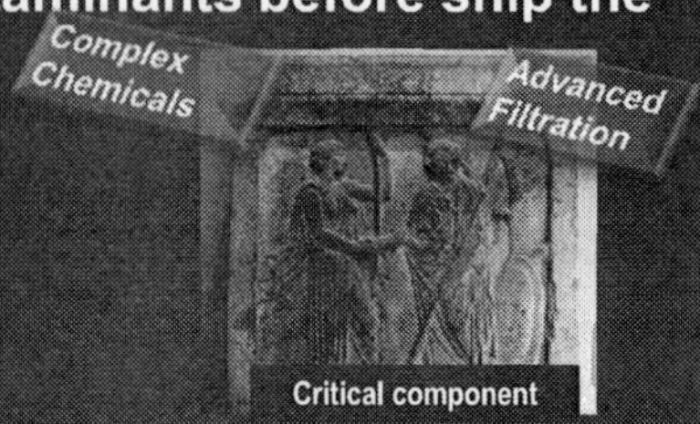

2nd Request: We Cannot Evaluate Advanced Node Chemistry/Process Without Aggressive "Super-Filtration And Purification"

- Need ultrapure materials, extreme retention, lowest possible interaction
- Optimized at chemical manufacturing and at factory processing/tools

3rd Request: Help us closing the Metrology/Analysis/Standard Gaps

4th Request: Work Together: Collaboration

Picture Ref: Hera & Athena handshaking, late 5th century BC, Acropolis Museum, Athens

archita.sengupta@intel.com SPCC 03/28/2017

Semiconductor Ecosystem Must Work Together

- Increasing interdependency with process complexity:
- Materials and Equipment Specs : **Process Control And Traceability**
- Need to re-write "Advanced Semi-Standards" for Measurement

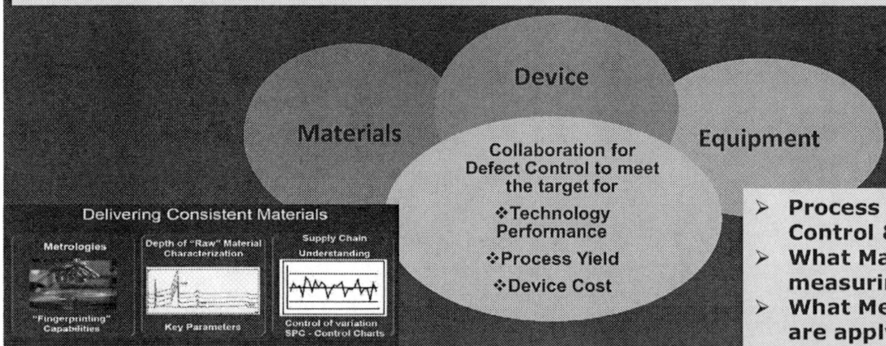

- Process is as good as Control & Baseline
- What Matrix are we measuring?
- What Methodology we are applying?

- Need Complete Supply Line Visibility and High Level of Trust
  - Develop Joint Analytical capabilities for Defect Detection and ID
- *Success requires collaboration across all supply chain partners*

Industry Must Collaborate on ONE PLATFORM to redefine Materials & Equipment "Specs & Metrology Standards" for <=14nm Node

archita.sengupta@intel.com SPCC 03/28/2017

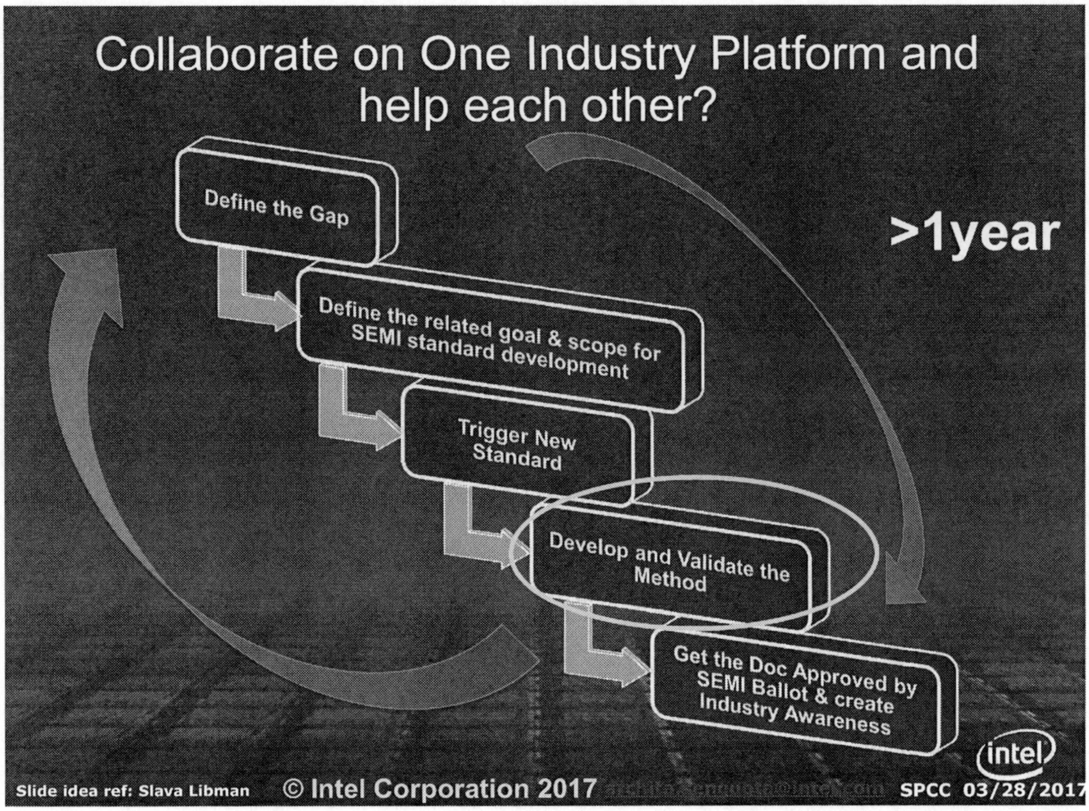
Collaborate on One Industry Platform and help each other?
Define the Gap
Define the related goal & scope for SEMI standard development
Trigger New Standard
Develop and Validate the Method
Get the Doc Approved by SEMI Ballot & create Industry Awareness
>1year
intel
Slide idea ref: Slava Libman
© Intel Corporation 2017
SPCC 03/28/2017

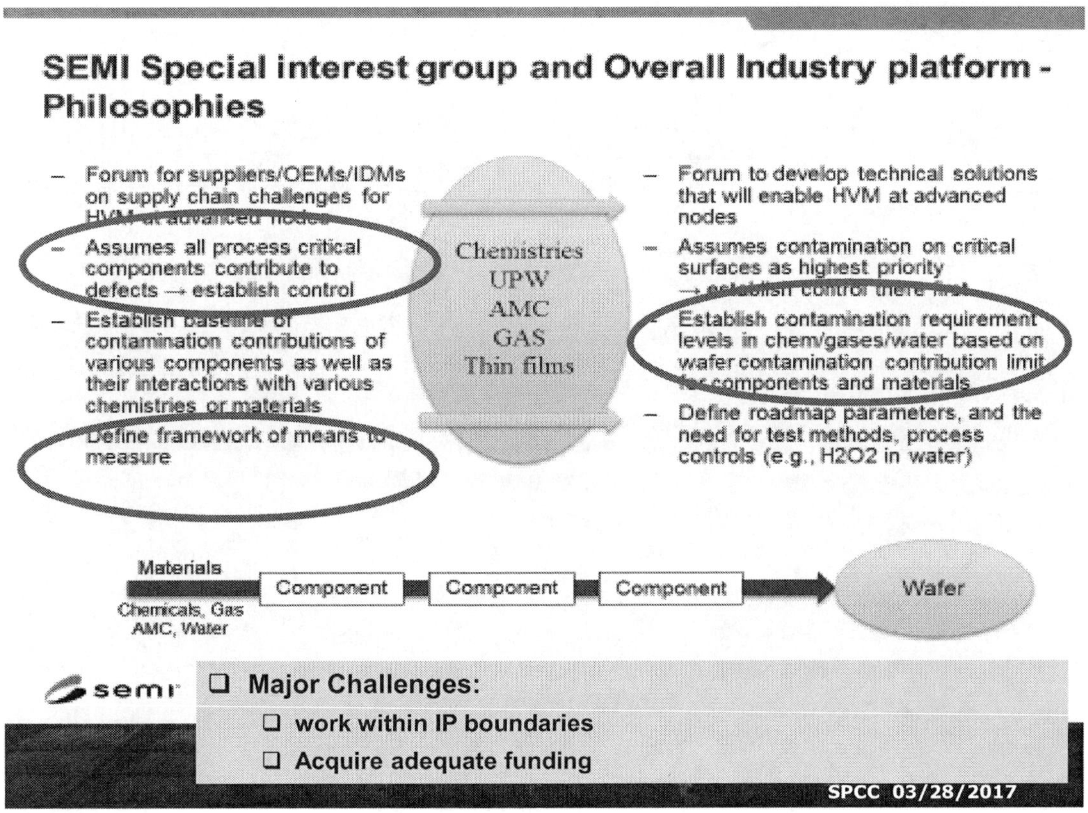
SEMI Special interest group and Overall Industry platform - Philosophies
– Forum for suppliers/OEMs/IDMs on supply chain challenges for HVM at advanced nodes
– Assumes all process critical components contribute to defects → establish control
– Establish baseline of contamination contributions of various components as well as their interactions with various chemistries or materials
– Define framework of means to measure
Chemistries
UPW
AMC
GAS
Thin films
– Forum to develop technical solutions that will enable HVM at advanced nodes
– Assumes contamination on critical surfaces as highest priority → establish control there first
– Establish contamination requirement levels in chem/gases/water based on wafer contamination contribution limit for components and materials
– Define roadmap parameters, and the need for test methods, process controls (e.g., H2O2 in water)
Materials
Chemicals, Gas
AMC, Water
Component
Component
Component
Wafer
semi
❑ Major Challenges:
❑ work within IP boundaries
❑ Acquire adequate funding
SPCC 03/28/2017

SEMI Component, Instrument and Subsystem (SCIS) Special Interest Group

- Goal:
- To Create Standard System of comparable Metrics which will be used to rate, compare and classify process-critical OEM components in order to reduce defects generated in Semiconductor manufacturing

- Establish a general "framework" to guide Industry partners to
  - Define observable/measurable defects
  - Define comparable category attributes specific to each category component
- Define methods for identifying defects related to each category
- Guide industry to utilize the framework

SCIS: Critical Component-OEM_End User Team

- **Components** such as seals, filters, mass flow controllers, valves, sensors, ion beam sources, etc.
- **Instruments** for in-line and off-line data measurement, collection, and monitoring
- **Sub-systems** that support process tools e.g. vacuum, robotics, power conversion, abatement, chillers, etc.

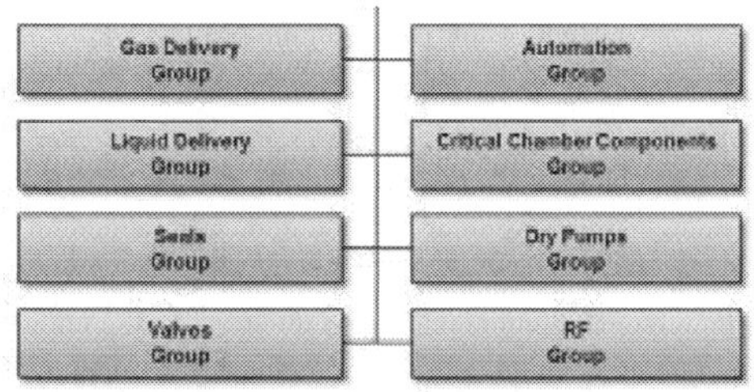

Slide Ref: Paul Trio, D. Rafferty, D. Vernikovsky

SPCC 03/28/2017

SCIS – SEMI – Other Industry Platforms Alignment
From Technology Gaps to SEMI Standards

SEMI SCIS Roles
- **Provide supply chain inputs** on critical components and related parameters that need measurement methods established
- **Define framework for component characterization**
- **Implement developed SEMI Standards** throughout supply chain, identify gaps and new requirements

SEMI SCIS

Provide regular updates, reporting on progress and gaps to maintain alignment

SEMI Standards ⟷ Other Industry platforms

- Other Industry Platforms
  - **Develop test methods** for measuring materials defect contribution as well as measuring performance of related components
  - **Obtain SCIS inputs on measurement methods** in development
  - **Conduct testing/experiments** to validate measurement methods

- **SEMI Standards Roles**
  - **Establish appropriate task force, develop/publish standards**

SPCC 03/28/2017

SCIS – SEMI TASK Force (TF)

• SEMI F57 was renamed in July 2016 to: High Purity Polymer Materials & Components TF

SEMI Standard TF	Final Goal	Contamination	Current Status
F57 – is the standard for contaminants in critical components as measured by F40	Revise Standard material to meet F63 standard - Include filtration	Ions (Fluoride first), metals, and TOC	- Testing is ongoing - Plans ready for Ballot late 2017
F104 – is the method for measuring particle contribution in critical components	Revise F104 to meet particle leaching requirements	Particles (static and dynamic testing) Considering TOC and NVR	SNARF will be submitted
F40 – is the method to measure extractable contaminants from critical components.	Rewrite adequate test methods for testing - Include filtration	Inorganic, Anions, TOC, NVR	Under discussion

semi

- **Definition of "Surface Contamination & High Surface Area components" – high purity polymers, high flow components**
- **Dynamic Rinse vs. Static Rinse contribution**
- **Definition of "Particles" (& Size to meet <=7nm Node requirement!) And Beyond Particles (TOC, NVR)**
- **F63: Gap in definition due to lack of metrology @ killer defects!**

SCIS Liquid Delivery Team Charter

Action	*Update and Next steps*
Complete OPC Technology considerations for particle contribution	Ongoing
Particle measurement for high performance polymer parts – Filters included	Revision of F57/104 for dynamic testing for all high performance polymer parts (update on F57) Update F104 for particle shedding of filters (reference to C79)
Ions, metals and TOC	Incorporation of Filters in F57 revision
Organic extractables	Team is discussing specific DOE

Alignment with SEMI F57, other SEMI Task Force and Other Industry Platforms

- Revision of F57 for Ions, metals and TOC (incorporation of Filters) filter performance for sub 15nm particle retention) *for leachable*
- Revision of F104 is being discussed – Static and dynamic particle shedding of *critical components* (Reference to C79 - Method to evaluate filters performance)
- Alignment with Chemical/UPW SEMI task forces

semi

intel

SPCC 03/28/2017

Success Story of Collaboration to Change SEMI Standard: SEMI C90-1015

- Changing the Industry – One Step at a time.
- Led by Intel - New industry standard for measuring/limiting the amount of Iron contamination in PFA materials (tube, valves, fittings, resin) used in the semiconductor industry for liquid chemical distribution has been published by the SEMI organization (2015)
- Next, defining the frequency of the test and the reporting medium for the results so as to monitor and manage the supply chain
- Many suppliers were involved with this new standard development and have adopted transition

semi

SEMI C90-1015
TEST METHOD AND SPECIFICATION FOR TESTING PERFLUOROALKOXY (PFA) MATERIALS USED IN LIQUID CHEMICAL DISTRIBUTION SYSTEMS

This Standard was technically approved by the Liquid Chemicals Global Technical Committee. This edition was approved for publication by the global Audits and Reviews Subcommittee on August 31, 2015. Available at www.semiviews.org and www.semi.org in October 2015.

1 Purpose

1.1 This Document defines perfluoroalkoxy (PFA) resin and component preparation and pretreatment procedures for chemical test methods used to evaluate liquid chemical distribution system components and resin.

intel

SPCC 03/28/2017

END REFLECTION

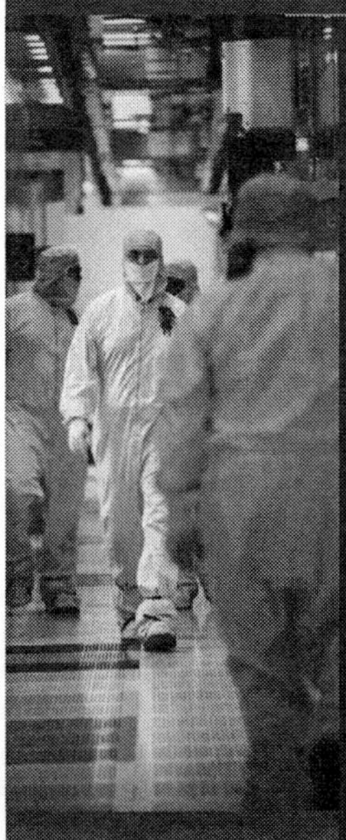

- Leadership means **Corporate Responsibility**
- 14nm and Beyond: We have **paradigm shift in process scaling/defectivity challenges: Process Variation can end up as Excursion**
- **Interdisciplinary expertise need to work together**
  - **Tool suppliers, Material suppliers, Filter Suppliers, Component suppliers, Metrology suppliers and IDMs need to work together to achieve required defectivity baseline and create new STANDARDS**
- Next Generation Quality Systems combined with **Sub-Supplier Management** helps to **achieve "Ship to Control Across Supply Chain"**, makes IDM "Quality Incident Free" and minimize costly learning during HVM ramp

intel

SPCC 03/28/2017

Thank YOU

- To My Co-Authors
- All My Intel colleagues for contributing to the content of the presentation
- To Abbas Rastegar as referenced and always providing valuable input
- To All Intel Suppliers: Thank You for Your Valued Partnership
- To SPCC Committee for the opportunity to present

SPCC 03/28/2017

Risk Factors

- The above statements and any others in this document that refer to plans and expectations for the first quarter, the year and the future are forward-looking statements that involve a number of risks and uncertainties. Words such as "anticipates," "expects," "intends," "plans," "believes," "seeks," "estimates," "may," "will," "should" and their variations identify forward-looking statements. Statements that refer to or are based on projections, uncertain events or assumptions also identify forward-looking statements. Many factors could affect Intel's actual results, and variances from Intel's current expectations regarding such factors could cause actual results to differ materially from those expressed in these forward-looking statements. Intel presently considers the following to be the important factors that could cause actual results to differ materially from the company's expectations. Demand could be different from Intel's expectations due to factors including changes in business and economic conditions, including supply constraints and other disruptions affecting customers; customer acceptance of Intel's and competitors' products; changes in customer order patterns including order cancellations; and changes in the level of inventory at customers. Uncertainty in global economic and financial conditions poses a risk that consumers and businesses may defer purchases in response to negative financial events, which could negatively affect product demand and other related matters. Intel operates in intensely competitive industries that are characterized by a high percentage of costs that are fixed or difficult to reduce in the short term and product demand that is highly variable and difficult to forecast. Revenue and the gross margin percentage are affected by the timing of Intel product introductions and the demand for and market acceptance of Intel's products; actions taken by Intel's competitors, including product offerings and introductions, marketing programs and pricing pressures and Intel's response to such actions; and Intel's ability to respond quickly to technological developments and to incorporate new features into its products. Intel is in the process of transitioning to its next generation of products on 22nm process technology, and there could be execution and timing issues associated with these changes, including products defects and errata and lower than anticipated manufacturing yields. The gross margin percentage could vary significantly from expectations based on capacity utilization; variations in inventory valuation, including variations related to the timing of qualifying products for sale; changes in revenue levels; product mix and pricing; the timing and execution of the manufacturing ramp and associated costs; start-up costs; excess or obsolete inventory; changes in unit costs; defects or disruptions in the supply of materials or resources; product manufacturing quality/yields; and impairments of long-lived assets, including manufacturing, assembly/test and intangible assets. The majority of Intel's non-marketable equity investment portfolio balance is concentrated in companies in the flash memory market segment, and declines in this market segment or changes in management's plans with respect to Intel's investments in this market segment could result in significant impairment charges, impacting restructuring charges as well as gains/losses on equity investments and interest and other. Intel's results could be affected by adverse economic, social, political and physical/infrastructure conditions in countries where Intel, its customers or its suppliers operate, including military conflict and other security risks, natural disasters, infrastructure disruptions, health concerns and fluctuations in currency exchange rates. Expenses, particularly certain marketing and compensation expenses, as well as restructuring and asset impairment charges, vary depending on the level of demand for Intel's products and the level of revenue and profits. Intel's results could be affected by the timing of closing of acquisitions and divestitures. Intel's results could be affected by adverse effects associated with product defects and errata (deviations from published specifications), and by litigation or regulatory matters involving intellectual property, stockholder, consumer, antitrust and other issues, such as the litigation and regulatory matters described in Intel's SEC reports. An unfavorable ruling could include monetary damages or an injunction prohibiting us from manufacturing or selling one or more products, precluding particular business practices, impacting Intel's ability to design its products, or requiring other remedies such as compulsory licensing of intellectual property. A detailed discussion of these and other factors that could affect Intel's results is included in Intel's SEC filings, including the annual report on Form 10-K for the fiscal year ended December 31, 2011.

intel

Impact of hydrogen peroxide in ultrapure water

KURITA

Kurita Water Industries Ltd.
○Yuichi Ogawa, Toru Masaoka, Nobuko Gan, Yu Fujimura, Yasuharu Minato, Yoichi Miyazaki

imec vzw
Kurt Wostyn, Antoine Pacco, Frank Holsteyns

SCREEN Semiconductor Solutions Co., Ltd.
Yukifumi Yoshida

SPCC 2017

Outline

1. Introduction
2. Experimental approach to assess the impact of H_2O_2 on SiGe epitaxial sequence using Si(100) wafers
3. Investigation of mechanism
   (1)Experiments to assess the H_2O_2 impact using Si(111) surface
   (2)Discussion
4. Summary

KURITA

1

H_2O_2 in ultrapure water (UPW)

Table. Impurities in General UPW

Impurities	Concentration
Particles	≦0.1 Pcs/mL
Organics	≦1.0 µg/L
Critical metals	≦0.0001 µg/L
Dissolved oxygen	≦1.0 µg/L
H2O2	5 to 25 µg/L

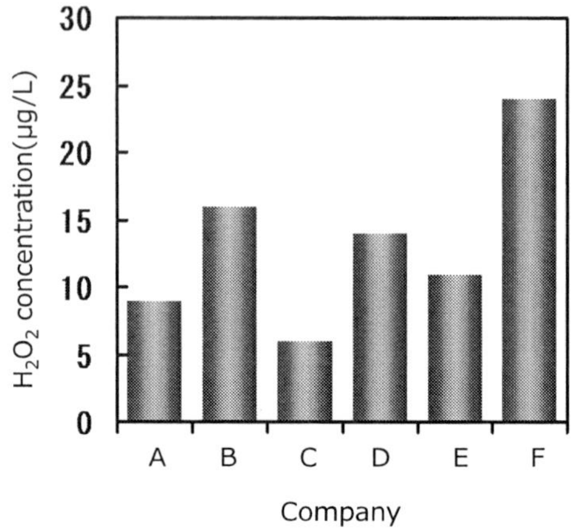

Fig. H_2O_2 concentration in UPW at a customer's Fab

H_2O_2 ,that is stronger oxidizer, exists in UPW and its concentration is ten times higher than these of other impurities

KURITA 2

Historical transition of the H_2O_2 issue in UPW

Year	Contents
2003	The first report, that mentioned the presence of H_2O_2 in UPW due to UV oxidation process was issued. [1]
2007	H_2O_2 decomposition catalyst for UPW was reported. [2]
2013	H_2O_2 in-line monitor for UPW was reported. [3]
2016	Item of H_2O_2 concentration was added to ITRS 2.0. [4]

<Reference>

1) Abe et al, *"The deterioration of Ultrapure Water Quality Under the Influence of Hydrogen Peroxide"*, Journal of Ion Exchange,Vol 14,p.273~276, July 2003

2) Kobayashi et al, *"Removal of Hydroxy Peroxide in an Ultra Pure Water System"*, The 54th JSAP Autumn meeting, 2007

3) Miyazaki et al, *"Advanced Hydrogen Peroxide Removal Technology Using Nano-sized Pt Particle Catalyst on Anion Exchange Resin"*, UPWMicro2013

4) Libman et al, *"UPW ITRS and SEMI: Synergy of Enabling Advanced Existing and Future Technologies"*, UPWMicro 2015

The issue of H_2O_2 has been widely known since more than ten years ago.

KURITA 3

Details of the target cleaning process

Logic semiconductor roadmap from imec

Year	2013	2014	2015	2016	2017	2018	2019	2020
Node (nm)	20～14			10			7	
Channel Material	Si			**Si, SiGe**			**Si, SiGe(Ge)**	
Structure	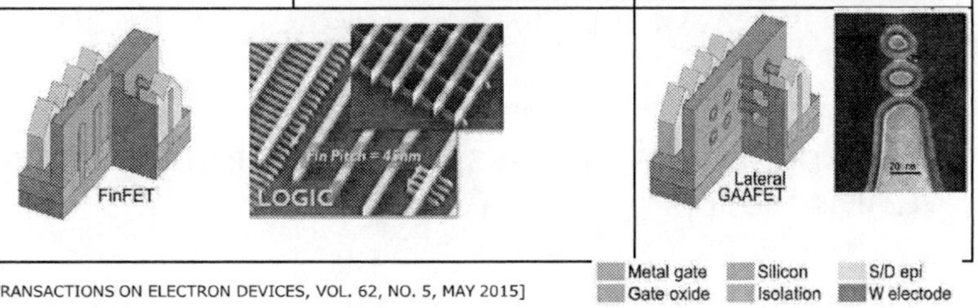							

[Ref.: Yakimets D. et al., IEEE TRANSACTIONS ON ELECTRON DEVICES, VOL. 62, NO. 5, MAY 2015]

SiGe epitaxial process on semiconductor and it's potential problem during cleaning processes

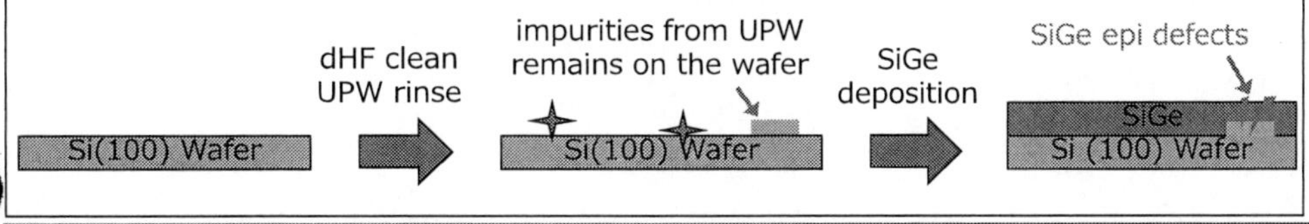

KURITA

Purpose of this study

Typical impact of impurities in UPW on a wafer during wet cleaning processes

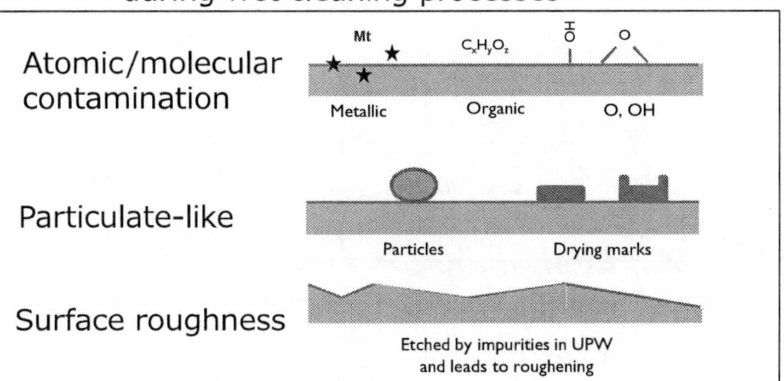

The impact of the dilute H_2O_2 in UPW on cleaning processes is unclear.

Evaluate the impact of the dilute H_2O_2 in the UPW using cleaning tools.

KURITA

Outline

1. Introduction
2. Experimental approach to assess the impact of H_2O_2 on SiGe epitaxial sequence using Si(100) wafers
3. Investigation of mechanism
   (1)Experiments to assess the H_2O_2 impact using Si(111) surface
   (2)Discussion
4. Summary

KURITA 6

Experimental method using single wafer cleaning tool

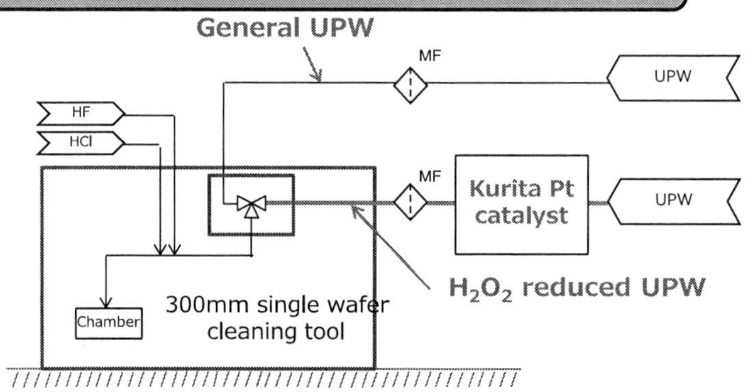

Fig: Single wafer cleaning tool

Table: H_2O_2 concentration in the UPW

General UPW	H_2O_2 reduced UPW (with Kurita Pt catalyst)
8-10 µg/L	<1 µg/L

A split test was conducted to clarify the impact of the dilute H_2O_2 during the SiGe pre-epi cleaning process using either H_2O_2 reduced UPW or general UPW.

KURITA 7

Feature of Kurita Pt catalyst <Decomposition>

SEM (cross-section Pt-resin)

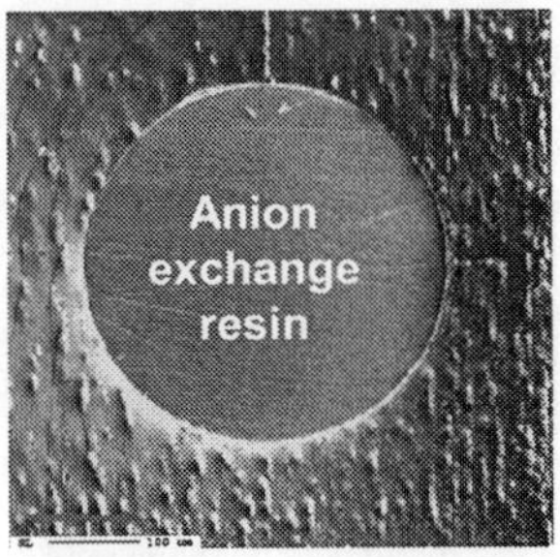

EDS (cross-section Pt-resin)

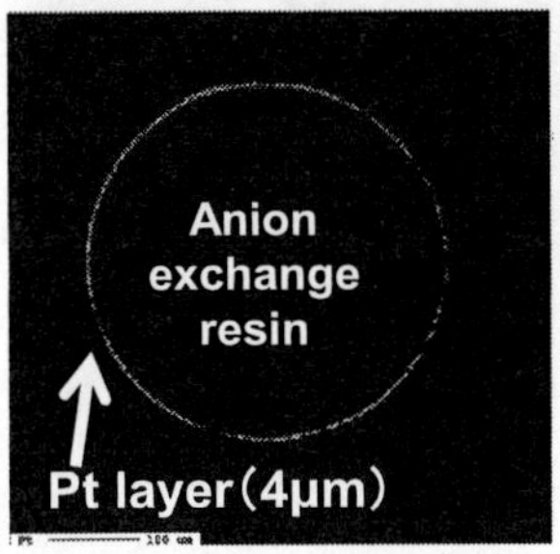

Spherically shaped anion exchange resin coated with Pt layer

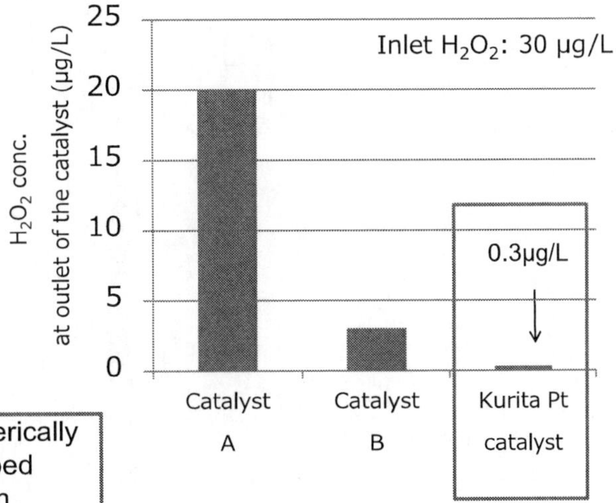

Fig. Decomposition performance compared with other catalysts

Kurita Pt catalyst decomposes H_2O_2 in UPW to less than 1 μg/L.

KURITA

8

Feature of Kurita Pt catalyst <Elution>

Table Elution amount from catalyst

Metal	Units	Kurita Pt catalyst	Catalyst B
Fe	ng/L	<0.1	<5
Cu		<0.1	<5
Al		<0.1	<5
Ni		<0.1	<5
Mn		<0.1	<5
Co		<0.1	<5
Ca		<0.1	<5
Zn		<0.1	<5
Mg		<0.1	<5
Li		<0.1	<5
K		<0.1	<5
Na		<0.1	10
Pd		<0.1	8
Pt		<0.1	<5

Condition
Amount : 200ml-R
SV : 300 1/hr

- Kurita Pt catalyst has less metal elution.

KURITA

9

Feature of Kurita Pt catalyst <Application>

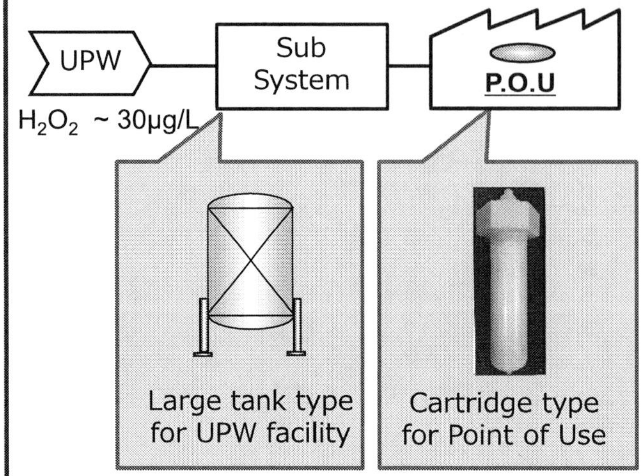

Fig Scope of application

Table performance of commercially installed Kurita Pt catalyst

	Number of years keeping H_2O_2 below 1μg/L
Fab A	8 years
Fab B	6 years
Fab C	4 years
Fab D	3 years

and so on ···

Kurita Pt catalyst has a lot of installation records in semiconductor Fabs.

KURITA

10

Process flow of experimental apparatus for SiGe epitaxial sequence

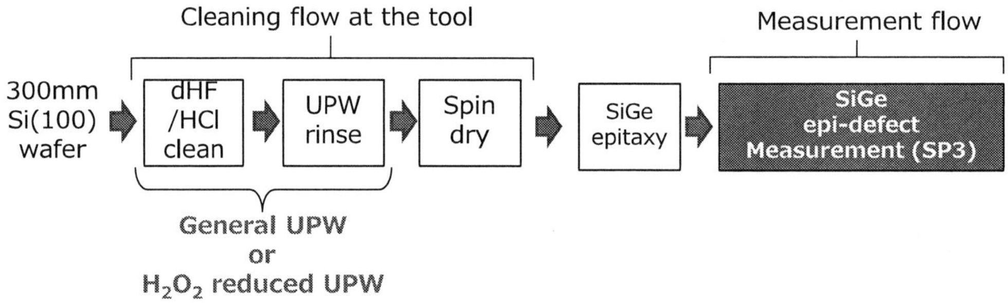

Contents	Condition
Wet cleaning	HF 0.7% clean diluted with General UPW or H_2O_2 reduced UPW
SiGe epi-defects measurement	- After wet clean, 80nm $Si_{0.8}Ge_{0.2}$ layer was epitaxially grown by an epi tool. - KLA-T Surfscan SP3 was used, [K. Wostyn et al ECS Trans (2014)]

KURITA

11

Result: SiGe epi-quality measurement

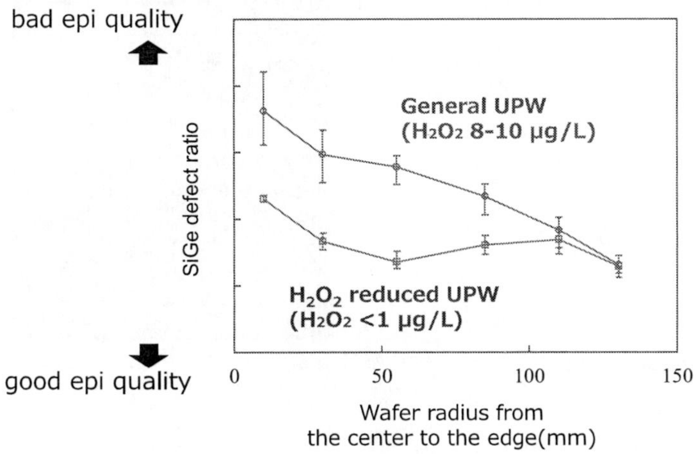

Fig. SiGe epi defect area on the wafer

SiGe epi quality can be improved by removing the dilute H_2O_2 in the UPW.

What has happened on the Si(100) surface by such a dilute H_2O_2?

KURITA

12

Outline

1. Introduction
2. Experimental approach to assess the impact of H_2O_2 on SiGe epitaxial sequence using Si(100) wafers
3. Investigation of mechanism
   (1)Experiments to assess the H_2O_2 impact using Si(111) surface
   (2)Discussion
4. Summary

KURITA

13

Why has Si(111) been chosen?

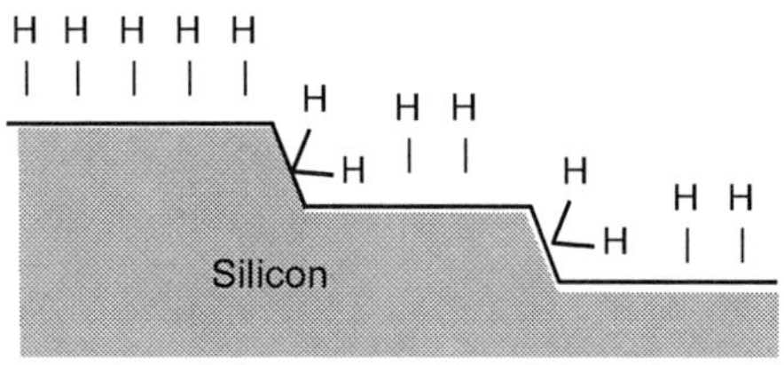

Fig. Surface image of Si(111)

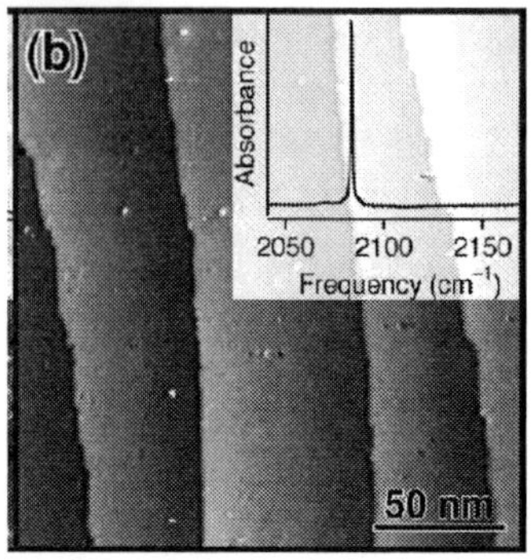

Fig. AFM image and FT-IR-ATR peak of Si(111)
[M.A. Hines, *Annu Rev Phys Chem.*, 54, 29-56 (2003)]

Si(111) surface: Terrace width with atomic steps, dominated by Mono-hydrogen termination

Si(111) will help to understand the impact of the dilute H_2O_2 on Si(100) due to its sensitivity for the state variation.

KURITA 14

Process flow of experiments using Si(111)

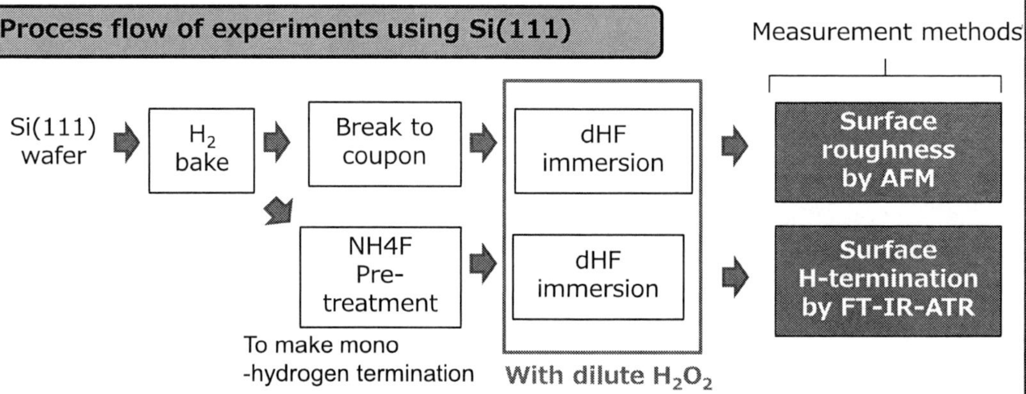

Contents	Test Conditions
Surface roughness measurement by AFM	H_2O_2 100 µg/L in 0.7%HF, N_2 bubbling, Immersion: 15, 30min
Surface H-termination measurement by FT-IR-ATR	H_2O_2 <1, 30, 100 µg/L in 0.5%HF, DO: 60ppb Immersion: 15min

AFM and FT-IR-ATR were used to clarify the impact of the dilute H_2O_2 in dHF on Si(111) surface.

KURITA 15

Result: AFM image on Si(111)

dHF with 100 µg/L H_2O_2

After H_2 bake

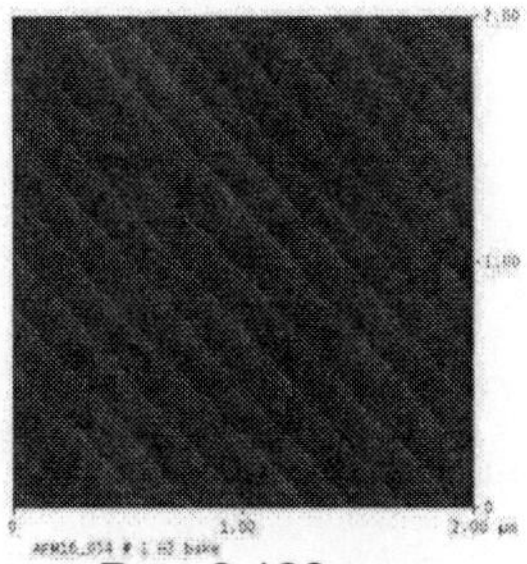

Rms 0.120nm
Ra 0.097nm

15min immersion

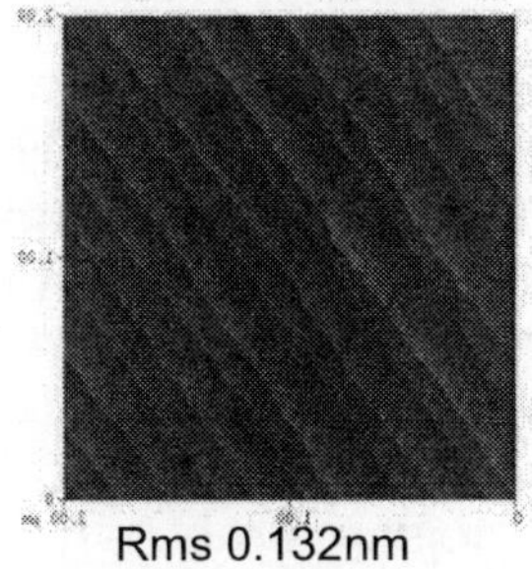

Rms 0.132nm
Ra 0.106nm

30min immersion

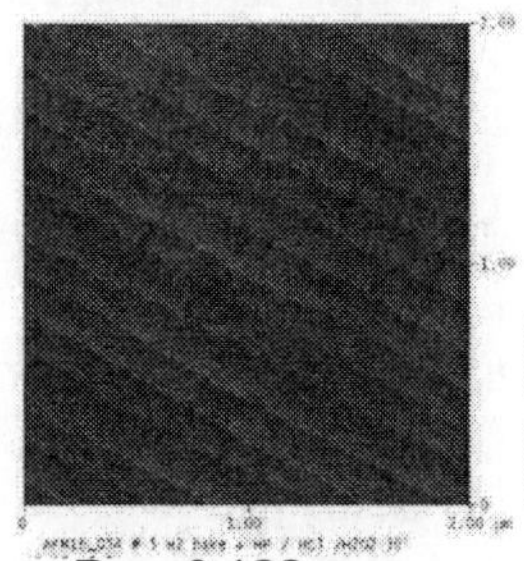

Rms 0.122nm
Ra 0.099nm

No clear difference between before and after immersion in dHF with 100 µg/L H_2O_2 was observed.

Si surface is not roughened by such a dilute H_2O_2.

KURITA

16

Result: FT-IR-ATR peak on Si(111)

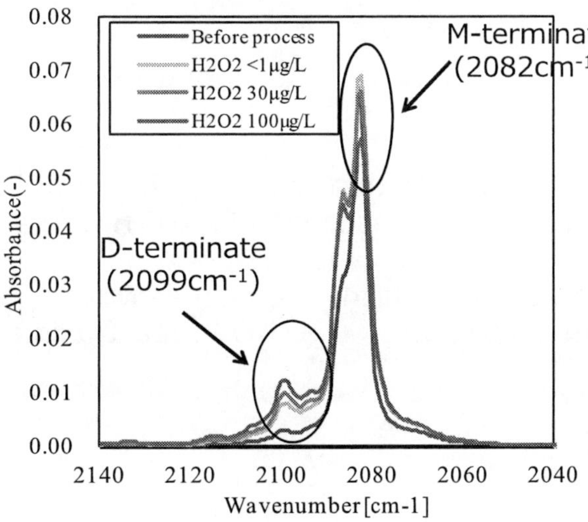

M-terminate is decreased while D-terminate is increased when the dilute H_2O_2 exists in the dHF .

Dilute H_2O_2 promotes Si oxidation, and the oxide is etched by dHF. So, the hydrogen–termination status has changed.

KURITA

17

Possible mechanism of the dilute H2O2 impact on Silicon

1. During dHF process, the dilute H_2O_2 promotes oxidizing the surface of Si(111).and the oxide is etched by dHF. By the oxidation and etching, the hydrogen termination status on Si(111) has changed.

2. Same reaction would happen on Si(100) by the dilute H_2O_2. These is a possibility that the same mechanism of the dilute H_2O_2 impact may happen on Si(100) also.

3. H_2O_2 may be one of the main factors of the SiGe epi defects.

Still work will be needed to support the above-mentioned possible mechanism

KURITA

18

Outline

1. Introduction
2. Experimental approach to assess the impact of H_2O_2 on SiGe epitaxial sequence using Si(100) wafers
3. Investigation of mechanism
   (1)Experiments to assess the impact of H_2O_2 impact using Si(111) surface
   (2)Discussion
4. Summary

KURITA

19

Summary

The SiGe epitaxial quality has been improved by H_2O_2 removed UPW compared with the dilute H_2O_2 containing UPW.

FT-IR-ATR measurements indicate an impact of the dilute H_2O_2 on the Si (111) surface morphology. Silicon can be etched by dHF solution involving dH_2O_2.

So it is assumed that H_2O_2 is one of the main factors which make the SiGe epi quality low.

H_2O_2 in UPW affects the production of semiconductors.

→Kurita Pt catalyst will solve this problem.

KURITA

20

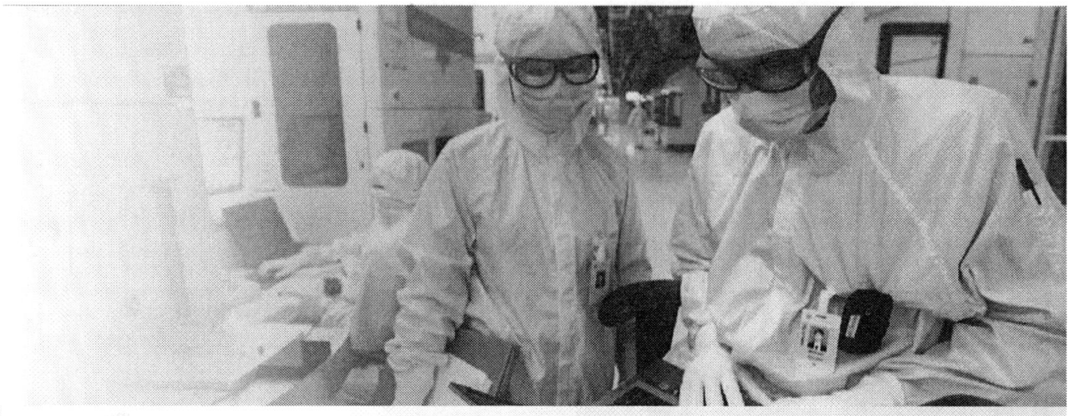

Effect of Incoming Quality of Sulfuric Acid on Inline Defects

Dhiman Bhattacharyya*, Sushil Patil, Mark Conrad, Hayley Manning, Fauzia Khatkhatay, Alexander Mena and Norberto DeOliveira

March 28, 2017

Dhiman Bhattacharyya, Ph.D.
GLOBALFOUNDRIES
400 Stonebreak Rd. Ext., Malta, NY - 12020
Mobile: 518-415-2311
Email: dhiman.bhattacharyya@globalfoundries.com

SPCC 2017
Surface Preparation and Cleaning Conference

Outline

- ❑ Background
- ❑ Problem Statement
- ❑ Root cause investigation and results
- ❑ Summary

GLOBALFOUNDRIES® 2

Outline

- ❑ **Background**
- ❑ Problem Statement
- ❑ Root cause investigation and results
- ❑ Summary

GLOBALFOUNDRIES 3

Background

- Hydrogen peroxide and sulfuric acid are the major consumables in Cleans.
- They are used in mixture in conventional cleans as Sulfuric and Peroxide Mixture (SPM or Piranha) which is then followed by Ammonia Peroxide Mixture (APM or SC1 or RCA1).
- Wet cleans using SPM is predominant in FEOL.
- High volume device manufacturers need to have multiple sources of materials for continuous supply chain.
- Any particle excursion in any chemical can potentially degrade the inline defects.

GLOBALFOUNDRIES 4

Background

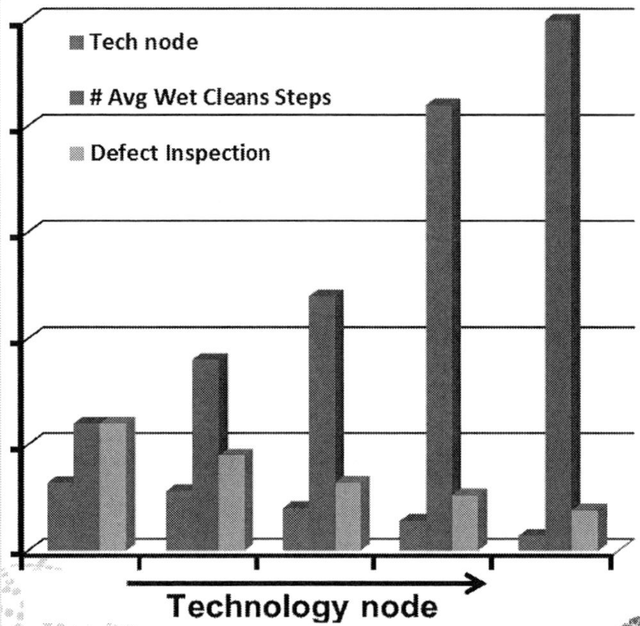

- Technology is shrinking,
- Wet cleans contribution to keep the wafers clean is significantly increasing,
- Sensitivity of Dark Field detection is increasing.

GLOBALFOUNDRIES 5

Background

Detection Limitation:

- Difficult to detect any excursion in incoming particles if the particles are smaller than the detection limit.
- Current state of the art detection limit is 45nm by liquid particle counter.
- Inline defect detection can also miss the defects on the wafer in similar situations where defects are smaller than the detection limit.

Variability in sulfuric acid manufacturing:

- Sulfuric acid is usually produced using either glass or Teflon-lined steel absorption tower in high volume manufacturing environment.
- Glass lined column may introduce silica like particles.
- Teflon lined column may introduce carbon particles.

GLOBALFOUNDRIES 6

Outline

- ❑ Background
- ❑ **Problem Statement**
- ❑ Root cause investigation and results
- ❑ Summary

GLOBALFOUNDRIES 7

Problem Statement:

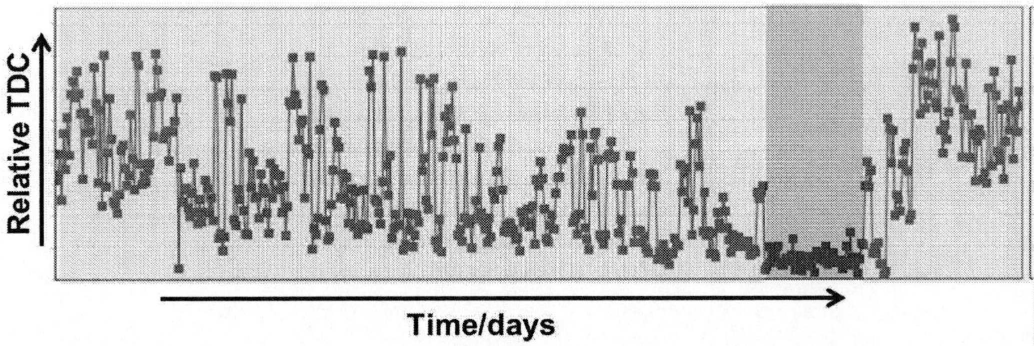

- The relative TDC (total defect counts) at a DF inspection step continuously improved over the time.
- Baseline defectivity significantly reduced.
- However, the TDCs jumped back to old baseline all of a sudden with no apparent correlation to any process or equipment or environmental changes.

GLOBALFOUNDRIES 8

Problem Statement:

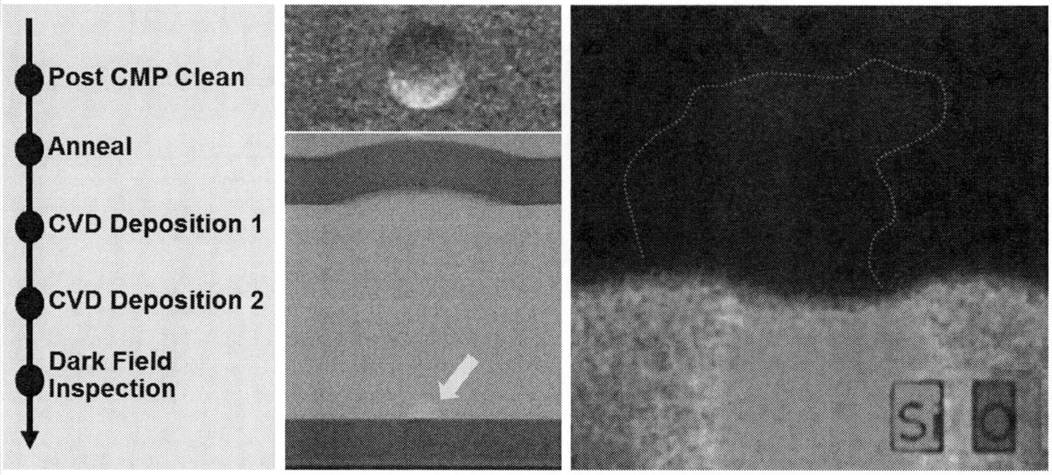

- This DF inspection detects any excursion in Post CMP Clean as embedded defects.
- Even particles as small as ~10nm are visible as embedded after being decorated by 2 CVD films.
- DA clearly showed that the particles were composed of Si & O.

GLOBALFOUNDRIES® 9

Outline

- ❑ Background
- ❑ Problem Statement
- ❑ **Root cause investigation and results**
- ❑ Summary

GLOBALFOUNDRIES® 10

Root cause investigation and results

Approaches to identify the root cause of these defects;

(1) Failure analysis of the particles,

(2) Comparison of defects from various components of the production recipe to conclude the defect source,

(3) Comparison of defects from cleans process recipes with different SPM dispense times, and

(4) Analysis of the defect trend versus the supply chain in real-time.

GLOBALFOUNDRIES 11

Root cause investigation and results

Comparison of defects from various components of the production recipe:

- The recipe was broken down into each chemistry.
- POR recipe, only which had SPM in it showed highest defect counts.

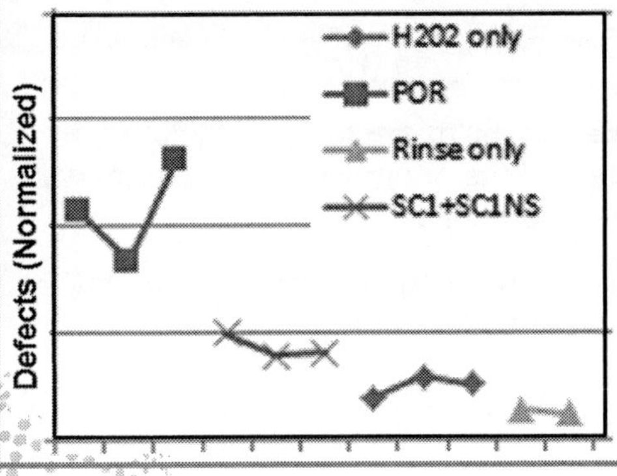

Comparison of defects from cleans process recipes with different SPM dispense times:

- An experiment with 50% reduced SPM dispense time at the same operation was performed.
- It helped in >75% reduction of the defects

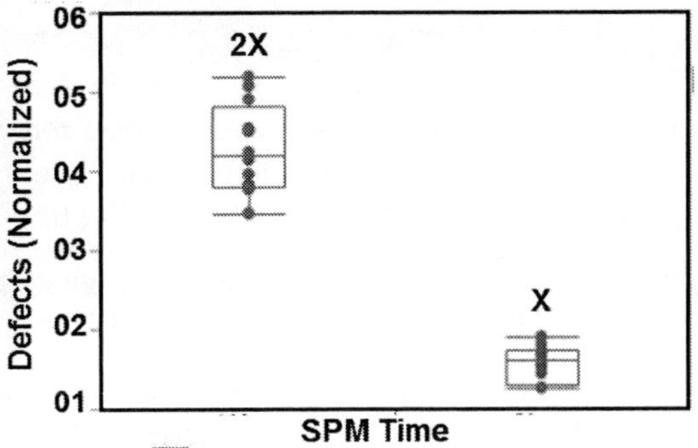

GLOBALFOUNDRIES 12

Root cause investigation and results

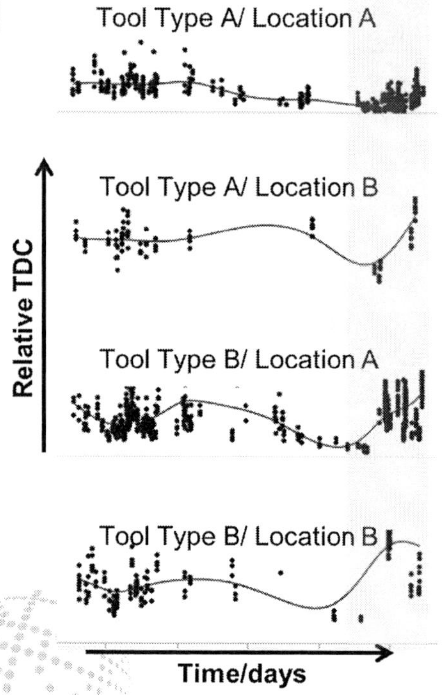

- Initial investigation was limited to very specific tools located to a specific location.
- The sensitivity of such tools in such location was in doubt.
- As time progressed, a clear uptrend in TDCs was noticed in all kinds of tools independent of where they were located.

GLOBALFOUNDRIES 13

Root cause investigation and results

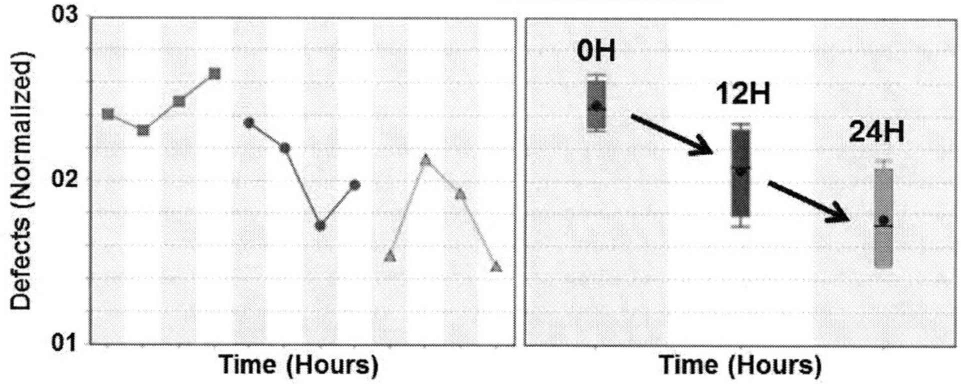

- A real-time experiment was carried out where the sulfuric acid manufactured in glass-lined steel absorption tower was stopped at the chemical feeder pump while the other source was continued to supply the operations.
- One experimental lot (12 wafers) was split in to three and each of the three sub-lots was processed at the post-CMP operation at a 12 hour interval.
- A very clear downtrend of defects in real-time was observed when the supply was flushing the silica contaminated material out.

GLOBALFOUNDRIES 14

Root cause investigation and results

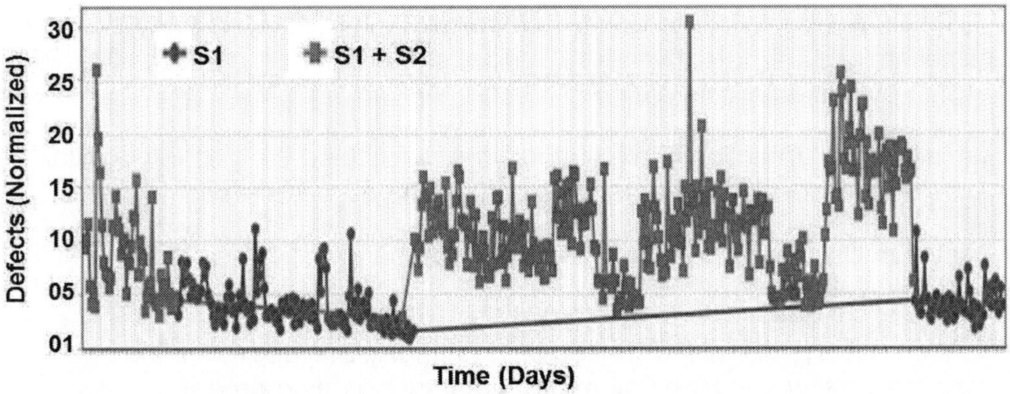

- Sulfuric acid has been used in bulk from both the suppliers (S1 and S2).
- The defect baseline came down when supply from S2 was interrupted for few months.
- Baseline trended up again when supply from S2 was restored.
- Defect level trended down after supply from S2 was stopped.
- S2 was found to have glass lined column while S1 had Teflon lined column.

GLOBALFOUNDRIES 15

Outline

- ☐ Background
- ☐ Problem Statement
- ☐ Evaluation of the supply
- ☐ **Summary**

GLOBALFOUNDRIES 16

Summary

- Difficult to detect any excursion in incoming particles if the particles are smaller than the detection limit. Current state of the art detection limit is 45nm by liquid particle counter.
- Inline defect detection can also miss the defects on the wafer in similar situations where defects are smaller than the detection limit.
- Any particle excursion in any chemical can potentially degrade the inline defects.
- Sulfuric acid is usually produced using either glass or Teflon-lined steel absorption tower in high volume manufacturing environment.
- The difference in manufacturing process introduces variation in the type of defects both in quality and quantity.

GLOBALFOUNDRIES® 17

Acknowledgment:

- *All colleagues in GLOBALFOUNDRIES*

GLOBALFOUNDRIES® 18

Thank You

GLOBALFOUNDRIES®

Blisters formation mechanism during High Dose Implanted Resist Stripping

Marion Croisy[a,b,c*], Cécile Jenny[a], Claire Richard[a] , Denis Guiheux[a], Sylvain Joblot[a], Alain Campo[b], Erwine Pargon[c] , Nicolas Posséme[b]

[a] STMicroelectronics, 850 rue Jean Monnet, 38926 Crolles Cedex, France

[b] CEA-Leti, 17 avenue des Martyrs, F-38054 Cedex 09, Grenoble, France

[c] CNRS-LTM, Université Grenoble Alpes, 17 rue des Martyrs, F-38054 Grenoble Cedex, France

* marion.croisy@st.com

life.augmented

cea leti

LTM LABORATOIRE DES TECHNOLOGIES DE LA MICROÉLECTRONIQUE

Outline 2

Context

Observation and characterization of blisters

- Blisters characterization (STEM, STEM-EDX and ToF-SIMS)
- Influence of implantation parameters on blisters formation
  - Dopants species (As vs P)
  - Implantation energy and dose

Solutions to avoid blisters formation

- Optimization of dry strip process
- Integration modification

Context
/High Dose Implantation Stripping (HDIS)

Dry strip process after implantation step

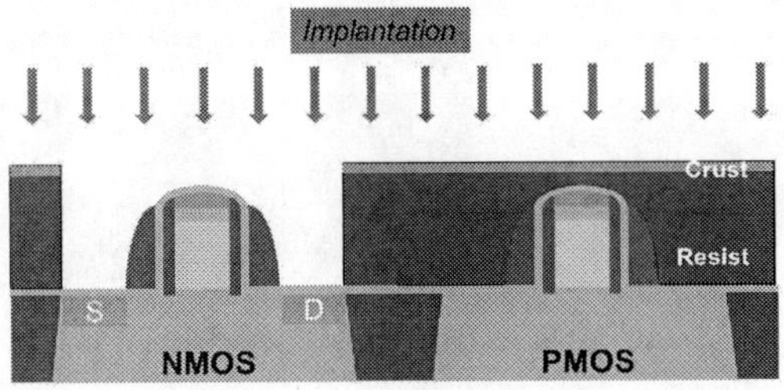

Dry strip challenges

- No remaining residues
- No added defectivity (blisters)
- Minimize substrate loss and oxidation
- No impact on dopant profile

N_2H_2 is the most suitable chemistry for implanted resist stripping in terms of residues removal and low substrate consumption

Limitation → blisters formation after several N_2H_2 stripping processes on implanted area

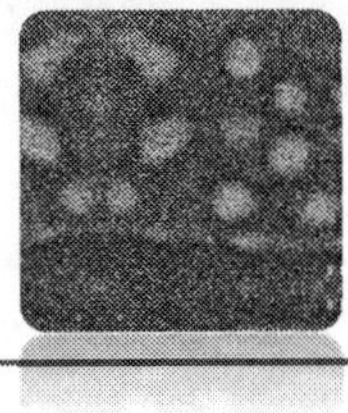

Context
/Integration

Example of CMOS integration from Spacer Z to SD formation

- Gate patterning
- SpacerZ
- Si epitaxy
- LDD x5
- Spacer2
- SD x4

After LDD and NSD → the dry strip chemistry used is forming gas (N2H2 with 3% of H2)

All lithographic rework are based on forming gas

Formation of blisters after several N2H2 processes

Example with P implantation (4keV $1.5^{E}15$ at/cm²)

After 4 N_2H_2 dry strip | ***5 N_2H_2 dry strip*** | ***6 N_2H_2 dry strip***

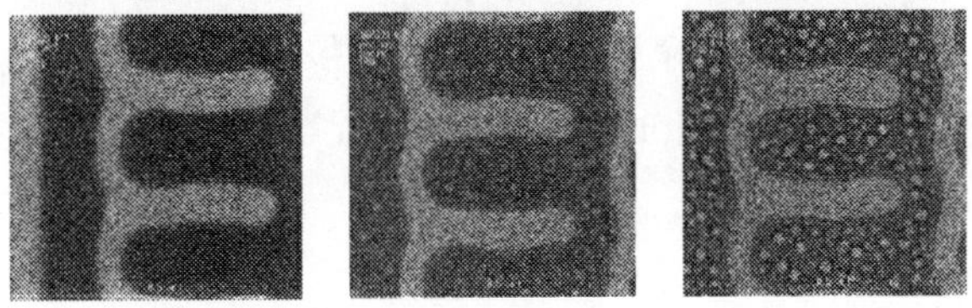

Experimental protocol

5

GAMMA tool (LAM Research)

- Downstream plasma
- 6 platens
    - 1 preheat station
    - 5 RF sources
- High temperature (285°C)

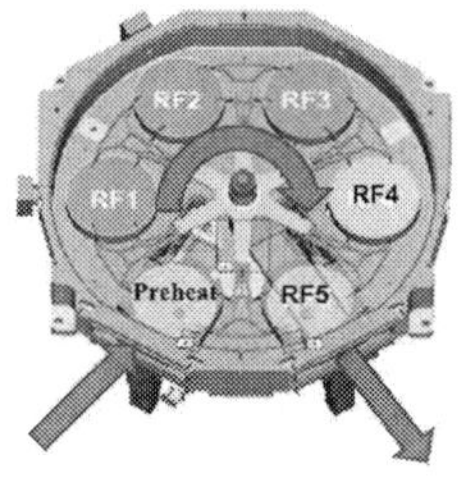

Wafer preparation

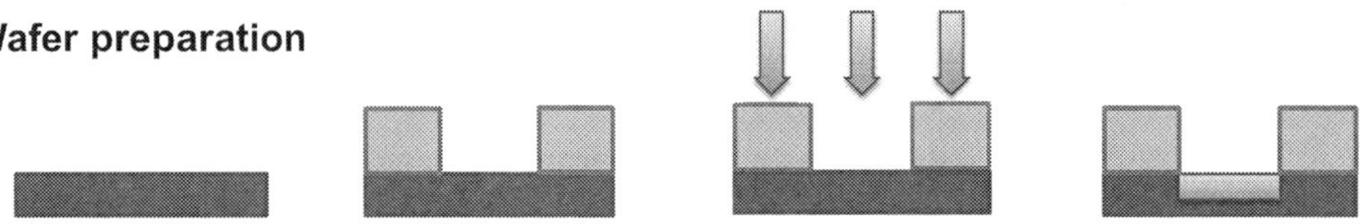

Characterizations

In-line : Top view SEM-CD observations

Off-line : Cross section STEM/EDX observation, chemical analysis by ToF-SIMS

Outline

6

Context

Observation and characterization of blisters

- Blisters characterization (STEM, STEM-EDX and ToF-SIMS)
- Influence of implantation parameters on blisters formation
    - Dopants species (As vs P)
    - Implantation energy and dose

Solutions to avoid blisters formation

- Optimization of dry strip process
- Integration modification

Blisters Characterization
/ STEM and EDX analysis

7

STEM cross section views

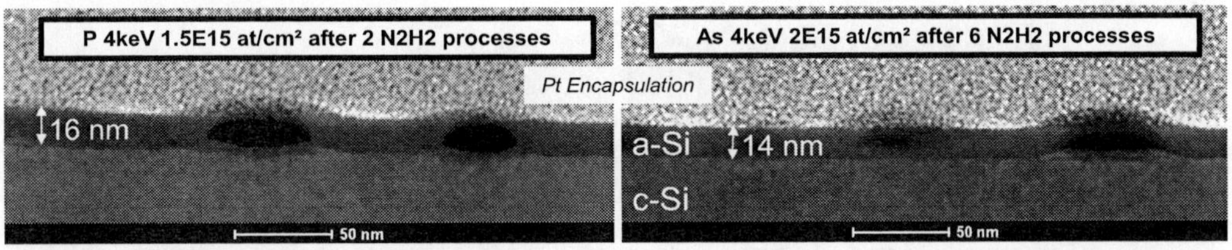

EDX analysis

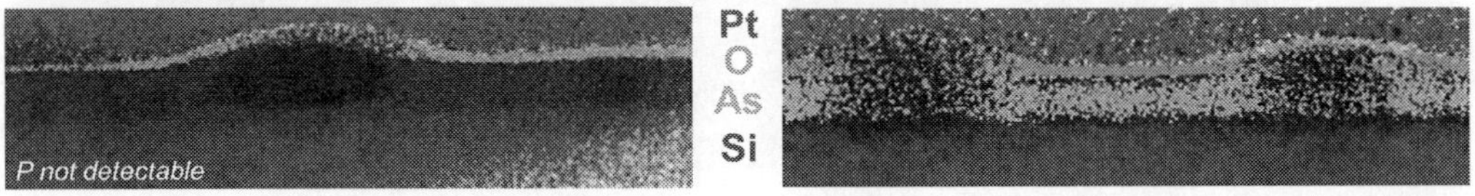

- Bubbles are formed near the Si surface in the implanted area corresponding to amorphous Si
- The surface is lifted by the bubbles
- No dopants are detected on the voids left by the bubbles

no implant 2 N2H2 processes
P 4keV 2N2H2 processes
P 4keV post implantation

Blisters Characterization
/ ToF-SIMS analysis

8

Evaluation of plasma diffusion in the Si substrate thanks to H^- and SiN^- profiles

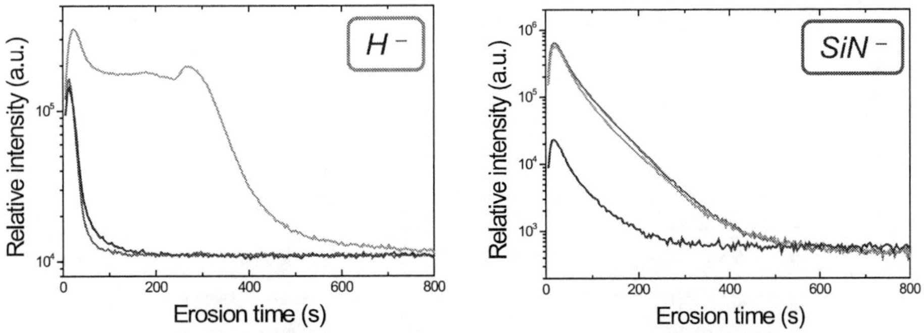

- No implantation → Diffusion of N but not H atoms

- P implantation → Similar N diffusion as for the non implanted substrate
  But H diffusion is greatly enhanced and an accumulation is observed

→ **H diffusion and accumulation in the amorphized Si created during the implant are responsible for blister formation**

- Photoresist coating and lithography
- Implant P *4keV 1.5E15 at/cm²*
  or
  Implant As *4keV 2E15 at/cm²*
- Dry strip N_2H_2

Blisters Observation
/ Influence of implanted specie (As vs P)

P 4keV

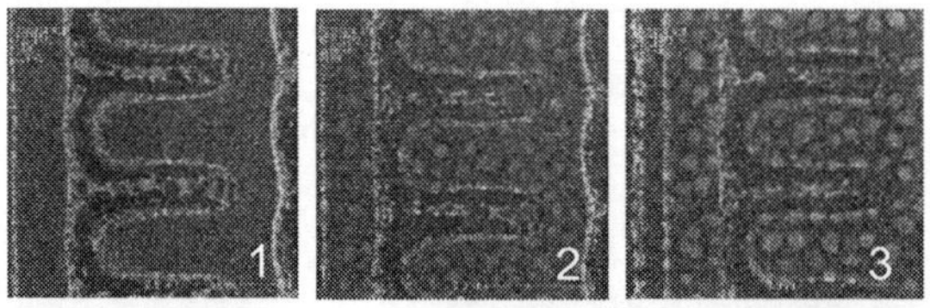

Blisters size 70 - 100nm

As 4keV

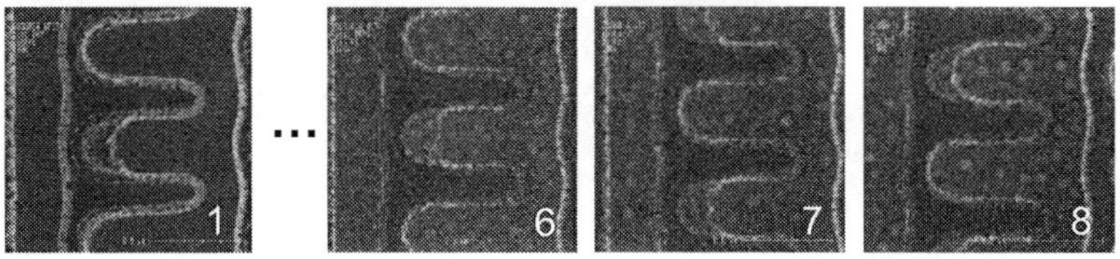

50 – 60nm

Influence of dopant specie
- blisters appear later with As
- larger blisters with P

	Number of dry strip							
	1	2	3	4	5	6	7	8
P 4keV 1.5^E15								
As 4keV 2^E15								

Blisters Characterization
/ Influence of implanted specie (As vs P)

What can explain the difference observed between As and P?

SRIM simulations

Dopant profile

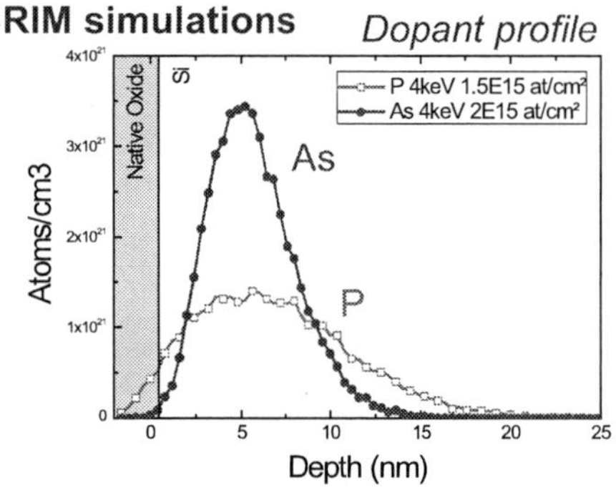

Si recoil profile

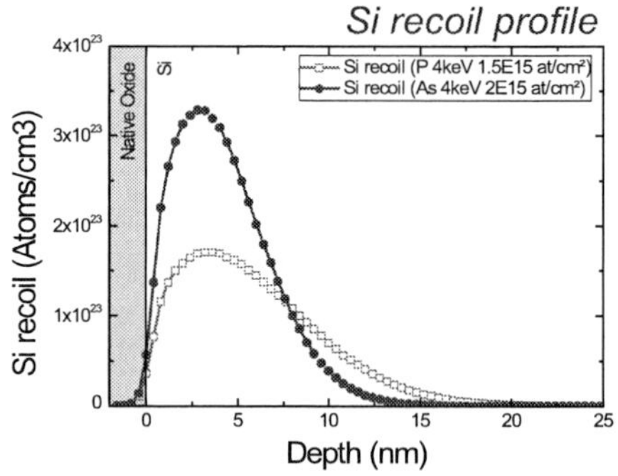

Si is considered amorphous when Si recoils > 3.6atoms/cm³

Amorphized Si thickness

	Simulated (SRIM) (nm)	Measured (STEM) (nm)
P	18	16
As	14	14

- Different dopants profiles but same Rp
- Higher concentration of Si recoils with As
  → higher amorphization rate
- Thinner amorphized layer for As

➡ **Effect of amorphization rate or amorphized layer thickness?**

- Photoresist coating and lithography
- Implant As *3keV*
- Dry strip N_2H_2

Blisters Post Dry Strip
/ Influence of implantation energy

11

	Nb of dry strip processes							
	1	2	3	4	5	6	7	8
As 5keV 2E15								
As 4keV 2E15								
As 3keV 2E15								

Amorphized Si thickness (SRIM)

As 5keV	16.4 nm
As 4keV	14 nm
As 3keV	11.4 nm

SRIM simulations

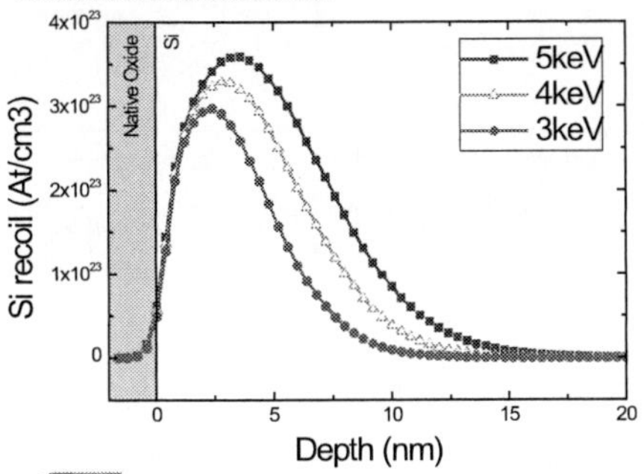

Influence of implantation energy
- Blisters are formed a little earlier with low implantation energies
- At lower energy, the amorphization rate and the amorphized thickness are lower

→ Influence of implantation energy on blisters is limited

- Photoresist coating and lithography
- Implant As *3keV*
- Dry strip N_2H_2

Blisters Post Dry Strip
/ Influence of implantation dose

12

	Nb of dry strip processes							
	1	2	3	4	5	6	7	8
As 3keV 2E15								
As 3keV 1E15								

Amorphized Si thickness (SRIM)

As 3keV 2E15	11.4 nm
As 3keV 1E15	10.4 nm

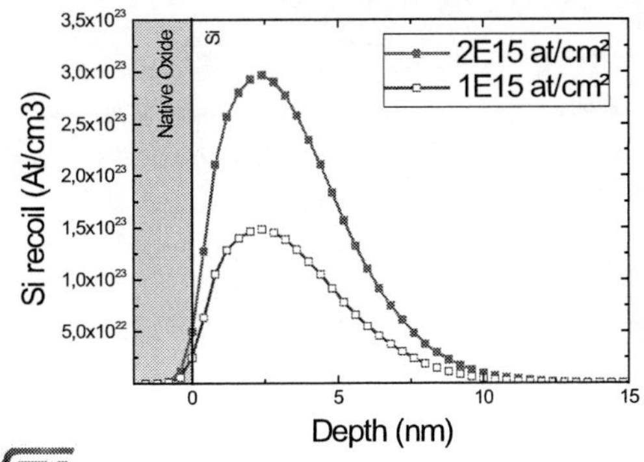

Influence of implantation dose
- Lower dose favors the blisters formation
- At lower dose, the amorphization rate is more than twice lower

→ Influence of amorphization rate on H diffusion and blisters formation

SPCC2017 | March 28-29 | Marion Croisy | Blisters formation mechanism during High Dose Implanted photoresist stripping

Outline

13

Context

Observation and characterization of blisters

- Blisters characterization (STEM, STEM-EDX and ToF-SIMS)
- Influence of implantation parameters on blisters formation
  - Dopants species (As vs P)
  - Implantation energy and dose

Solutions to avoid blisters formation

- Optimization of dry strip process
- Integration modification

- Photoresist coating and lithography
- Implant As *3keV 1^E15 at/cm²*
- Dry strip N_2H_2

Blisters Post Dry Strip
/ Influence of stripping recipe

14

Recipe	Number of dry strip processes				
	1	2	3	4	5
Reference (30s/platen)					
T° ramp + Low P (45s/platen)					
Low P (30s/platen)					

Low process pressure and temperature ramp → the blisters formation is delayed compared to reference recipe

Limitation → remaining residues after dry and wet processes
The stripping is less efficient because of temperature ramping

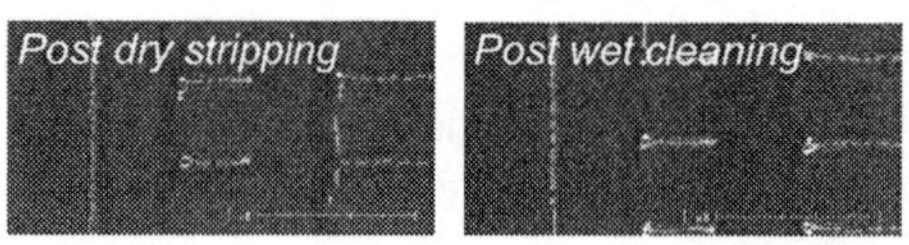

With low process pressure only→ the blisters formation is delayed and no residues remain

- Thermal oxide growth (55A)
- Implantation P
- Dry strip N_2H_2

Blisters Post Dry Strip
/ Influence of oxide layer

15

Modification of the stack→
- Implantation is made through an thermal oxide layer (55A)
- Implantation step is adapted to have similar dopant profiles

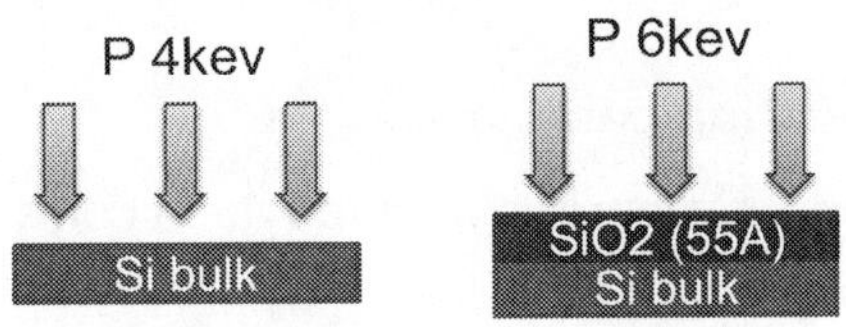

SRIM simulations

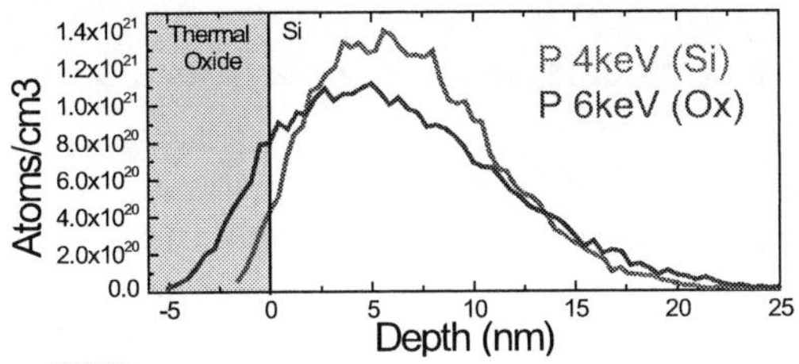

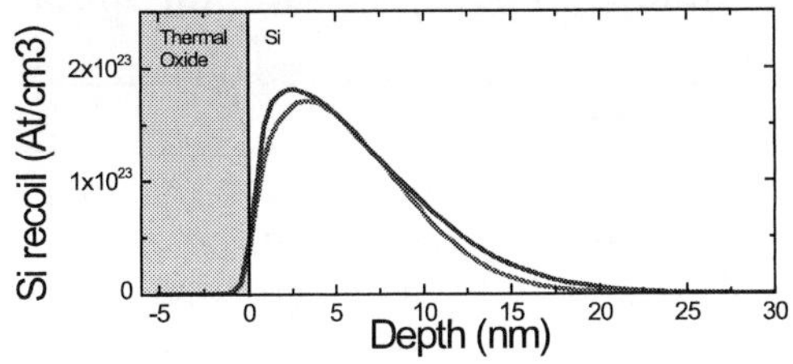

- Thermal oxide growth (55A)
- Implantation P 1.5E15 at/cm² 4keV or 6keV
- Dry strip N_2H_2

Blisters Post Dry Strip
/ Influence of oxide layer

16

		Nb of dry strip				
		1	2	3	4	5
Si	P 4keV					
Oxide	P 6keV					

The oxide layer allows to delay the formation of blisters
→ **What is the effect of the oxide layer?**

		Nb of dry strip				
		1	2	3	4	5
Oxide	P 6keV					
Oxide removed before strip	P 6keV					

→ *Effect on H diffusion*

The blisters are formed sooner if the oxide is removed before stripping

The growth of an oxide layer before implantation allows to delay blisters formation by limiting H diffusion in Si substrate

SPCC2017 | March 28-29 | Marion Croisy | Blisters formation mechanism during High Dose Implanted photoresist stripping

Conclusions

- **Blisters characterization**
  - During N2H2 stripping, H atoms diffuse though the amorphous Si layer and accumulate to form a bubble
  - N2 also diffuses in the substrate but is not responsible for blistering
- **Influence of implantation parameters**
  - Blisters are favored by low implantation dose corresponding to less amorphized Si conditions
- **Solutions against blistering effect**
  - Dry stripping process optimization → low pressure allows to delay blisters formation
  - Integration modification → oxide layer growth before the implantation is beneficial in particular because it decreases H diffusion in the substrate during the plasma process

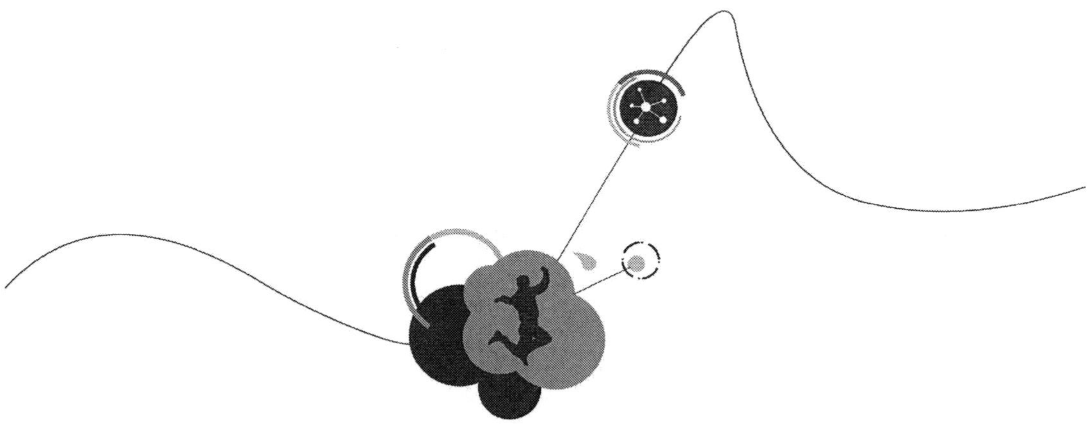

Thank you for your attention

Entegris

50 YEARS OF PURE ADVANTAGE

AUSTIN 28 & 29th MARCH 2017

Gas Purge or Wet Cleaning: Decontamination Solutions to Control AMCs in FOUPs

Surface Preparation and Cleaning Conference

Paola González-Aguirre, Entegris, *paola.gonzalez@entegris.com*
Hervé Fontaine, CEA-Leti, *herve.fontaine@cea.fr*
Carlos Beitia, CEA-Leti, *carlos.beitia@cea.fr*
Jim Ohlsen, Entegris, *jim.ohlsen@entegris.com*
Jorgen Lundgren, Entegris, *jorgen.lundgren@entegris.com*

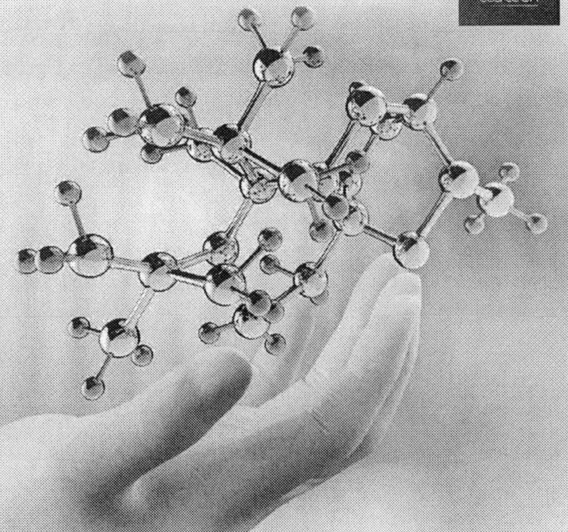

OUTLINE

- Introduction
- Goals
- Gas purge
- Wet clean
- Numerical simulation
- Summary

INTRODUCTION: AMC

- Airborne molecular contamination (AMC) may be responsible for severe yield losses
- Among AMCs, HF is identified as root cause of defectivity, where moisture and time play a critical role:

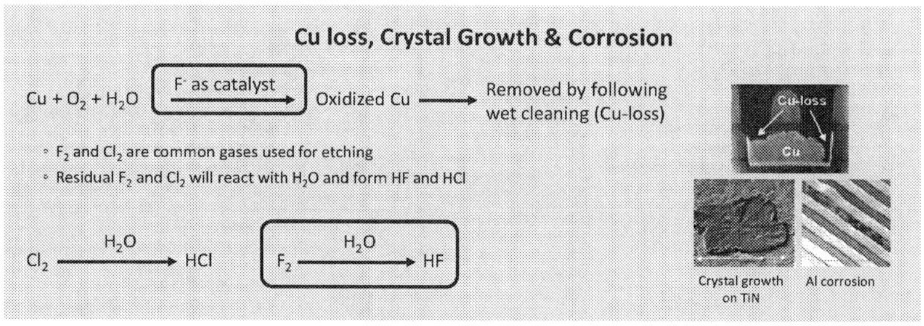

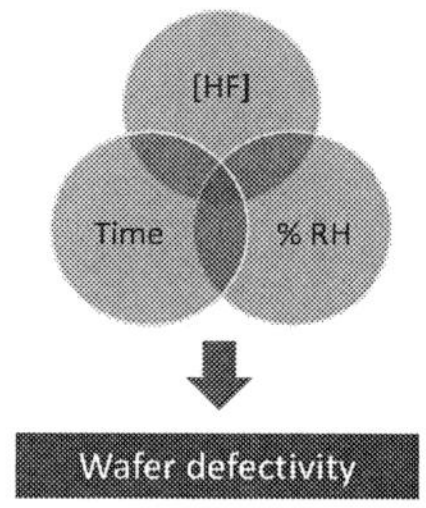

Jun B., Bai H., Lin K.I., Lui S.S., AL-Cu Pattern wafer study on metal corrosin due to chloride ion contaminants. IEEE Transactions on Semiconductor Manufacturing., Vol 23 (2010), 553-558

Luo V., Ding J., Zhan W., Peng L., Wang A., Liu B., Song Y., Yin D., Method to redice crystal defects in AlCu bond pad. ECS Transactions. 27 (2010). 321-325

T. Kamoshima, Fujii Y., Noguchi T., Saeki T., Takata Y., Ochi H., Koiwa A., "Controlling ambient gas in slot-to-slot space inside FOUP to suppress Cu-loss after dual damascene patterning," IEEE Trans. Semicon. Manufact., vol. 21. (2008), 573-577

INTRODUCTION: FOUP TO WAFER CROSS CONTAMINATION

Demonstration of a cross-contamination chain between FOUP and wafers

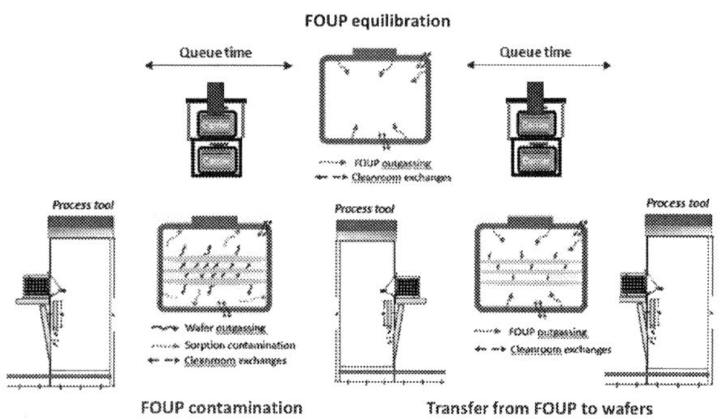

- Solution-diffusion model (polymer membranes): molecular transfer governed by gas **solubility** and **diffusion** in polymers
- New generation FOUPs must minimize the impact of the AMCs (not only particles) onto the wafers
- Air FOUP control or FOUP decontamination must be implemented to limit AMC contamination
  - Gas Purge
  - Wet Clean

GOALS

- Evaluate gas purge & wet clean efficiency for critical HF contamination in terms of contaminant transfer to wafer
  - Quantitative assessment of cross-contamination phenomena depending on decontamination solution → direct relevant data at the FOUP scale
  - Comparison of different FOUP materials in terms of contamination reduction

☑ SPECTRA PC

Polycarbonate

☑ SPECTRA PC/CP

Polycarbonate/C-powder

☑A300 EBM/CNT

EBM/C-nanotubes

GAS PURGE TEST: EXPERIMENTAL

FOUP Conditioning
Clean room equilibration (21°C 40% RH)
7 days

Intentional Gaseous HF Contamination
10 µl- Droplet of HF 2%
(9.2 ppmv after total evap.) 2 hours

25 300mm Si wafers
6 200 mm Cu Wafer Storage
LPE-IC 2 hours, 25 hours

Standard Purge 5L/min
door closed 24 hours

No Purge
door closed 24 hours

GAS PURGE TEST: RESULTS OF THE HF TRANSFER ON STORED CU-WAFERS

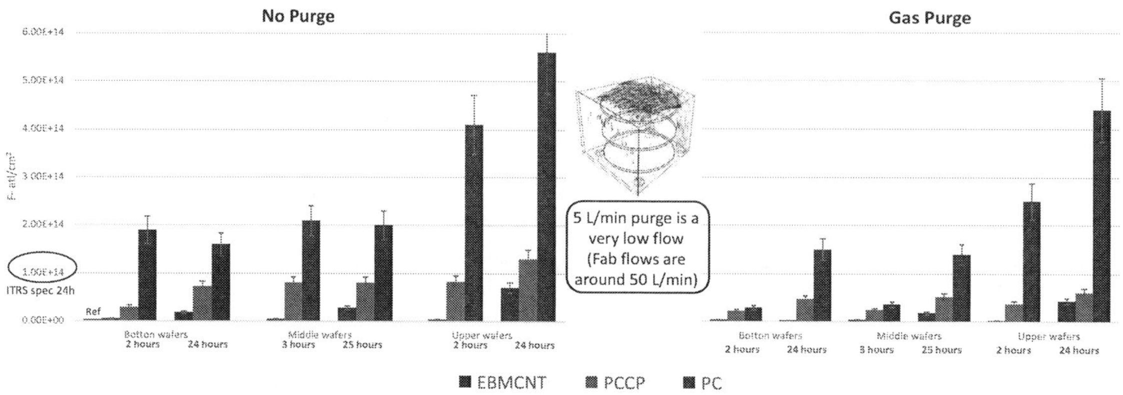

- HF transfer: EBMCNT < PCCP < PC
- Remarkable 25 slot effect
- Main transfer during first hours of exposition
- EBMCNT respect ITRS limits

- HF transfer: EBMCNT < PCCP < PC
- 25 slot effect
- Significant HF transfer reduction (40-80% short time and 20-40% long time)
- EBMCNT & PCCP respect ITRS limits

WET CLEAN TEST: EXPERIMENTAL

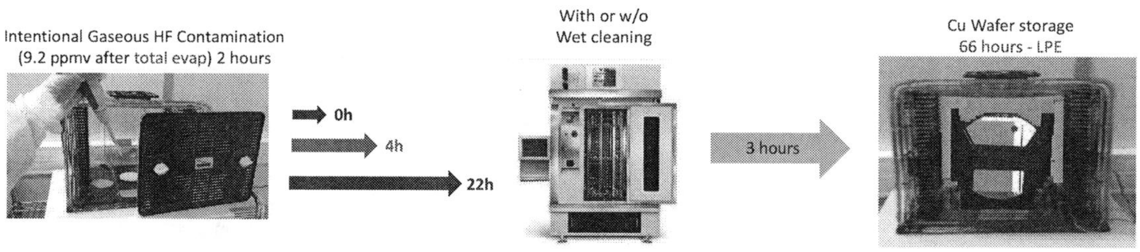

DMS M300, Spray surfactant 1 min, total rinse 3 min, total dry (70°C) 29 min

WET CLEAN TEST: RESULTS OF THE HF TRANSFER ON STORED CU-WAFERS

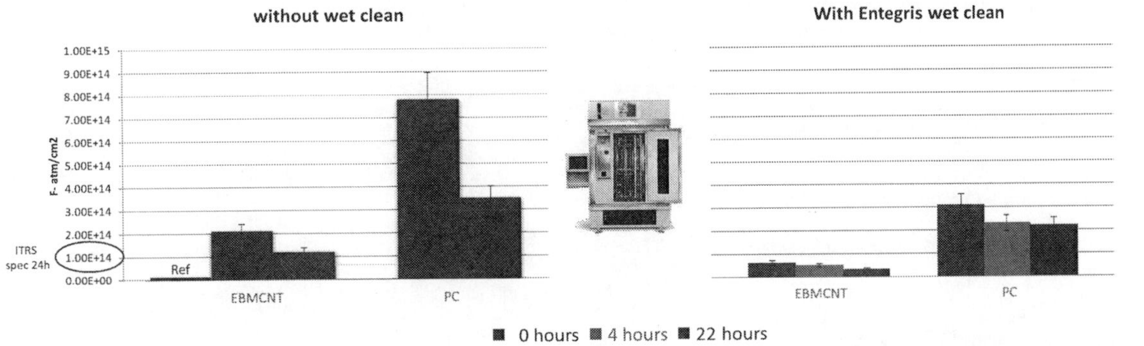

- HF cross contamination: EBMCNT < PC
- Contamination transfer decreases with the time: HF diffusion in bulk

- Wet clean decreases HF transfer between 40 & 60% (depending on time) in PC and around 70% in EBMCNT
- EBMCNT respect ITRS limits after clean

WET CLEAN TEST: HF TRANSFER VS WAITING TIME

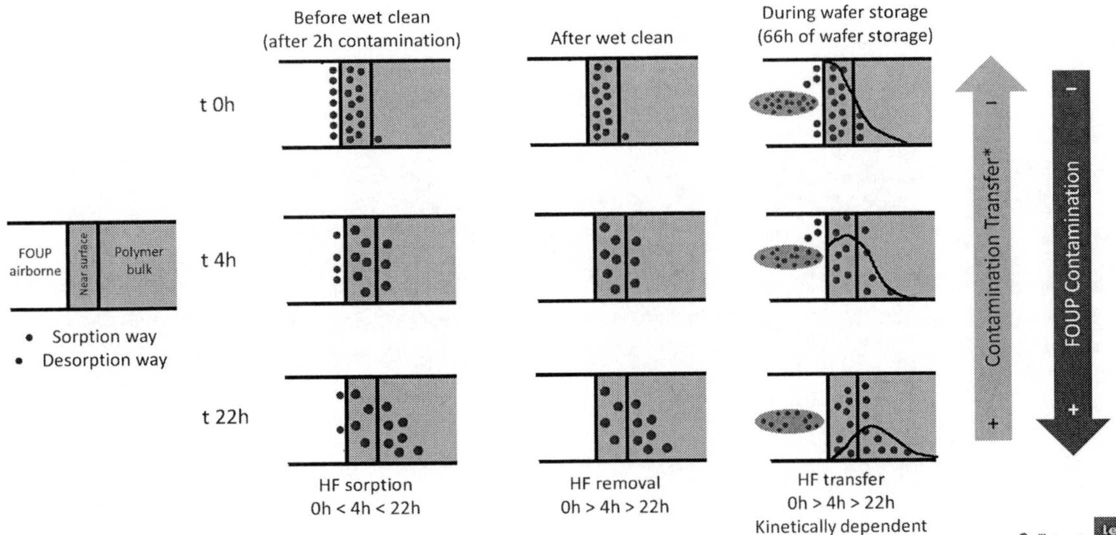

NUMERICAL SIMULATION: MODEL IMPLEMENTED

Knowledge of basic HF transport properties [1]

Polymer	PC		EBM
D (m^2/s) 10^{15}	1.70 ± 54%	>	0.37 ± 50%
S ($m^3(STP)/m^3pol\ Pa$)	0,344 ± 21%	≈	0.386 ± 19%

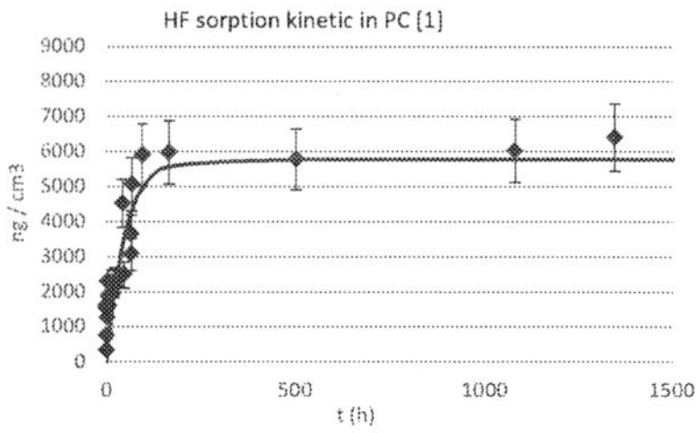

Simple membrane model using Comsol multiphysics tool [2]

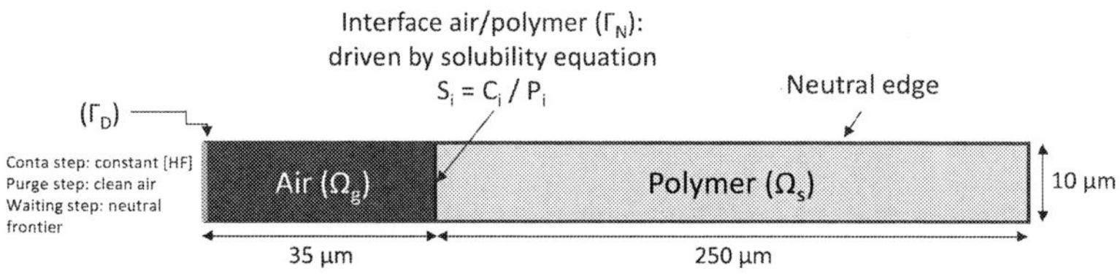

- Only diffusion phenomenon is considered in Ω_g and Ω_s
- Mesh element size: max 1.10^{-2} µm (422 400 elements)

[1]: HF transport coefficients in polymers used for microelectronic applications.
Gonzalez-Aguirre P., Fontaine H.; Beitia C.; Pastorello R.; Ohlsen J., Lundgren J. Defect and Diffusion Forum, Vol 367 (2016), 68-76
[2]: Mathematical Modeling of the AMCs Cross-Contamination Removal in the FOUPs: Finite Element Formulation and Application in FOUP's Decontamination.
N. Santatriniaina, J. Deseure, T. Q. Nguyen, H. Fontaine, C. Beitia, L. Rakotomanana. International Journal of Mathematical, Computational Science and Engineering, Vol 8, No 4, (2014), 30-35

NUMERICAL SIMULATION: EVOLUTION OF [HF] PROFILES IN POLYMER (WET CLEAN SCENARIO)

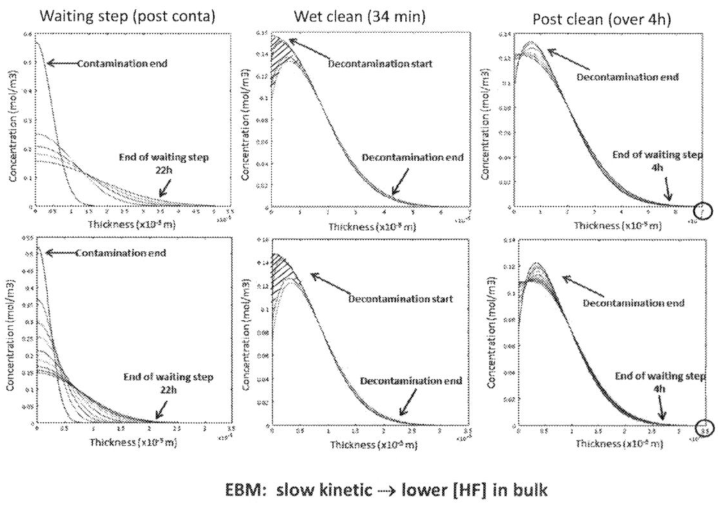

Contamination reduction: -5%

Contaminant diffusion depth : 60 µm

Contamination reduction: -7%

Contaminant diffusion depth: 27 µm

EBM: slow kinetic → lower [HF] in bulk

NUMERICAL SIMULATION VS. EXPERIMENTAL RESULTS

Intentional Gaseous HF Contamination

0h

4h

22h

Wet cleaning

3 hours

Cu Wafer storage

HF cross contamination
EBMCNT < PC

Waiting time effect
0h > 4h > 22h

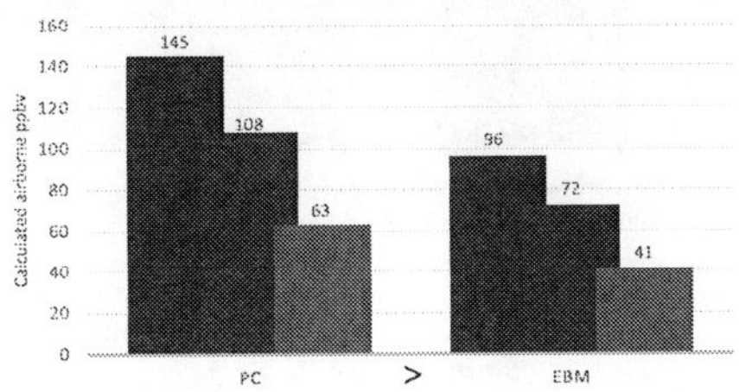

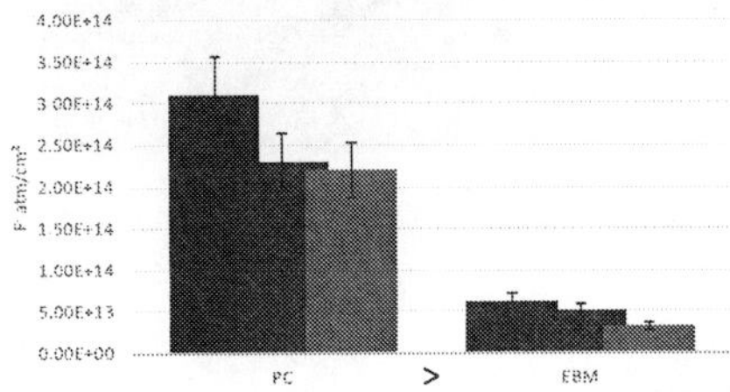

Entegris leti

SUMMARY

- HF transfer from FOUP to Cu-wafers:
  - In purge solution: PC > PC/CP > EBM/CNT
  - In wet clean solution: PC > EBM/CNT

  } Mainly diffusivity-dependent
- Gas purge
  - Limits (control) HF cross contamination by airborne removal during wafers storage
- Wet clean
  - Removes (partially) HF contamination in near polymer surface
  - More efficient closer to the contaminant event/process
- Use of obtained D,S coefficients allow HF behavior estimations as a contaminant into the polymer by simulation based on a membrane model
- Numerical estimations presents the same trend as experimental results

Among the tested FOUPs, EBM/CNT is the most efficient to limit HF contamination transfer to wafers → reduced wafer defectiveness is expected

Entegris leti

ADVANCED WAFER HANDLING: SOLUTIONS FOR CONTAMINATION CONTROL

Cross contamination is a synergic phenomena, decontamination solution as well

- No single solution can avoid cross contamination

Barrier FOUP

Moisture, HF content remains low longer

Barrier material and/or coatings

+

Advanced FOUP Purge

Moisture content decreases faster

Flow distribution diffusers to improve purge near to wafer critical position (edge/top)

+

Wet clean

Very efficient for particles removal

Wet clean by ECP

FOUP material, purge and wet cleaning are a set of variables that have great impact on yield, until now better strategy to improve yield and minimize the wafer loss is to combine these three

50 YEARS OF PURE ADVANTAGE

Non-uniform contamination results in sub-ppm fail rates

analytical challenges

Martin Knotter
SPCC2017, Austin

March 28, 2017

SECURE CONNECTIONS
FOR A SMARTER WORLD

History & Future

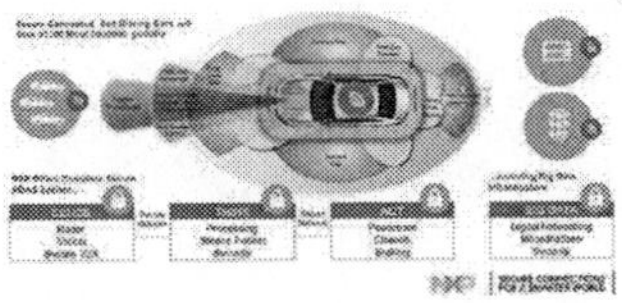

Expected end of 2017.
One of the reason is **Automotive Quality**

What is automotive quality for surface preparation

• ITRS FEP road map was also based on 99% yield	10000 ppm
• Most customer applications have 99.98%	200 ppm
• A few years ago we had 1-ppm mind set: 99.9999%	1 ppm
• Today we have zero-ppm: 99.999998%	20ppb
• Goal is to have per failure mode > 99.9999997%	<< 3 ppb!

With 1 ppm and 1000 IC's in a car 1 of 1000 will fail due to one IC failing
With 200 ppm 18% of the cars will fail

3. March 24, 2017

ITRS Roadmap surface preparation

Road map was base on 99% and in ITRS 2.0 99.9% calculation are given

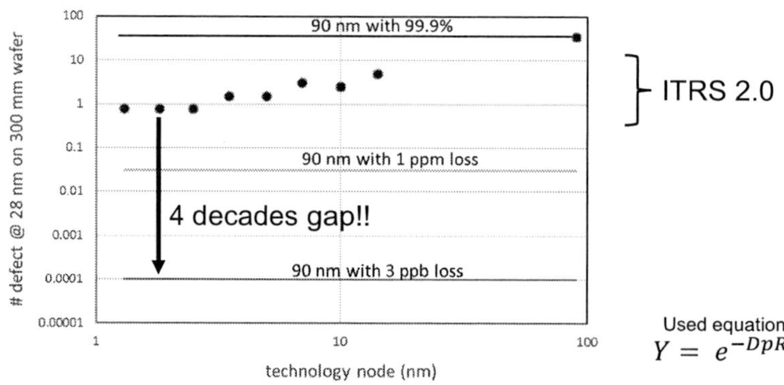

Defects on critical gate oxide area

Used equations:
$Y = e^{-DpRpAeff}$

4. March 24, 2017

Yield: failures in the field

- Many opportunities to fail with a chance, p_x

$$Y_{loss} = \sum_{x=1}^{n} p_x$$

- For many failure modes there is a detection with a chance to detect (d_x)

$$Y_{loss\ at\ customer} = \sum_{x=1}^{n} d_x p_x$$

What is going to the customer?

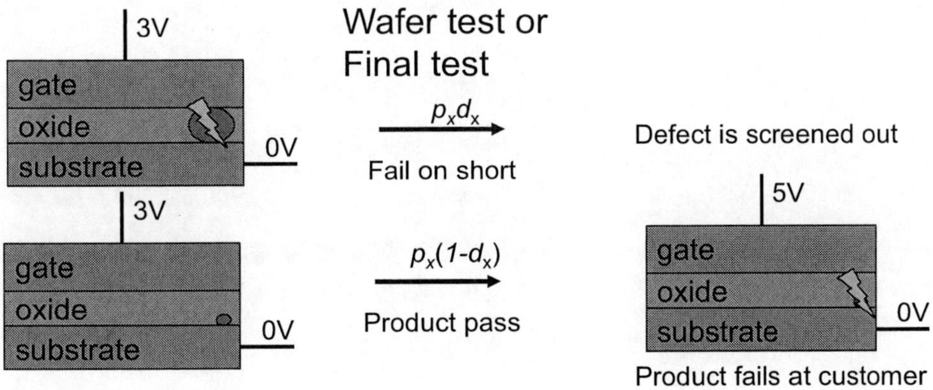

What is going to the customer?

- Gap is large, but is bridged by:
- Improving processes
  - Reduce defect addition
  - Improve cleaning
- Filter out by advanced detection methods

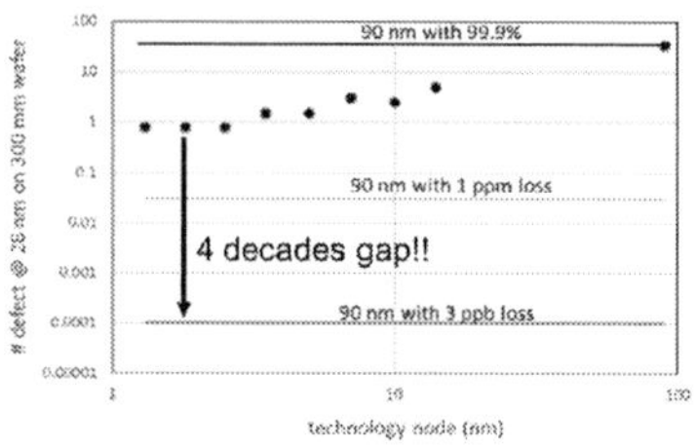

What are we doing to focus on the problems…

7. March 24, 2017

OUR TOOLS

Manufacturing to field failure

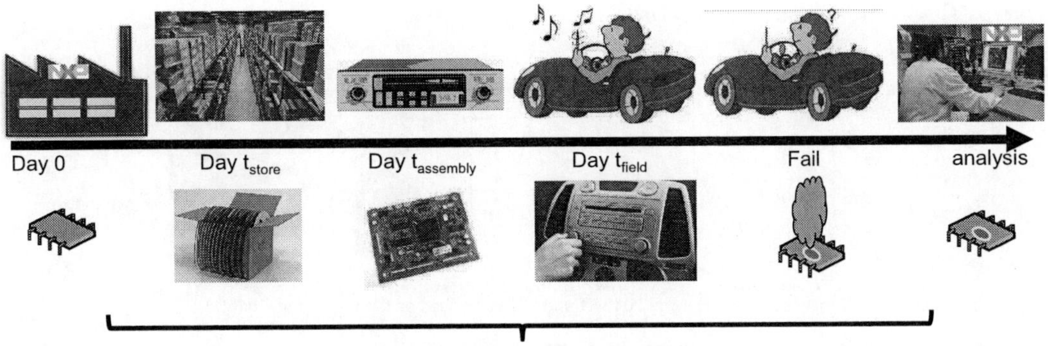

Return Time has lot of variation

9. March 24, 2017

Forecast of severity: crawl chart

Return profile of products manufactured in 2013 Q1

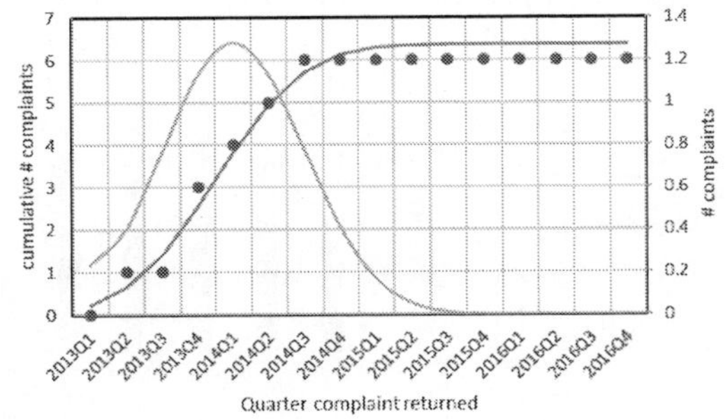

No real data

After Failure Analysis, the root cause is known and such charts can be made for, i.e. gate oxide fails

10. March 24, 2017

Crawl Charts: extrapolation

- Forecast using Crawl Chart

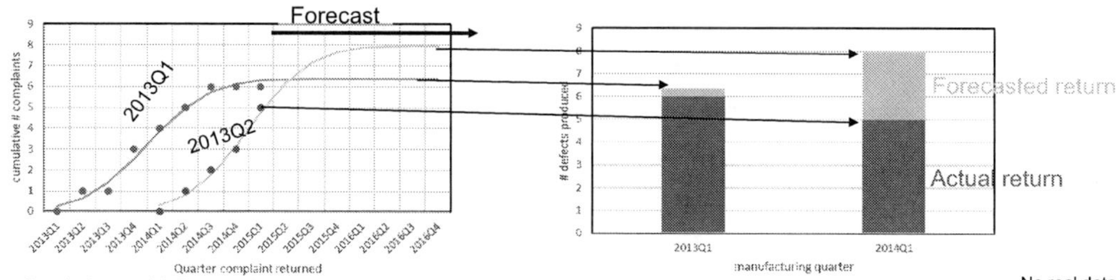

Crawl charts of failing products produced in 2013Q1 and 2014Q1 based on the return date 2015Q1

No real data

Since the shape of crawl chart is assumed to be constant, the maximum CQC can be forecasted by scaling the curve to fit the known points

NXP

Crawl Charts: trends and extrapolation

- Multiple Crawl Charts information gets mixed up

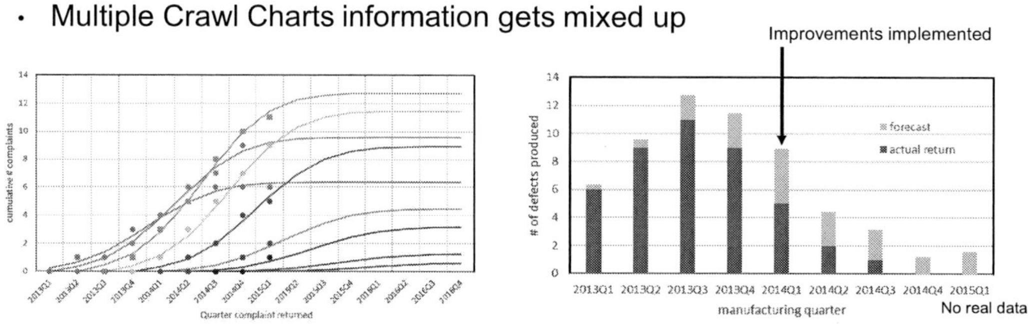

No real data

Summary of multiple crawl charts

Problem is solved

NXP

Failure Analysis (FA) with full traceability

- Samples are rare
- Physical FA: Many tricks to investigate reason and root cause of fail preferably non destructive.
- Process reason investigation:
  - Die ID or batch ID
  - Test-yield check (e.g. escaped product, $p_x(1-d_x)$, of large group found in test $p_x d_x$)
  - Trace back the full process history.
    - On which tool it has been on which time including rare events.
    - If more defective products returned: Communality studies: e.g. all 3 defects were processed on the same wet bench (of the four that are in use for this process step)

13. March 24, 2017

#1 failure mode in FE

- Gate oxides blown up (root cause evidence is gone)
- There are many reasons for gate oxide fails
- In some cases there are communalities such as with clean, wet etch,or dryer. Containment:
  - Golden flow
  - Make changes in process
- What is the root cause?
- Do we have defects outside our vision?

14. March 24, 2017

POTENTIAL FAILURE MODE

Non uniform contamination: hypothesis

In spec:

But potential defect

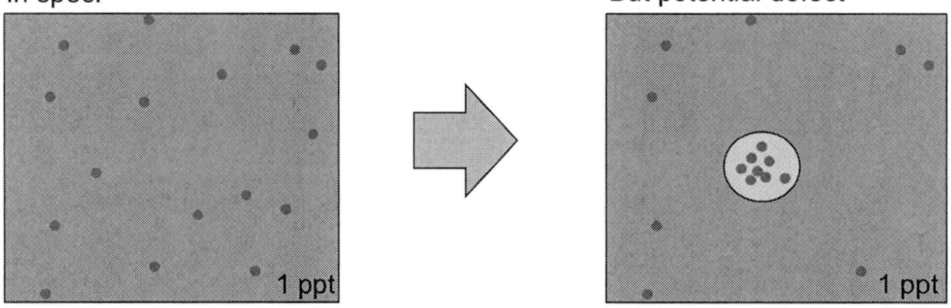

Deposition by adsorption and carry-over layer

Deposition by adsorption, particle deposition, and carry-over layer

Known reaction

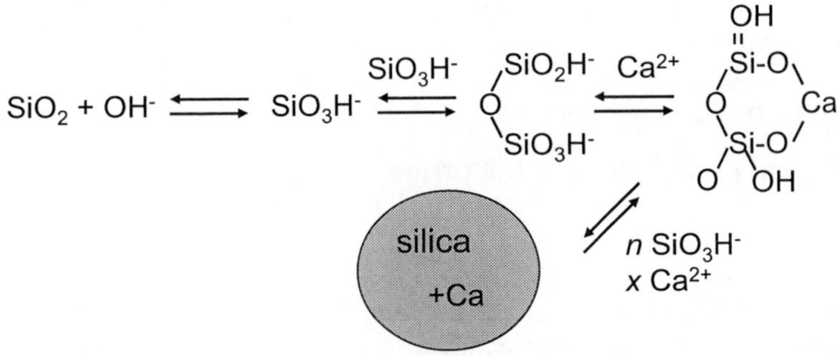

We are able to make these particles (but >> 50 nm)
These cause defects, while similar Ca-free silica particles cause no defects.
What determines kill ratio of particle?

Impact of Metal-Ion Contaminated Silica Particles on Gate Oxide Integrity by I. Rink, F. Wali, D.M. Knotter in Solid State Phenomena 145-146:131-134

17. March 24, 2017

Weak responses: particles in water

$p_x d_x$

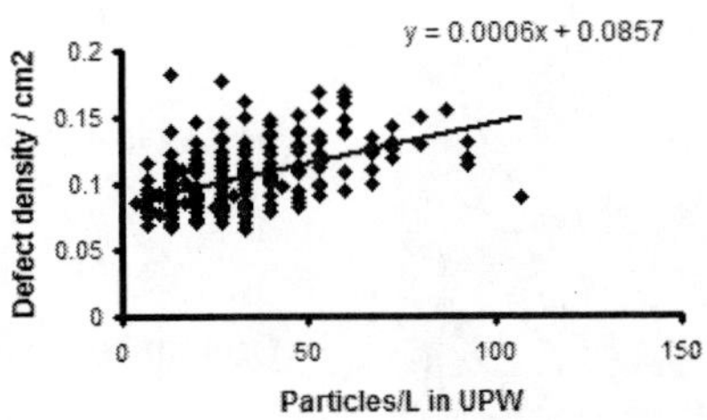

??? → $p_x(1-d_x)$

Particles do not account for all defects, but slope is significantly different from 0.

Impact of particles in ultra pure water on random yield loss in IC production F. Wali, D. M. Knotter, A. Mud, F. Kuper Microelectronic Engineering 86(2):140-144

18. March 24, 2017

Challenge FE analysis

- We have circumstantial evidence for particles having high M contamination can result in fails
- We lack analytical methods for a full prove.

- By the way: Solution is <u>not</u>:
  - Reduce silica in water (it will etch wafers)

- In BE we have an example….

19. March 24, 2017

Contamination in mold compound

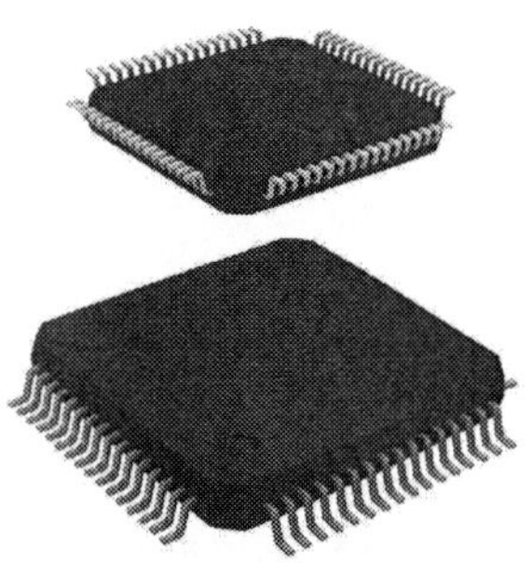

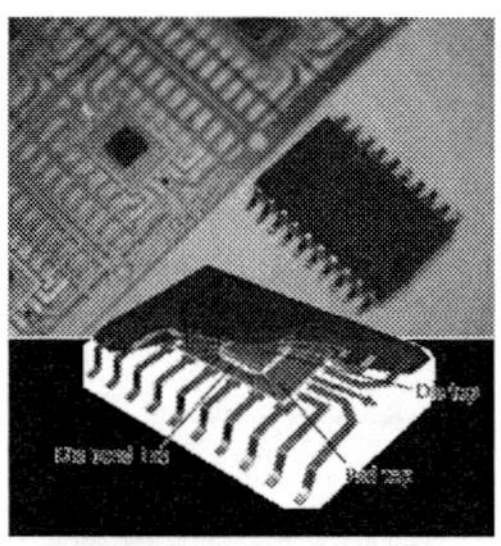

AlCu + Cl^- + xH_2O ⟶

AlO_xH_y + Cu + Cl^- + z H_2

⇩

Ball lift and defective product

⇩

Specification Cl in mold compound
< 35 ppm

20. March 24, 2017

Contamination in mold compound: particle

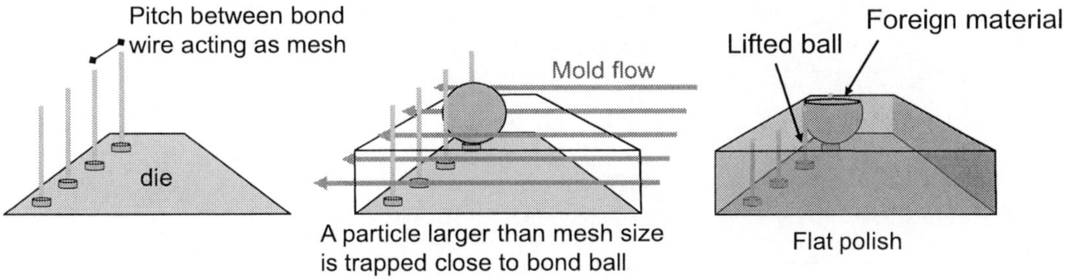

Evidence of root cause of failure is still present

21. March 24, 2017

Cl-containing particle in package

SEM-EDX

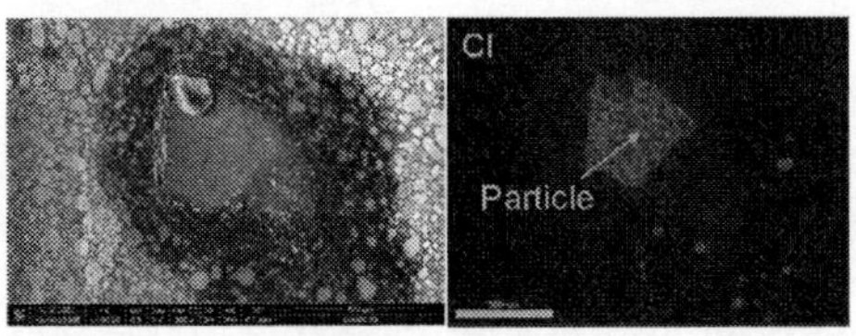

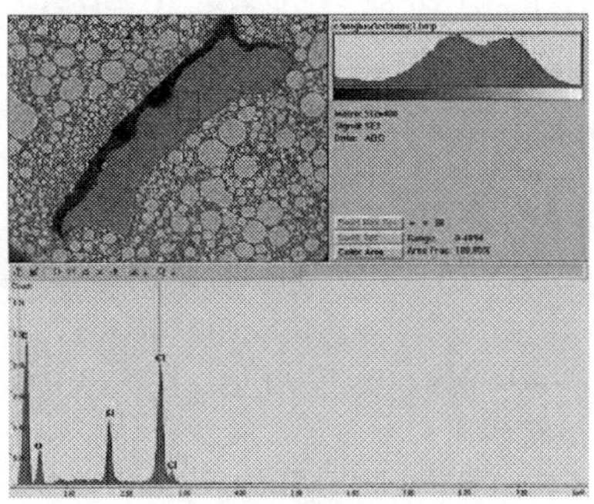

Rik Otte, Rob Fonville and Martin Knotter ": Identification of Foreign Particles in Packages of Failed Products" ESREF 2017 submitted to : E - Packaging and Assembly Reliability and Failure Analysis

22. March 24, 2017

Bond pad corrosion / PDC / Martin Knotter

Confirmation of PVC particle

Note that XPS would also differentiate between $CaCl_2$, NaCl

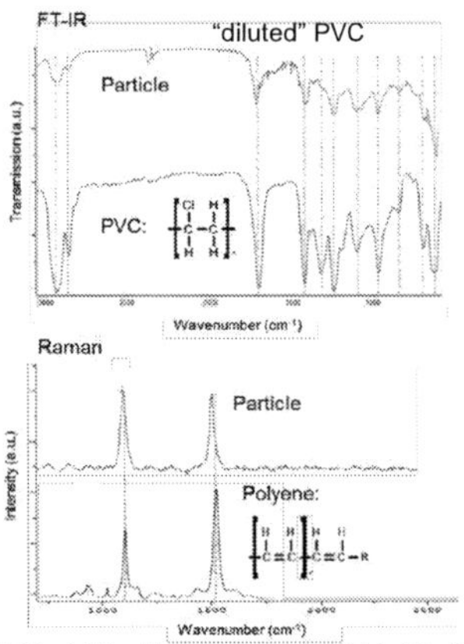

XPS:
Organically bonded chloride is at 200eV and chloride salts typically 198.5eV.

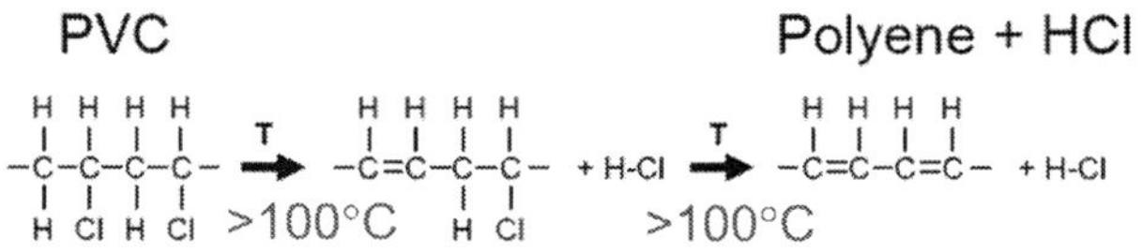

Hjertberg T and Sörvik EM. Thermal degradation of PVC. Degradation and Stabilisation of PVC. Springer Netherlands, 1984, pp 21-79

Rik Otte, Rob Fonville and Martin Knotter ": Identification of Foreign Particles in Packages of Failed Products" ESREF 2017 submitted to : E - Packaging and Assembly Reliability and Failure Analysis

Summary of BE problem

- Bulk material is well within specification for Cl content and particles
- Product fails because of a particle generating locally a contamination level higher than specification.
- In this case we succeeded in the development of
  1. An analytical method to measure local high contamination levels and the composition of this material
  2. An Analytical method (not in this presentation) to determine this in the bulk material
- This enables us to find root causes and improve our processes.

Conclusion

- We need to develop analytical methods to determine the distribution of contaminants within a bulk material.
- We need analytical methods to determine contamination levels at the location of failure.

SECURE CONNECTIONS
FOR A SMARTER WORLD

ENGINEERING
TEXAS A&M UNIVERSITY

HIPER LAB

HIPER Lab
Harris Integrative Photonics and Electronics
Research Laboratory

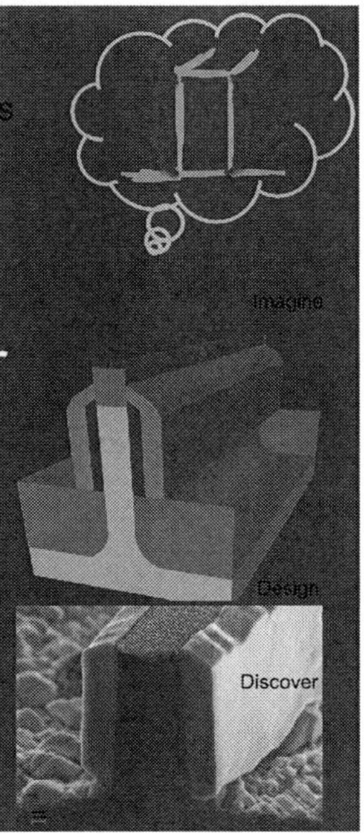

Nitride Device Surface Processes: A Review of Surface Science for Power and RF Technologies with Nitrides

H. Rusty Harris - Associate Professor
Department of Electrical and Computer Engineering
Department of Physics and Astronomy
Texas A&M University
College Station, TX

ELECTRICAL & COMPUTER ENGINEERING

email: rusty.harris@tamu.edu

ELECTRICAL & COMPUTER ENGINEERING

Outline

- Gallium Nitride Power and RF Market
- What's so unique about Nitrides?
  - POLARIZATION!
  - Formation of 2D Electron Gas
  - Surface Defects – the good with the bad
  - Threshold voltage
- Surface and Interface case studies
  - ALD High k dipole formation
  - Surface and Heterostructure interface charge trapping
  - Ohmic Contacts
- Conclusion

Why GaN?

Johnson Figure of Merit

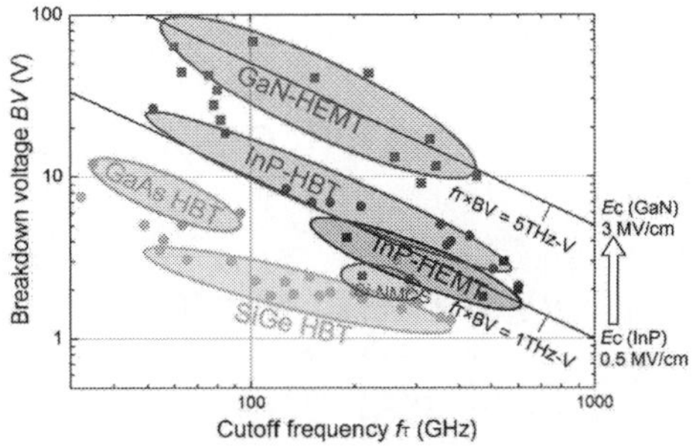

HIGH SCALABILITY OF GAN/ALGAN HEMTS

Scaling	Material properties	Advantage
Vertical scaling	·High DoS in GaN (5× InGaAs) ·Polarization doping ·High potential barrier of AlGaN	·High vertical scalability ·Thin top barrier w/ high n_s ·Low gate leakage
Lateral scaling	·High peak saturation velocity ·High breakdown field	·High *JFoM* ($JFoM = f_T \times BV$)
Parasitic reduction	·High n_s ($1\text{-}2\times10^{13}$ cm^{-2}) ·High μ (1500 cm^2/V·s)	·Low parasitic resistance ($\sigma = n_s \cdot q \cdot \mu$)

$$\tau = \frac{1}{2\pi f_T} = \frac{C_{gs}+C_{gd}}{g_m} + C_{gd}\cdot(R_s+R_d)\cdot\left[1+\left(1+\frac{C_{gs}}{C_{gd}}\right)\frac{g_d}{g_m}\right]$$

K. Shinohara, et al. IEEE Transactions on Electron Devices doi:10.1109/TED. 2013.2268160

Baliga Figure of Merit

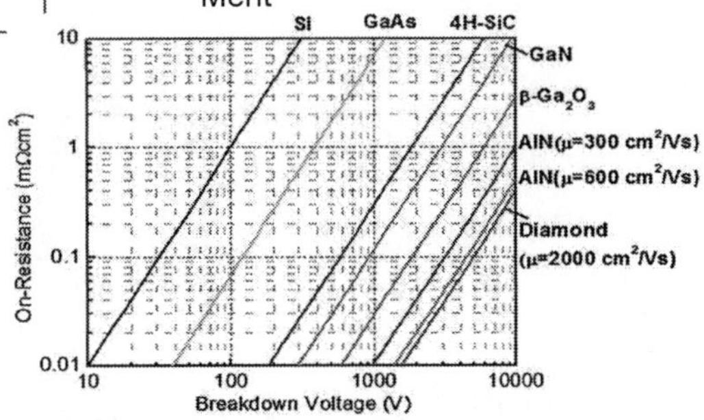

Smaller, tougher, faster, & more efficient power systems

2015 projection: GaN state-of-the-art device (V & Amp) vs. application requests

(Source: Power GaN Market 2014, Yole Développement, June 2014)

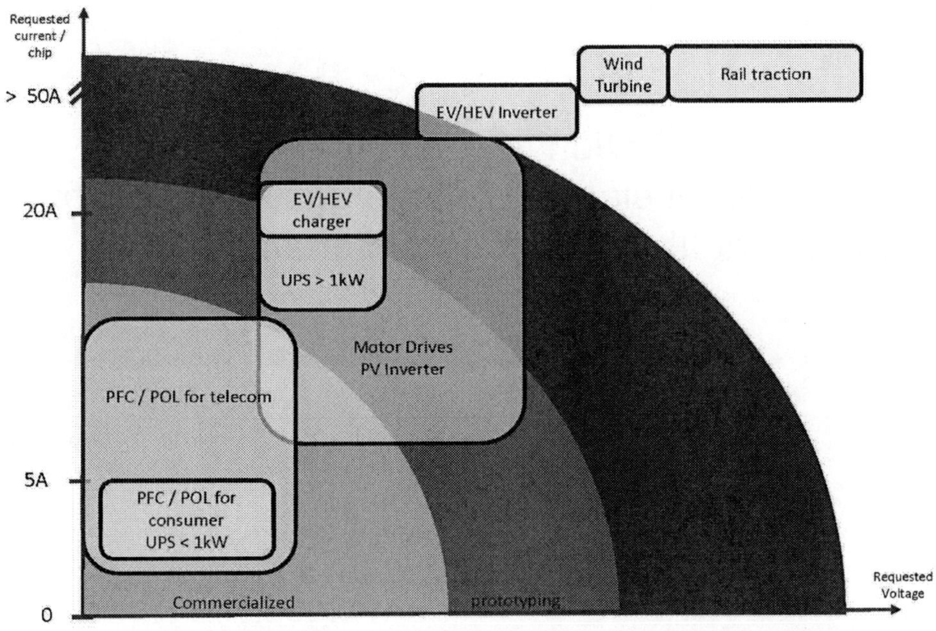

ELECTRICAL & COMPUTER ENGINEERING

Background - GaN

- Wurzite Structure
- Spontaneous Polarization
  - Electronegativity
  - Asymmetric
- Piezoelectric Polarization
  - Strain
- Wide Band gap

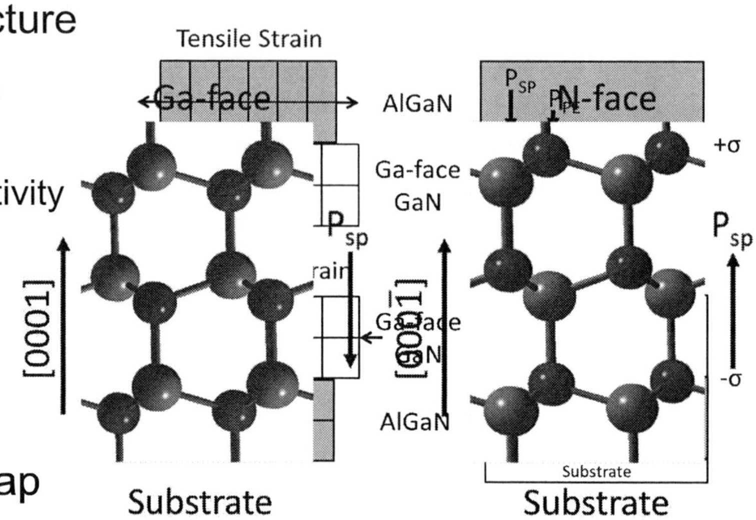

ELECTRICAL & COMPUTER ENGINEERING

Fundamental Physics

- These polarization vectors can lead to two dimensional electron or hole gases (2DEG or 2DHG) at III-Nitride heterointerfaces

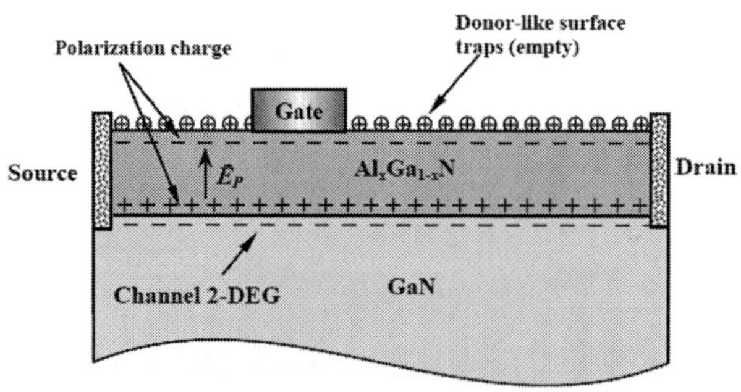

Outline

- Gallium Nitride Power and RF Market
- What's so unique about Nitrides?
  - POLARIZATION!
  - Formation of 2D Electron Gas
  - Surface Defects – the good with the bad
  - Threshold voltage
- Surface and Interface case studies
  - ALD High k dipole formation
  - Surface and Heterostructure interface charge trapping
  - Ohmic Contacts
- Conclusion

Surface States and interface problems with GaN

JOURNAL OF APPLIED PHYSICS 101, 081704 (2007)

Microscopic origins of surface states on nitride surfaces*

Chris G. Van de Walle[a] and David Segev
Materials Department, University of California, Santa Barbara, California 93106

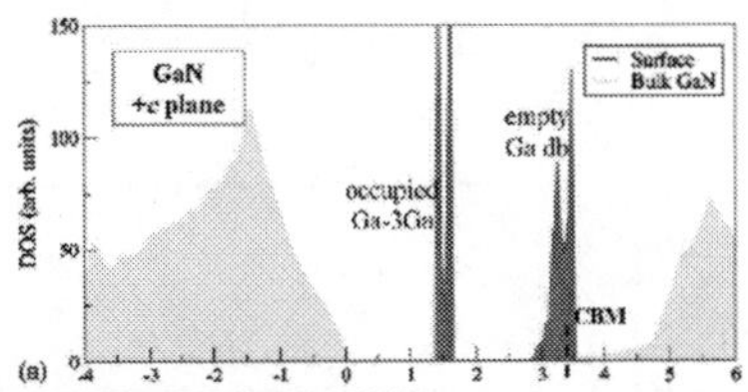

IEEE TRANSACTIONS ON ELECTRON DEVICES, VOL. 48, NO. 3, MARCH 2001

Polarization effects, surface states, and the source of electrons in AlGaN/GaN heterostructure field effect transistors

J. P. Ibbetson,[a] P. T. Fini, K. D. Ness, S. P. DenBaars, J. S. Speck, and U. K. Mishra[b]
Department of Electrical and Computer Engineering and Materials Department, University of California, Santa Barbara, California 93106

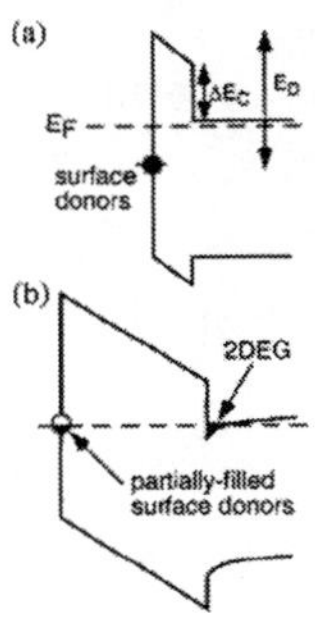

The Impact of Surface States on the DC and RF Characteristics of AlGaN/GaN HFETs

Ramakrishna Vetury, *Member, IEEE*, Naiqain Q. Zhang, Stacia Keller, and Umesh K. Mishra, *Fellow, IEEE*

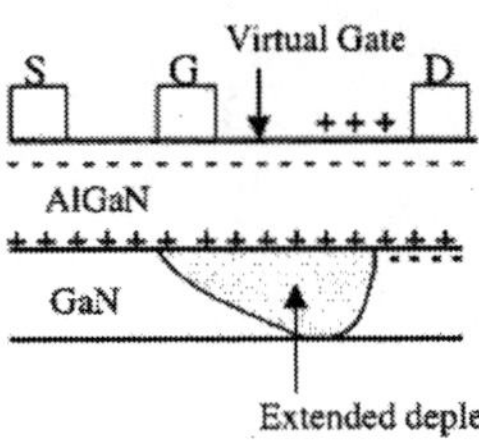

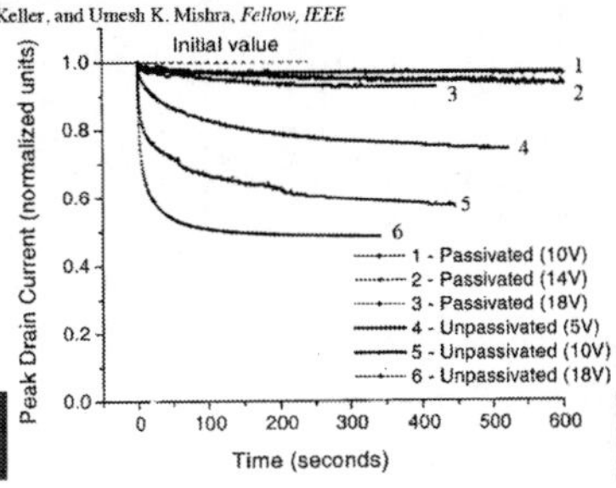

Because 2DEG exists at equilibrium…

…already on!

IEEE TRANSACTIONS ON ELECTRON DEVICES, VOL. 53, NO. 9, SEPTEMBER 2006

Control of Threshold Voltage of AlGaN/GaN HEMTs by Fluoride-Based Plasma Treatment: From Depletion Mode to Enhancement Mode

Yong Cai, Yugang Zhou, Kei May Lau, *Fellow, IEEE*, and Kevin J. Chen, *Member, IEEE*

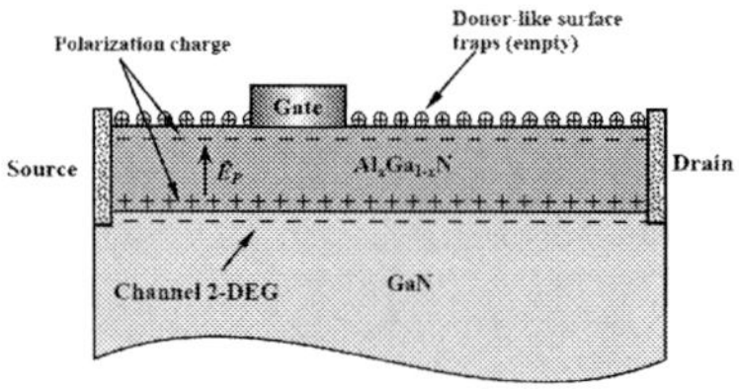

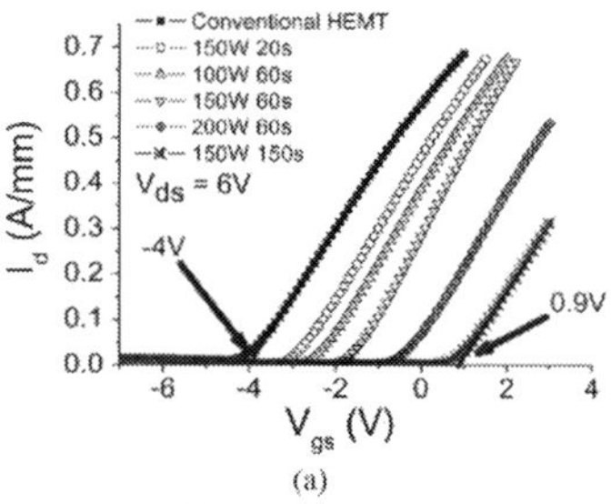

(a)

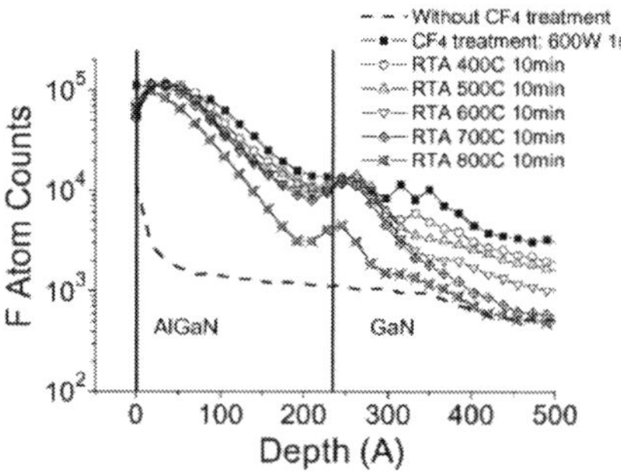

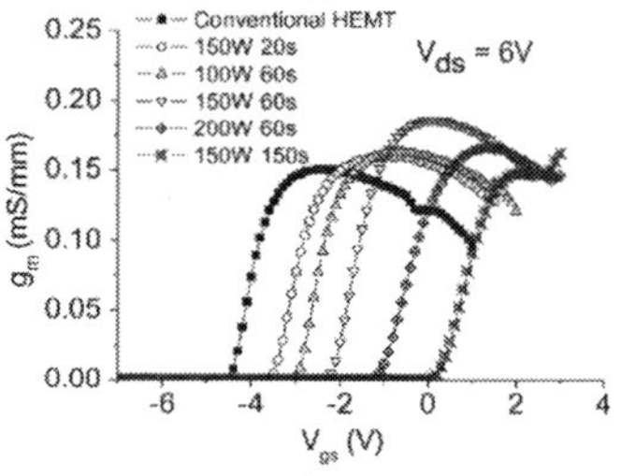

(b)

These imp

As the channel is t… …terface decrease the virtual gate pot…

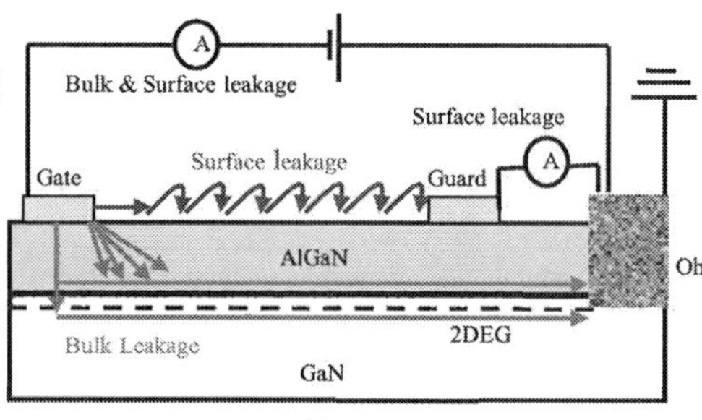

(a)

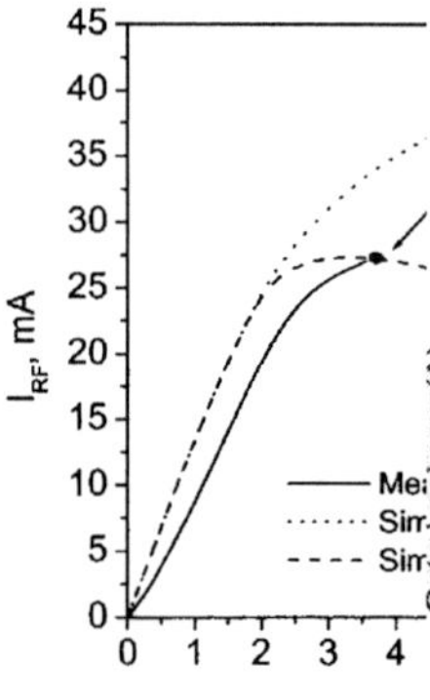

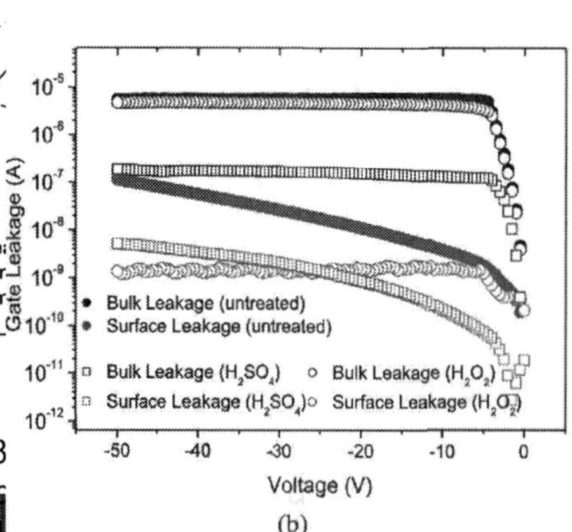

(b)

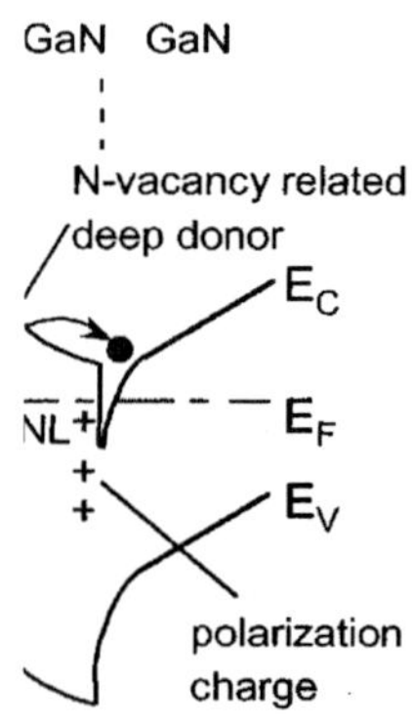

(a)

Appl. Phys. Lett. 78
http://dx.doi.org.ez
10.1063/1.1363694

Outline

- Gallium Nitride Power and RF Market
- What's so unique about Nitrides?
  - POLARIZATION!
  - Formation of 2D Electron Gas
  - Surface Defects – the good with the bad
  - Threshold voltage
- Surface and Interface case studies
  - ALD High k dipole formation
  - Surface and Heterostructure interface charge trapping
  - Ohmic Contacts
- Conclusion

Background – Cleaning GaN

- Cleaning Procedures
  - aq-HCl with or without anneal
    - N-deficiency and Chemisorbed Cl^-
    - Annealing – lower concentrations of O, Cl, and C
  - aq-HF with or without anneal
    - Chemisorbed F^-
    - Annealing – lower concentrations of O, F, and C
  - Chemical Vapor Cleaning (CVC) using NH_3 at ≈ 900°C
    - "Atomically" clean surface

Band Offset Measurements of GaN (0001) – dielectric interfaces

- Experimental Set-up
  - aq-HCl Clean GaN
    - Cleaned using HCl, rinsed with water and dried
  - HfO_2/GaN
    - 3 nm ALD HfO_2 with RTP
  - Al_2O_3/GaN
    - 3 nm ALD Al_2O_3 with RTP
  - XPS, UPS and TEM

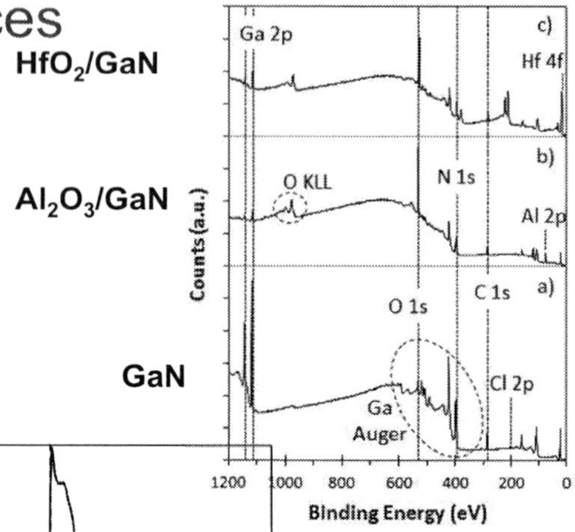

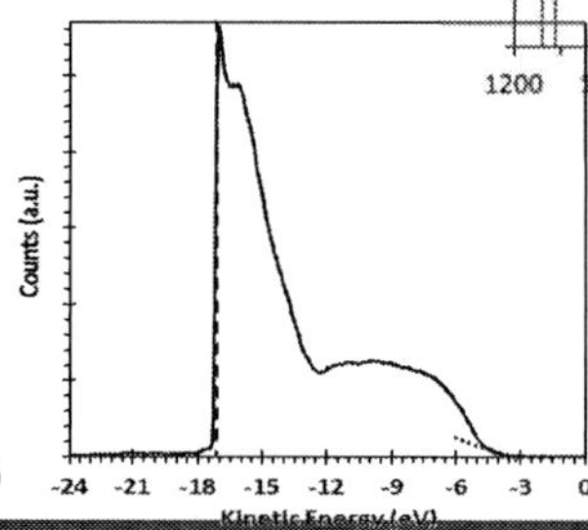

$$\chi = h\nu - W - E_g$$

Electron Affinity, Photon Energy, Spectra Width, Bandgap

And…

XPS analysis of GaN (0001) – *find overall shift in N and/or Ga with additional of each layer

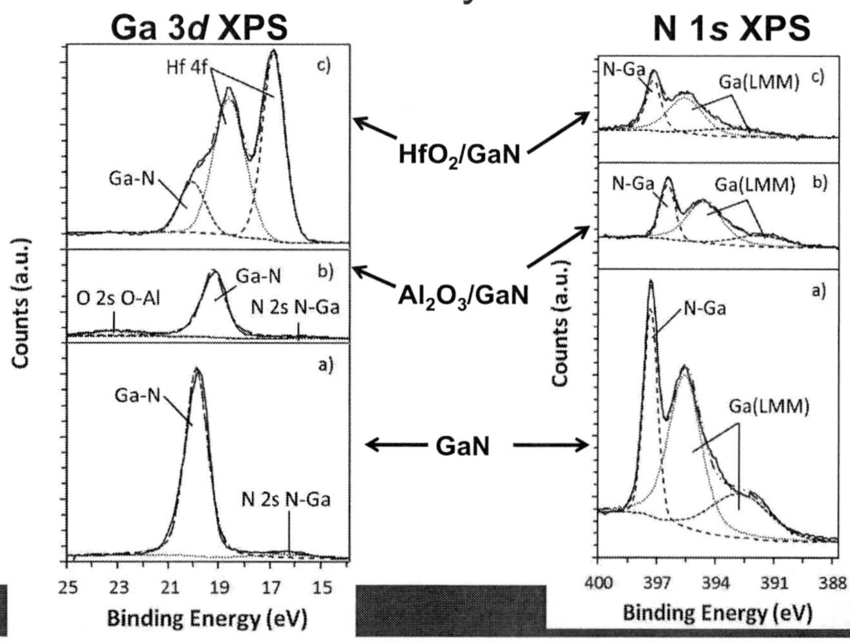

ELECTRICAL & COMPUTER ENGINEERING

Band Offset Measurements of aq-HCl cleaned GaN (0001)

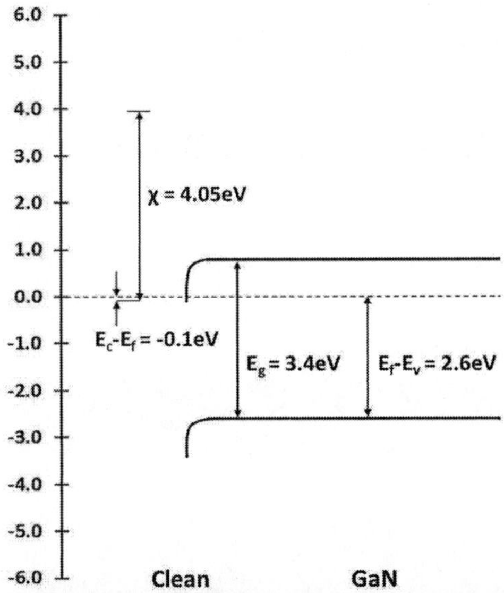

- VBM = 3.5 eV, X = 4.1 eV
- Downward band bending ~ 0.9 eV
  - Ga dangling bonds or N-vacancies

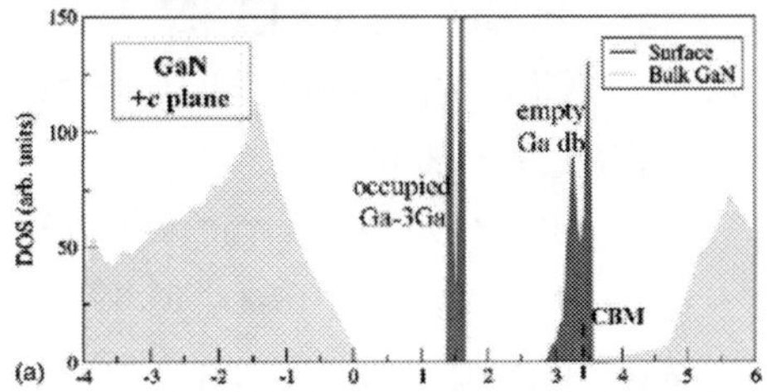

ELECTRICAL & COMPUTER ENGINEERING

Band Offset Measurements of GaN (0001) – HfO_2

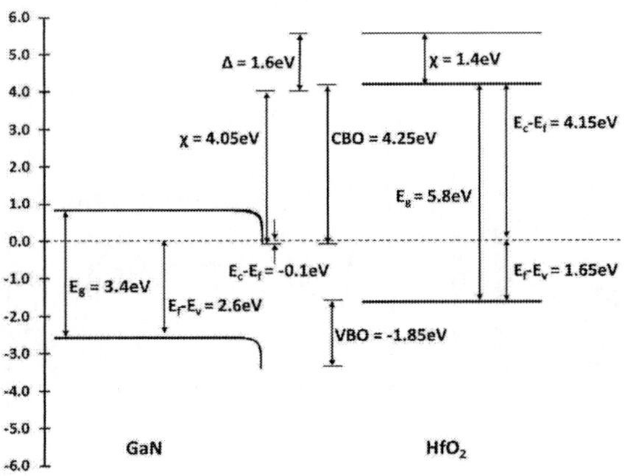

- No additional band bending
  - ΔGa 3*d* ≈ ΔN 1*s* ≈ 0 eV
    - Within error
  - $E_g(HfO_2)$ = 5.8 eV
  - CBO = 4.25 eV
  - VBO = -1.85 eV
- Δ = 1.6 eV

Band Offset Measurements of GaN (0001) – GaON

- Does an interfacial layer effect the calculated dipole at the GaN/HfO_2 interface?
  - Band offset measurements of thermally grown GaON on GaN
- Experimental Set-up
  - aq-HCl Clean GaN
    - Cleaned using HCl, rinsed with water and dried
  - GaON/GaN
    - 800 °C for 30 min with Dry O_2
    - ~ 3 nm thermal GaON
- XPS/UPS analysis

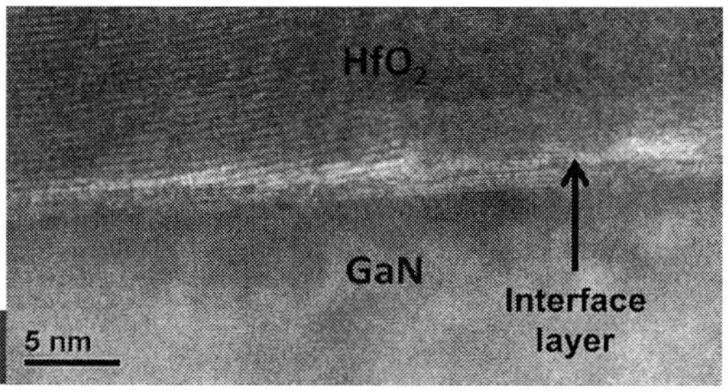

Band Offset Measurements of GaN (0001) – GaON

- aq-HCl GaN
  - VBM ~ 3.5 eV, Ionization Energy = 7.45 eV
- GaN/GaON
  - VBM ~ 7.4 eV below GaN, Ionization Energy = 9.55 eV
    - Similar shape and width to previously published work
    - Increase in Ionization Energy is similar to previously published work on the oxidation of AlGaN

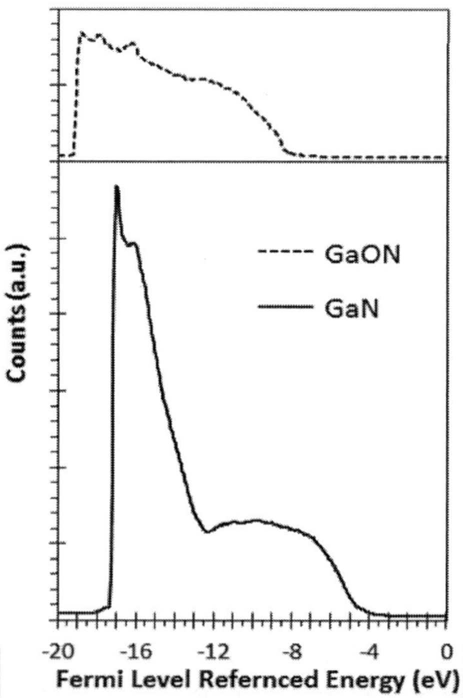

ELECTRICAL & COMPUTER ENGINEERING

Band Offset Measurements of GaN (0001) – GaON

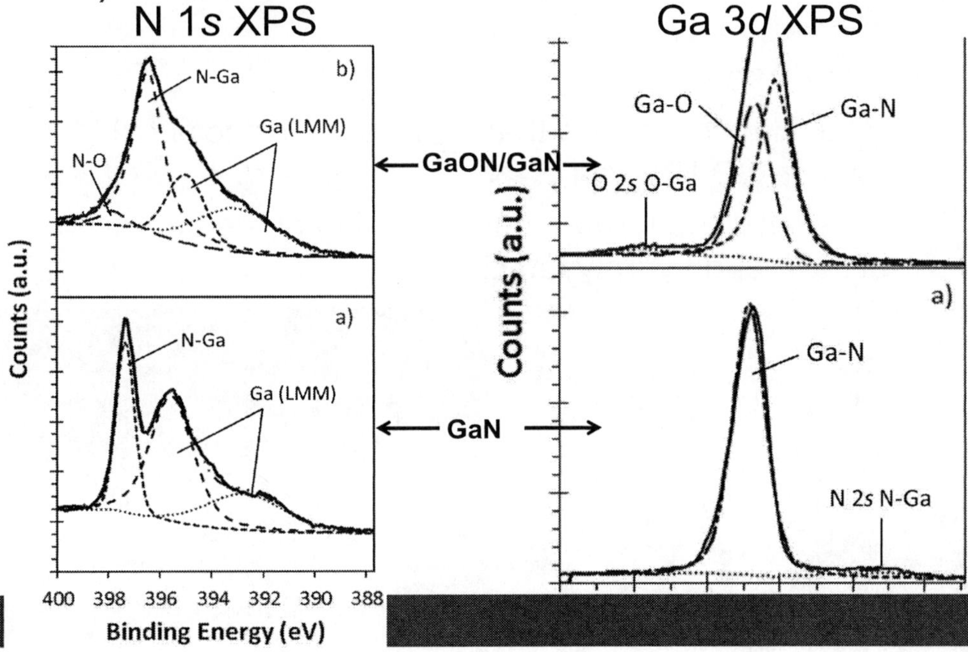

ELECTRICAL & COMPUTER ENGINEERING

Band Offset Measurements of GaN (0001) – GaON

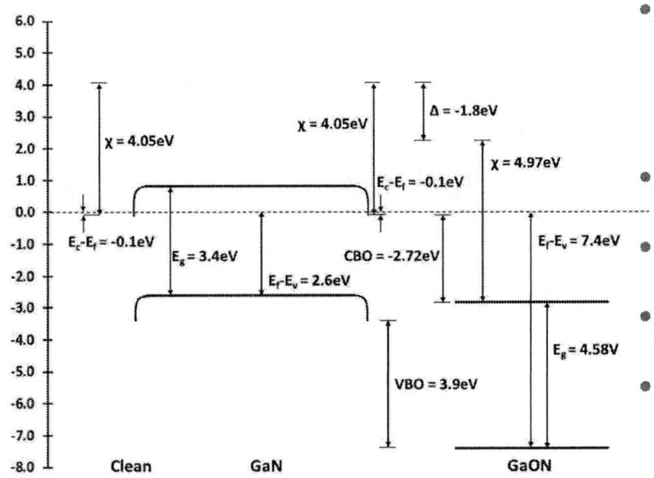

- Assume all shifts are due to changes in bonding
- E_g(GaON) = 4.6 eV
- CBO = -2.7 eV
- VBO = 3.9 eV
- Δ = -1.8 eV

Ab initio Analysis of the Interactions of Hydrolyzed GaN Clusters with TEMAH

ELECTRICAL & COMPUTER ENGINEERING

- In two cases a Ga-N(CH_2CH_3)(CH_3) bond formed
  - Possible mechanism that forms an interfacial oxide during ALD HfO_2 on GaN
  - No such bond formed during testing of the interactions of TMA and hydrolyzed GaN

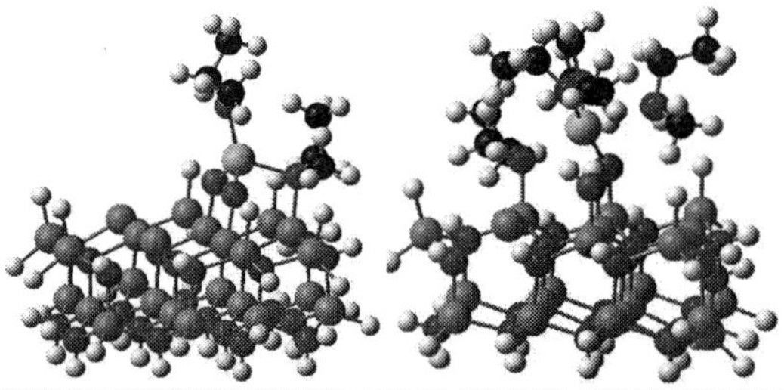

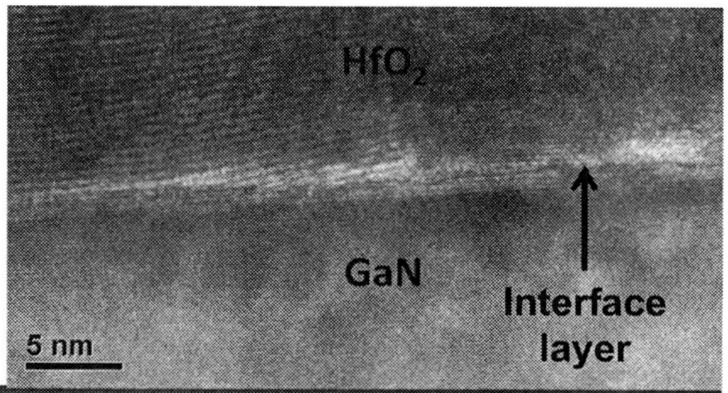

Outline

ELECTRICAL & COMPUTER ENGINEERING

- Gallium Nitride Power and RF Market
- What's so unique about Nitrides?
  - POLARIZATION!
  - Formation of 2D Electron Gas
  - Surface Defects – the good with the bad
  - Threshold voltage
- Surface and Interface case studies
  - ALD High k dipole formation
  - Surface and Heterostructure interface charge trapping
  - Ohmic Contacts
- Conclusion

C-V of dielectric-gated AlGaN/GaN

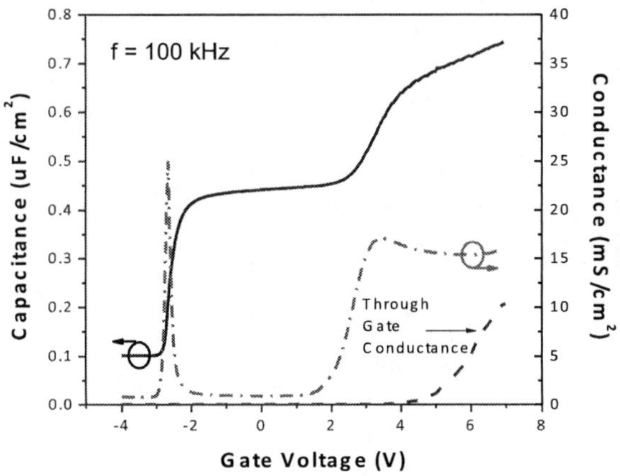

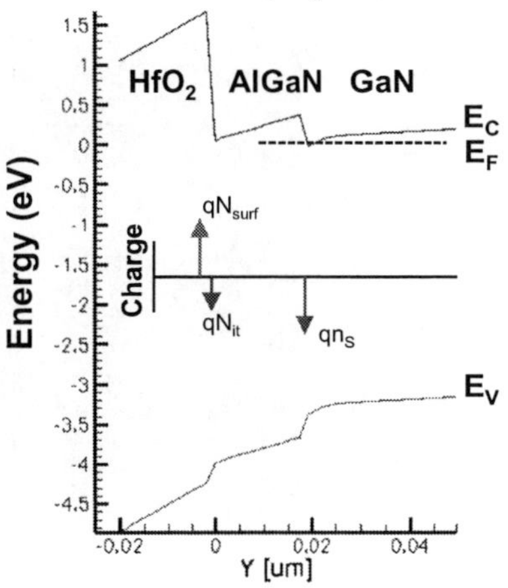

Secondary conductance peak onset at 2V indicative of Oxide / AlGaN interface trapping and / or accumulation

Interesting Shift in C-V

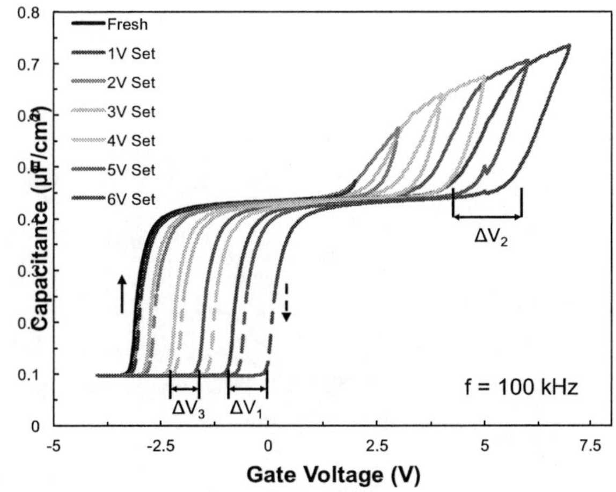

Charge trapping in the interfacial layer results in voltage shift of both 2DEG and oxide/barrier accumulation thresholds.

Baseline Performance (5/6)

Physical origin of trapping behavior

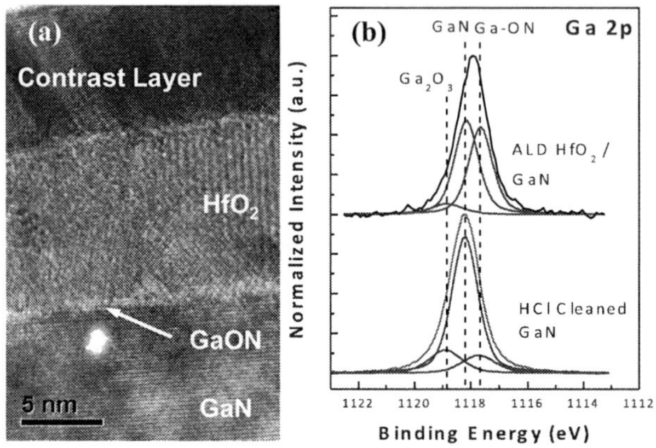

Ga^{1+} Oxynitride State -0.5 eV from core

Gate Stack – Interface Traps *(1/5)*

- Methodology

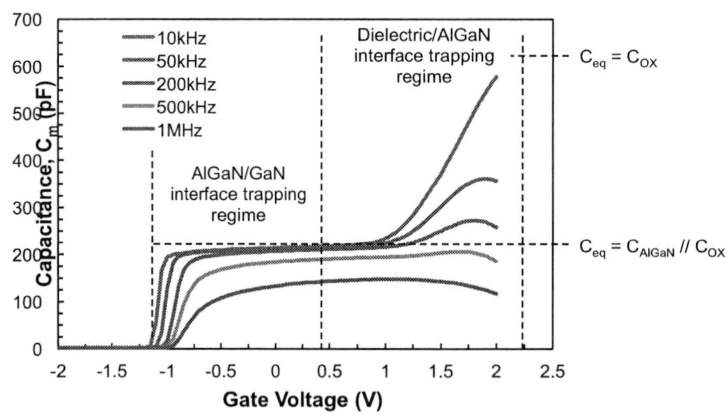

Careful analysis required in Dielectric/AlGaN region due to leakage current

Gate Stack – Interface Traps *(2/5)*

- Methodology – Conductance Method

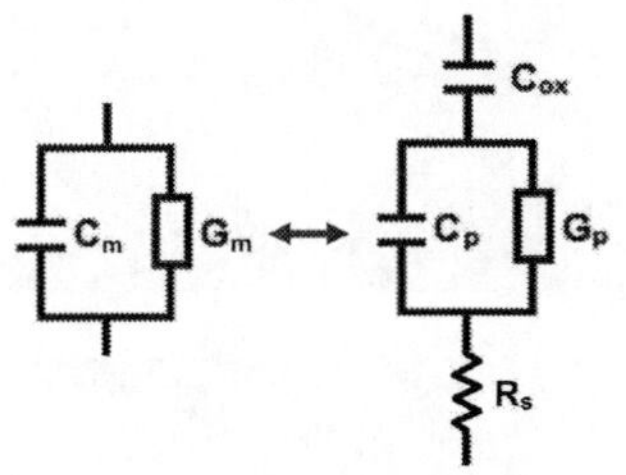

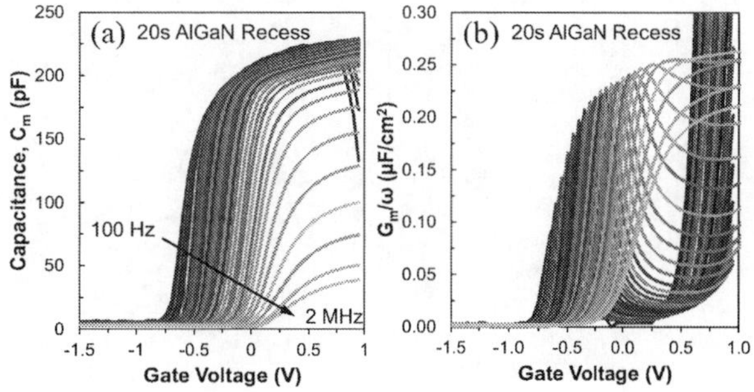

$$R_S = \frac{G_{m,a}}{G_{m,a}^2 + \omega^2 C_{m,a}^2} \quad (1)$$

$$C_{ox} = C_{m,a}\left[1 + \left(\frac{G_{m,a}}{\omega C_{m,a}}\right)^2\right] \quad (2)$$

27

Gate Stack – Interface Traps *(3/5)*

- Methodology – Conductance Method

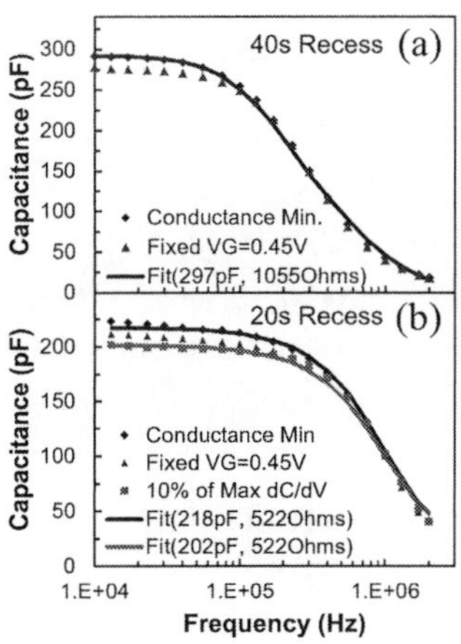

(c) $$C_m = \frac{C_{ox}[G^2 + (\omega C_{ox})^2]}{\{G + R_S[G^2 + (\omega C_{ox})^2]\}^2 + (\omega C_{ox})^2}$$

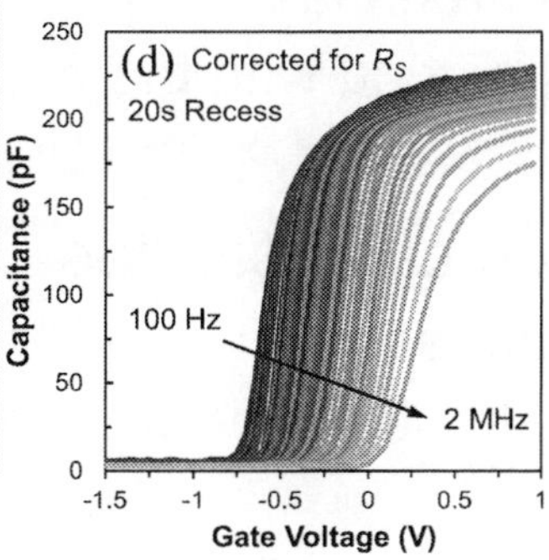

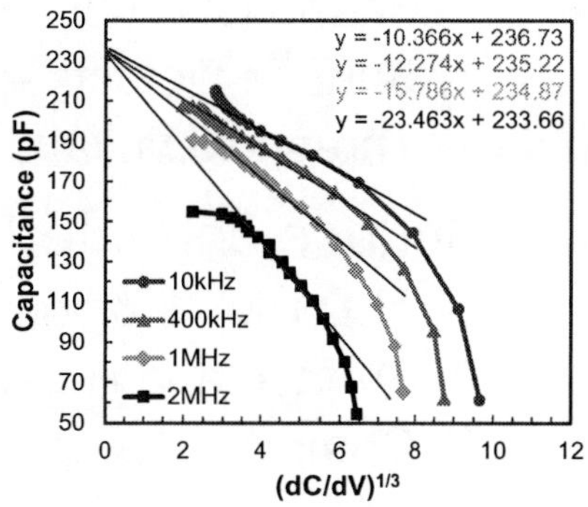

Series resistance determination

Barrier capacitance determination

28

Gate Stack – Interface Traps *(4/5)*

ELECTRICAL & COMPUTER ENGINEERING

- Methodology – Conductance Method

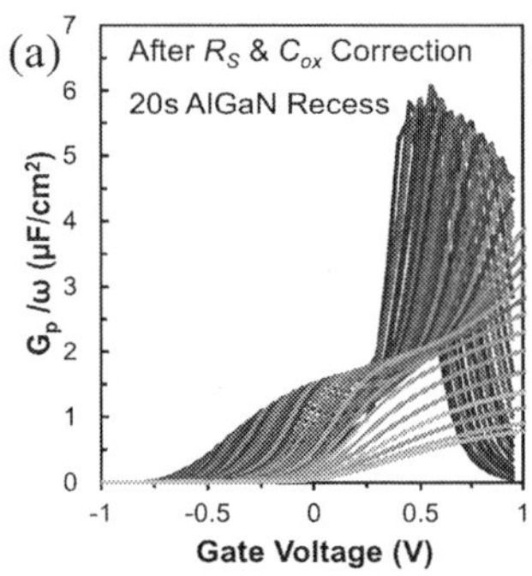

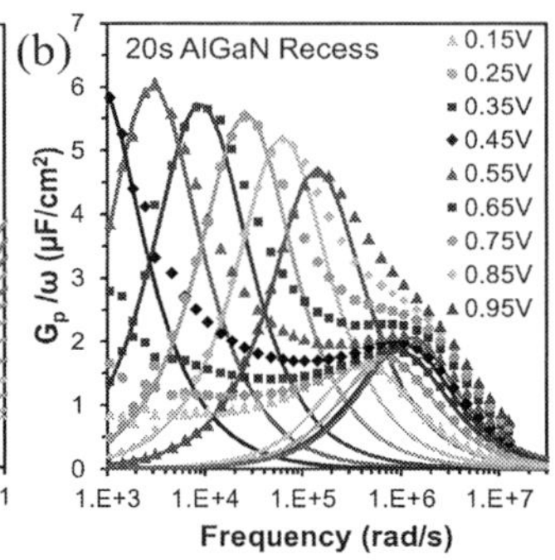

$$\frac{G_p}{\omega} = \frac{qD_{it}\omega\tau}{1+(\omega\tau)^2}. \qquad (4)$$

$$D_{it} \approx \frac{2.5}{A*q}\left(\frac{G_p}{\omega}\right)_{max} \qquad (5)$$

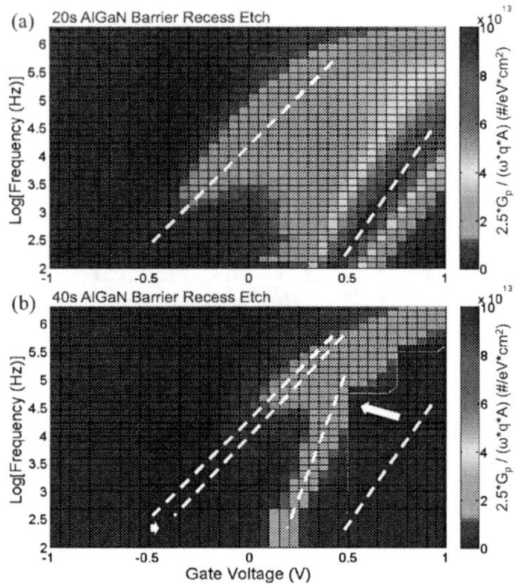

29

Outline

ELECTRICAL & COMPUTER ENGINEERING

- Gallium Nitride Power and RF Market
- What's so unique about Nitrides?
  - POLARIZATION!
  - Formation of 2D Electron Gas
  - Surface Defects – the good with the bad
  - Threshold voltage
- Surface and Interface case studies
  - ALD High k dipole formation
  - Surface and Heterostructure interface charge trapping
  - Ohmic Contacts
- Conclusion

ELECTRICAL & COMPUTER ENGINEERING

Ohmic contact to the 2DEG… with AlGaN in the way!

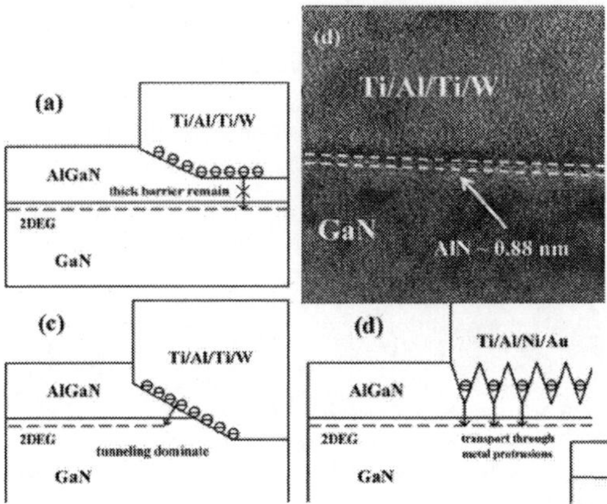

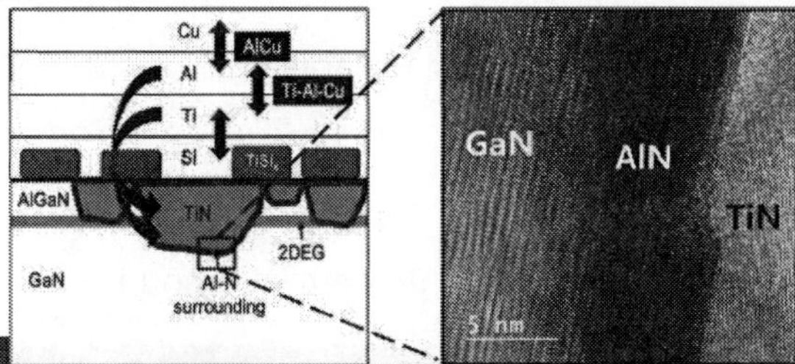

ELECTRICAL & COMPUTER ENGINEERING

Selective removal of unreacted Ti

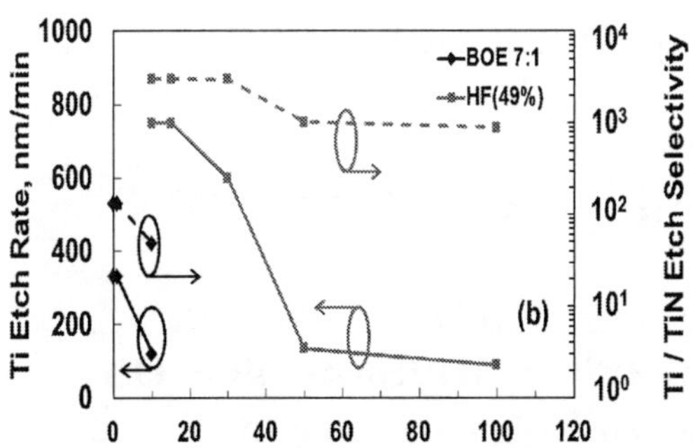

Titanium etch rate and selectivity relative to TiN for different dilutions of HF(49%) in DI-H2O and BOE (7:1).

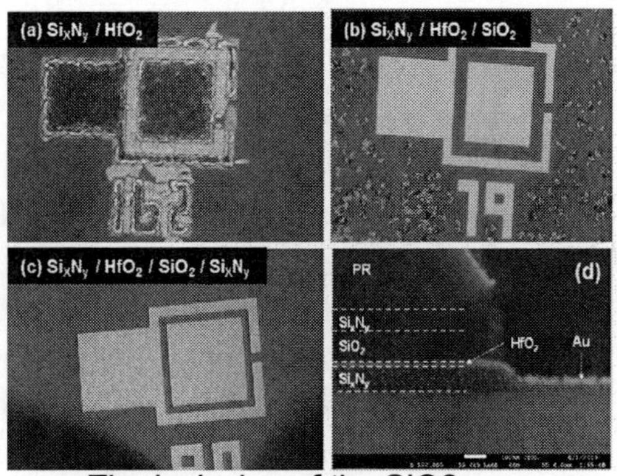

The inclusion of the SiO2 delamination layer and 15nm of Si3N4 fully removes residual Ti in the field region and small windows (d) SEM of final field dielectric stack.

And self-aligned ohmic contact transistors to GaN HEMTs!

ELECTRICAL & COMPUTER ENGINEERING

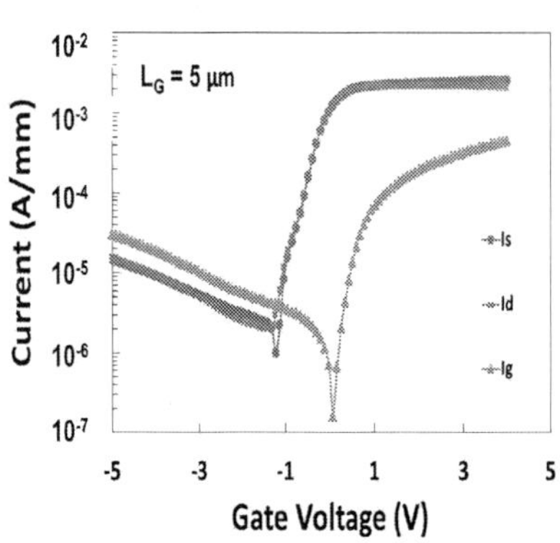

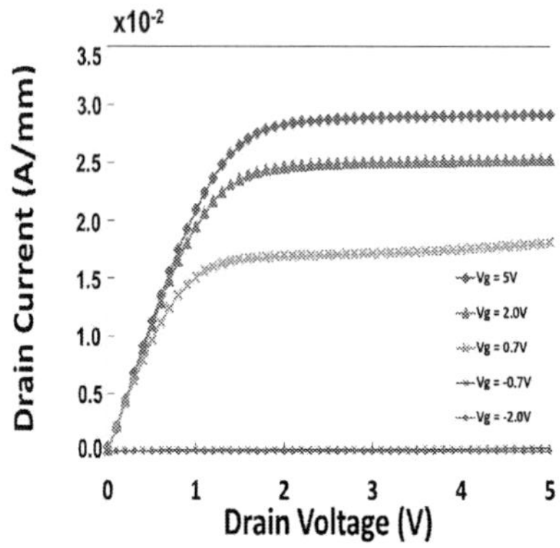

Left: Id-Vg at V_D = 0.1V with SS = 196 mV/dec, I_{ON}/I_{OFF} = 1100, V_{th} = -0.48 V and $g_{m,max}$ = 2.41 mS/mm

Right: R_{ON} = 16.98 ohm*mm and $I_{D,sat}$ = 29.85 mA/mm.

Summary

ELECTRICAL & COMPUTER ENGINEERING

- Nitrides will become more prevalent in Power and RF technologies
- Ga-N polar bond plays a major role in surface properties
- Clever use of cleans and device physics give us clear pathways to nitride success

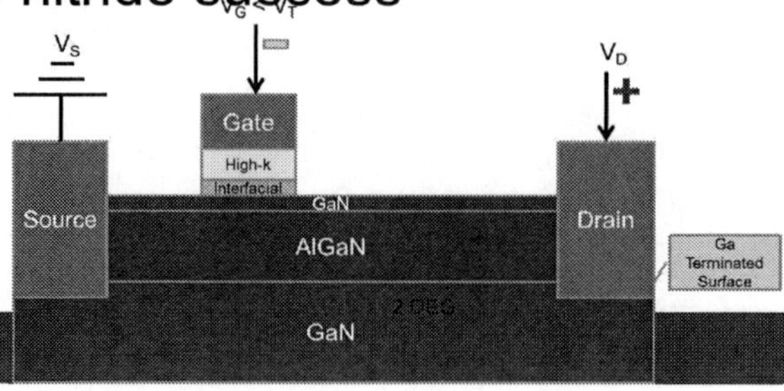

Aqueous Ammonium Sulfide Passivation and $Si_{1-x}Ge_x$ MOSCaps

Lauren Peckler[1], Stacy Heslop[2], and Anthony Muscat[1]

[1]Department of Chemical & Environmental Engineering, University of Arizona

[2]Department of Chemistry, University of Arizona

1

Motivation

- Ge and SiGe are alternatives to Si for CMOS devices owing to their smaller bandgap and increased hole mobility.
- Incorporating Ge into SiGe films can be used to tune material properties.
- Ge oxides form readily and the interface contains a high density of defects.
- SiGe with < 50% Ge has been integrated into current Si semiconductor manufacturing.
- SiGe with > 50% Ge has been difficult to integrate.
- Sulfur chemistry is accepted by industry for passivation.

2

Chemistry on the SiGe surface lead to poor MOSCap performance

Surface Chemistry

Dangling Bonds → Electronic states appear between band gap

Native Oxides → Poor nucleation of oxide layer; Poor control and repeatability in device manufacture; Multiple dielectrics

Electrical/Device Defects

- Electronic states appear between band gap
- Poor nucleation of oxide layer
- Poor control and repeatability in device manufacture
- Multiple dielectrics

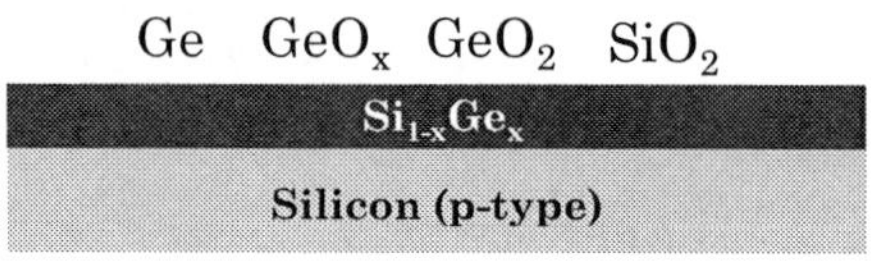

3

Goals

1. Remove oxides from SiGe surface and passivate Ge dangling bonds with $(NH_4)_2S$ chemistries to:
   - Reduce interface defects between SiGe and high k dielectric
   - Minimize defects in bulk layer of dielectric
   - Minimize electric oxide thickness
2. Compare 25% and 75% Ge substrates

$Si_{1-x}Ge_x$ (x = 0.25)

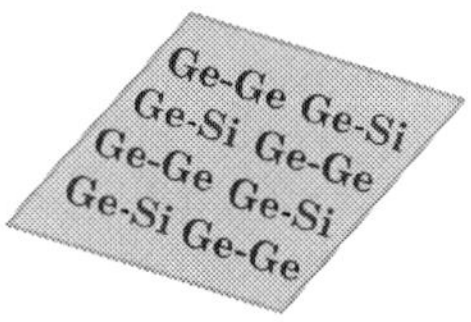

$Si_{1-x}Ge_x$ (x = 0.75)

4

Removing Oxides and Forming Ge-S

Remove oxides through wet chemistry clean*

Control Treatment

SC-1: 1:1:500, RT, 2 min
UPW, 1 min,
Slow dry with N_2 (~30 s)

↓

HF:HCl:H_2O (1:3:300)
5 min

Bond Ge dangling bond to S with $(NH_4)_2S$*

$(NH_4)_2S$ Treatment

(Control)

↓

$(NH_4)_2S$:H_2O
(1:100)
20 min

Acidic Treatment

(Control)

↓

$(NH_4)_2S$:HF:HCl:
H_2O
(1:0.15:0.15:100)
20 min

*Coupon Size Samples: 1 x 1 cm^2

5

A Study of Two Surfaces and Two MOSCaps

$Si_{1-x}Ge_x$ (x = 0.25)

Surface

~40 nm ↕ $Si_{1-x}Ge_x$ (x = 0.25)
Silicon (p-type)

MOSCap

Au (25 nm)
Ni (50 nm)
Al_2O_3 (10 nm)
$Si_{1-x}Ge_x$ (x = 0.25)
Silicon (p-type)

$Si_{1-x}Ge_x$ (x = 0.75)

Surface

~180 nm ↕ $Si_{1-x}Ge_x$ (x = 0.75)
Silicon (p-type)

MOSCap

Au (25 nm)
Ni (50 nm)
Al_2O_3 (10 nm)
$Si_{1-x}Ge_x$ (x = 0.75)
Silicon (p-type)

6

$Si_{1-x}Ge_x$ (x = 0.25)

Chemical & Electrical Characterization

7

On $Si_{1-x}Ge_x$ (x = 0.25) Ge-S not detected with $(NH_4)_2S$ treatment.

$(NH_4)_2S$ Treatment

Control Treatment

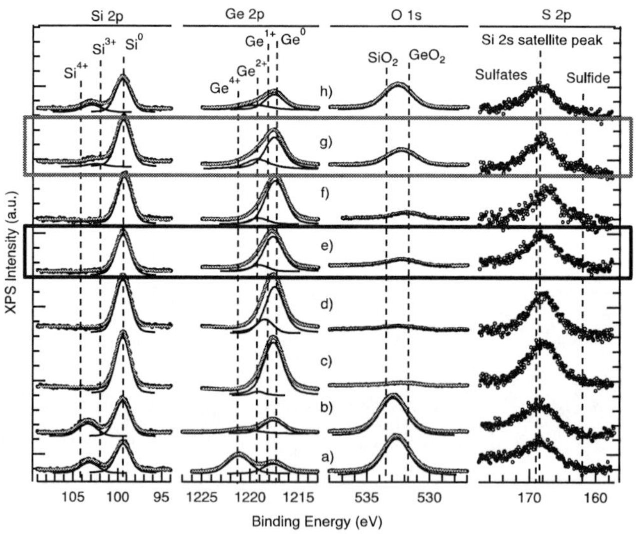

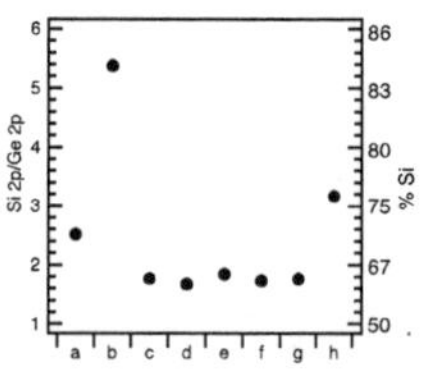

8

Ge-S is detected with acidic $(NH_4)_2S$ (1:100 v/v) treatment, as well as oxides.

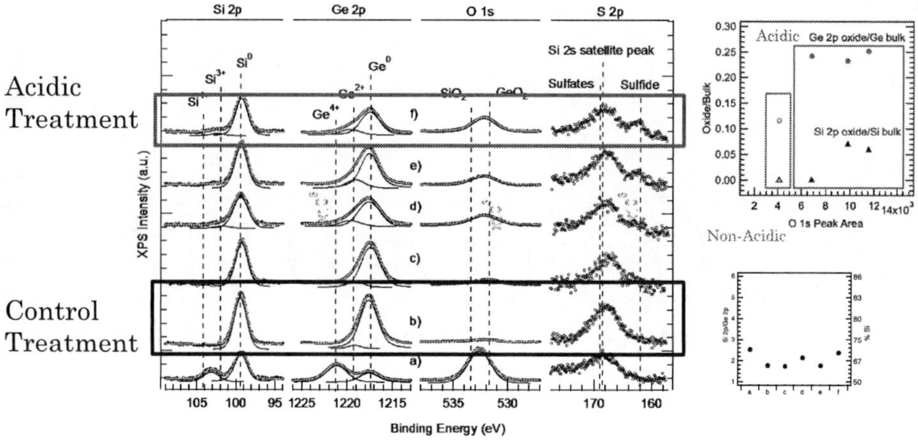

Acidic treatment reduces defects and capacitance.

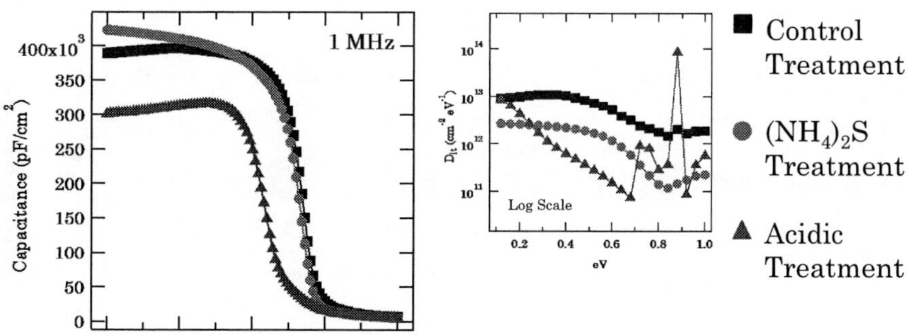

Flatband shift possibly due to thickness of dielectric layer.

$Si_{1-x}Ge_x$ (x = 0.75)

Chemical & Electrical Characterization

11

On $Si_{1-x}Ge_x$ (x = 0.75) Ge-S and oxides detected after $(NH_4)_2S$ treatment.

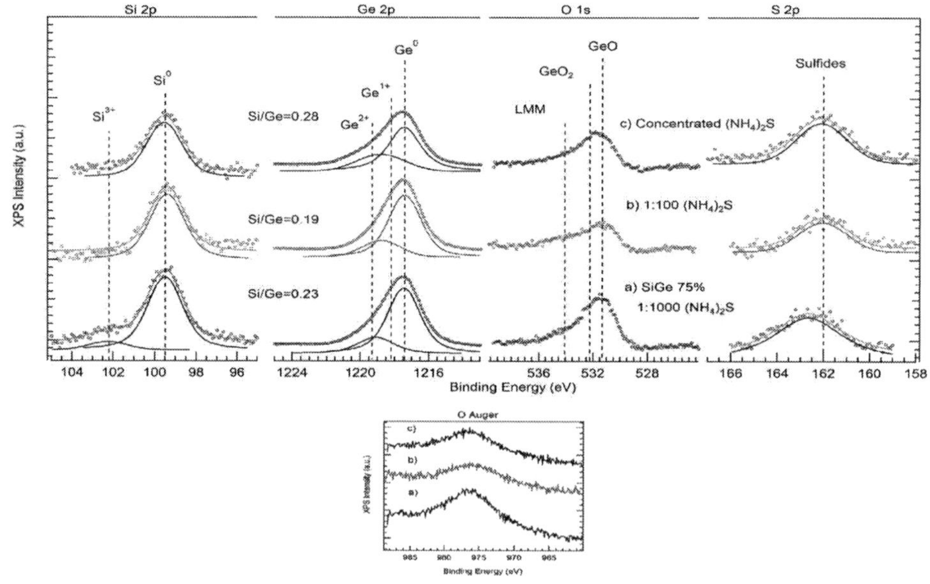

12

Ge-S and oxides correlate to less capacitance.

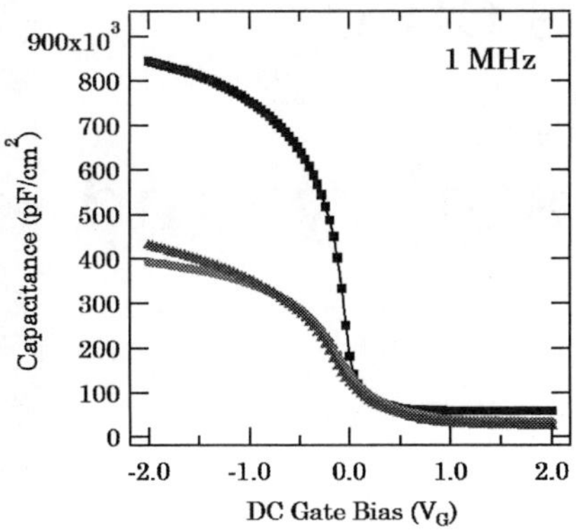

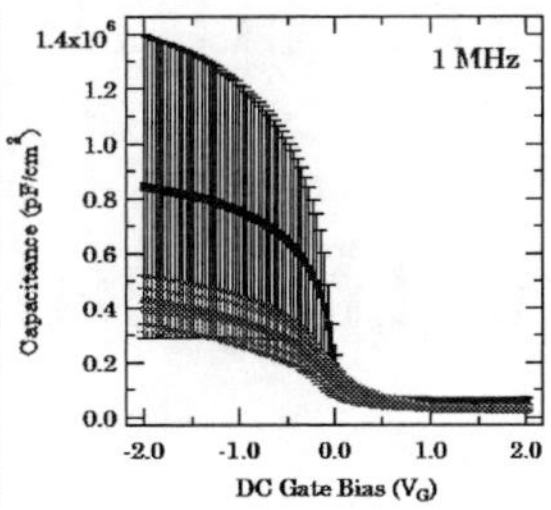

- ■ Control Treatment, 75% Ge
- ● $(NH_4)_2S$ Treatment, 75% Ge
- ▲ Acidic Treatment, 75% Ge

13

$(NH_4)_2S$ treatments decrease interface defects at valence band edge for both substrates.

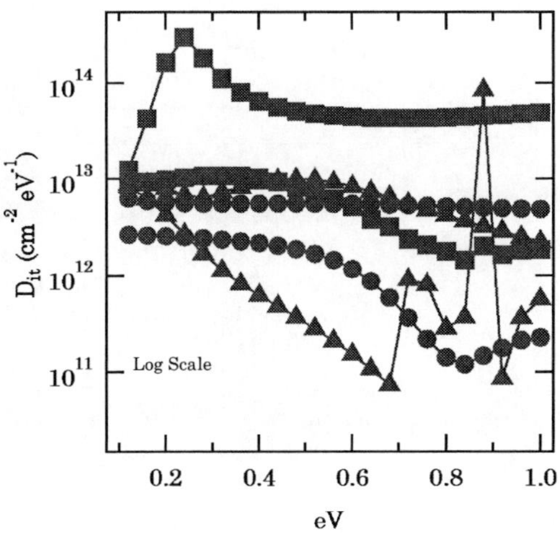

- ■ Control Treatment, 75% Ge
- ● $(NH_4)_2S$ Treatment, 75% Ge
- ▲ Acidic Treatment, 75% Ge
- ■ Control Treatment, 25% Ge
- ● $(NH_4)_2S$ Treatment, 25% Ge
- ▲ Acidic Treatment, 25% Ge

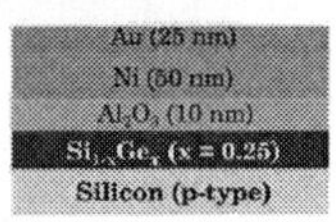

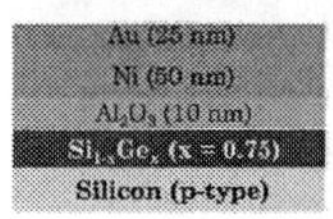

14

Additional Studies on S Desorption, Surface Variation, Repeatability, and Time

15

Checking S desorption after ALD

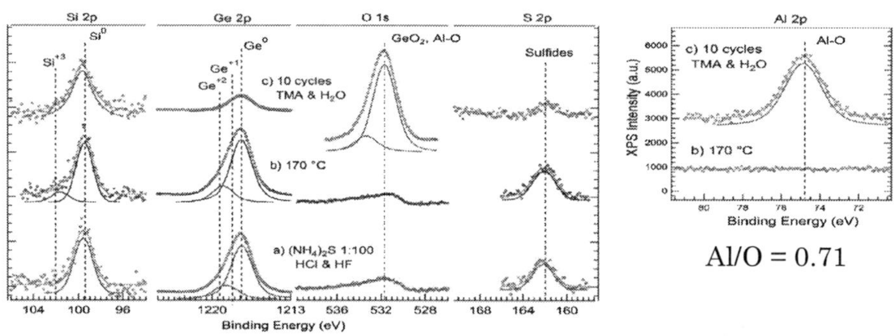

Al/O = 0.71

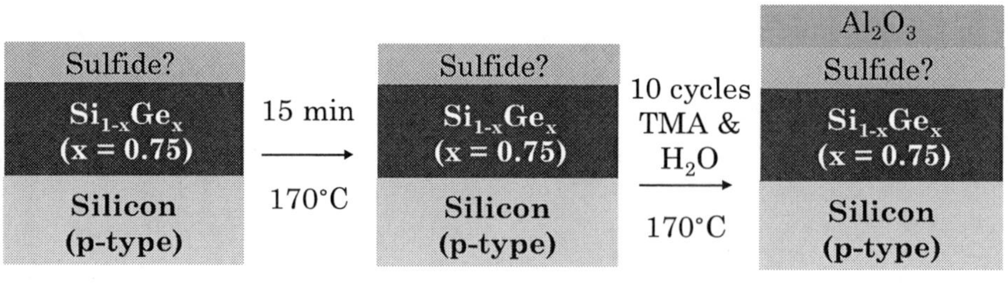

16

Surface variation on $Si_{1-x}Ge_x$ (x = 0.25) observed through D_{it}

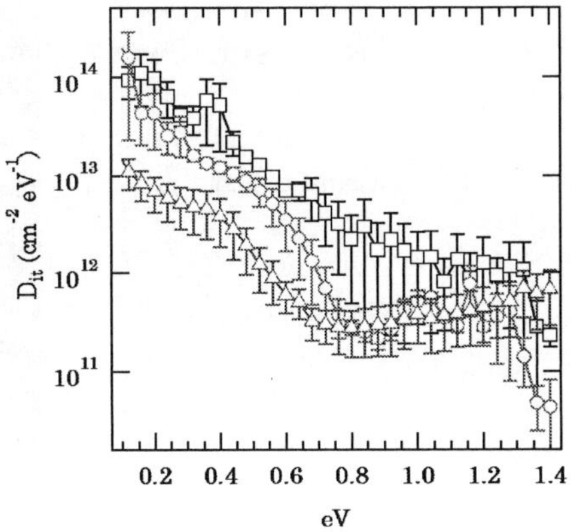

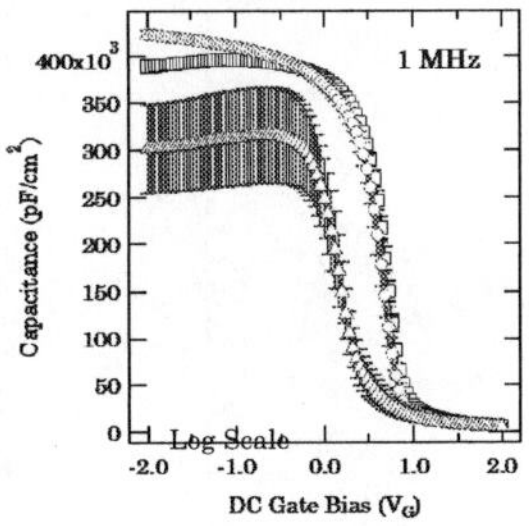

□ Control Treatment

○ $(NH_4)_2S$ Treatment

△ Acidic Treatment

Micron scale capacitors on coupon

17

Between three sets of $Si_{1-x}Ge_x$ (x = 0.25) MOSCaps D_{it} trend does not hold.

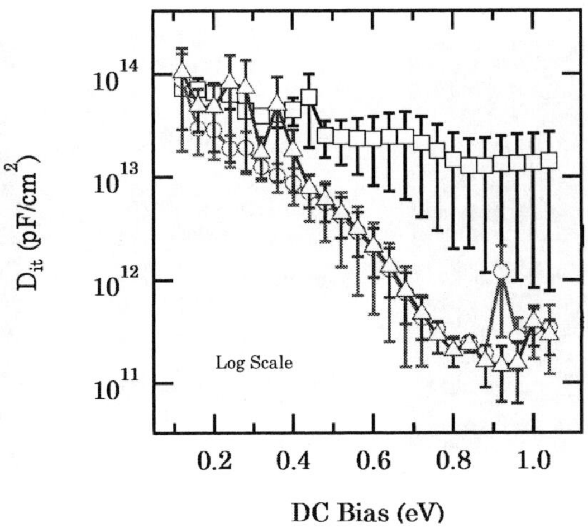

□ Control Treatment

○ $(NH_4)_2S$ Treatment

△ Acidic Treatment

Three sets of coupons

18

Interface defects increase with time.

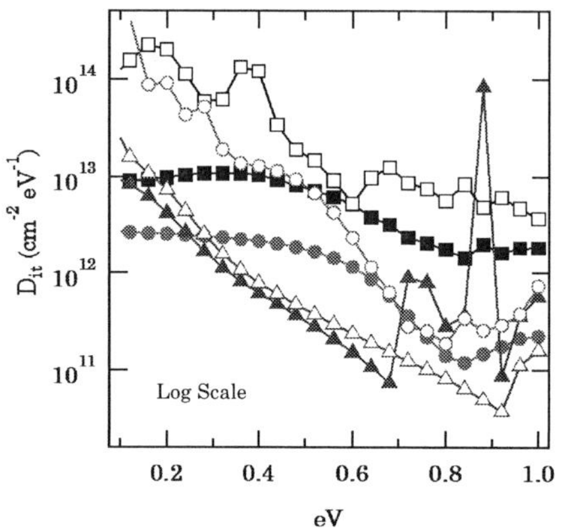

■ Control Treatment, Time 1
□ Control Treatment, Time 2

● $(NH_4)_2S$ Treatment, Time 1
○ $(NH_4)_2S$ Treatment, Time 2

▲ Acidic Treatment, Time 1
△ Acidic Treatment, Time 2

Time 2 – Time 1 = 5 months

19

SEM Image of $Si_{1-x}Ge_x$ (x = 0.25) MOSCap

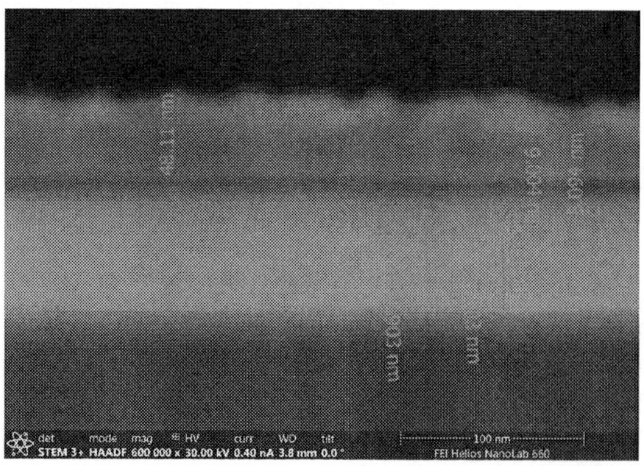

Control treatment

Pre annealing

Thicknesses for metal layers are not expected.

Metal layers are indistinguishable.

20

Conclusions

- Ge-S forms on both $Si_{1-x}Ge_x$ (x = 0.25, 0.75) in either the $(NH_4)_2S$ or the $(NH_4)_2S$ + Acid solution.
- Oxides regrow faster on both $Si_{1-x}Ge_x$ (x = 0.25, 0.75) when S is present on the surface.
- D_{it} reduction could be caused by Ge-S bond and/or oxide formation.
- MOSCap repeatability and surface homogeneity varies significantly between the two SiGe substrates.
- Interface defects increase with age on the $Si_{1-x}Ge_x$ (x = 0.25) surface.

21

Future Work

- Form Ge-S without oxide regrowth by non-aqueous solutions and gas phase deposition.
- Improve D_{it} analysis with series resistance correction.
- Improve metal contact deposition with calibration and switch to thermal deposition.
- Deposit thinner layers of high k to reduce flat band shift.

22

Surface Preparation and Wet Cleaning for Germanium Surface

J. Snow[2*], M. Otsuji[1], Y. Yoshida[1], H. Takahashi[1], F.Sebaai[3], F. Holsteyns[3], M. Sato[1] and H. Shirakawa[1]

[1] SCREEN Semiconductor Solutions Co., Ltd.
480-1 Takamiya-cho, Hikone, Shiga, 522-0292 Japan
[2] SCREEN SPE USA, LLC
820 Kifer Road, Suite B, Sunnyvale, CA 94086 USA
[3] IMEC vzw
Kapeldreef 75, B-3001 Leuven, Belgium

Outline

- **Introduction**
  - Key Cleaning Steps
  - Challenges on Wet Cleaning of SiGe and Ge
- **Ge Surface Preparation**
  - PRE (Particle Removal Efficiency)
  - MRE (Metal Removal Efficiency)
- **Material Removal Selective to SiGe/Ge**
  - Unreacted Ni Removal
- **Controlled SiGe/Ge Wet Etch**
  - SiGe Wet Etch Selective to Ge
- **Summary**

- **Introduction**
  - Key Cleaning Steps
  - Challenges on Wet Cleaning of SiGe and Ge
- **Ge Surface Preparation**
  - PRE (Particle Removal Efficiency)
  - MRE (Metal Removal Efficiency)
- **Material Removal Selective to SiGe/Ge**
  - Unreacted Ni Removal
- **Controlled SiGe/Ge Wet Etch**
  - SiGe Wet Etch Selective to Ge
- **Summary**

Assumed Logic Device Fabrication

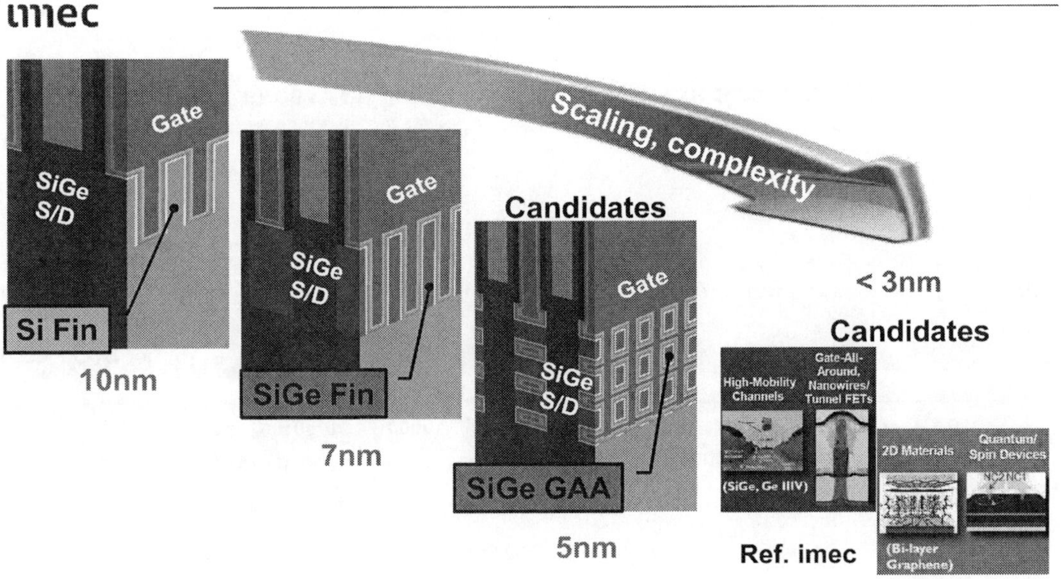

- **Structure becomes more complex while scaling down**
- **New materials as SiGe, Ge and III-V will be introduced**

Key Cleaning Steps on Ge

Surface preparation
- Post etch clean
- Pre-epi clean

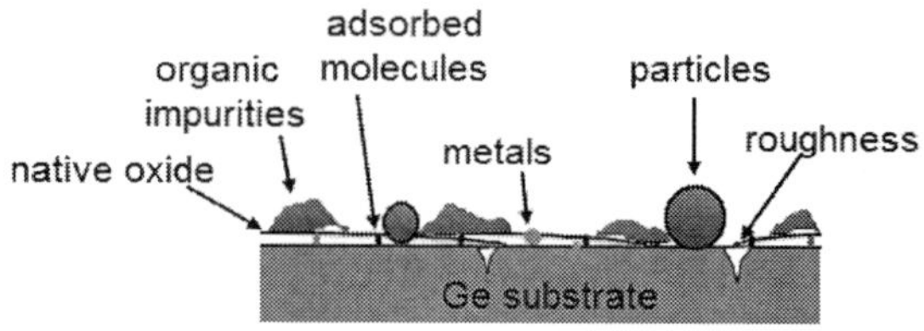

Material removal
- PR strip
- Unreacted Ni removal

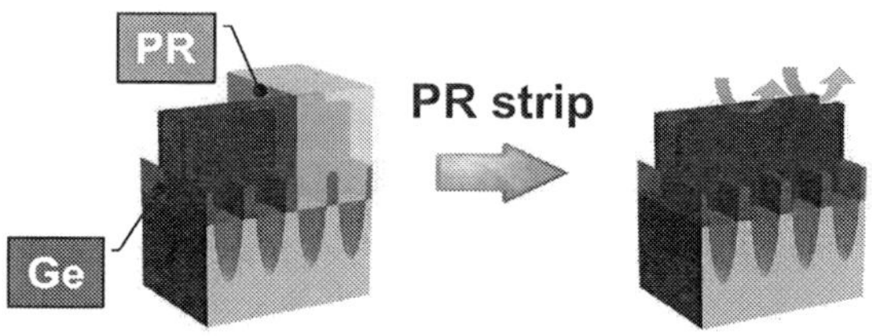

Controlled SiGe/Ge etch
- Controlled Ge-Fin Trimming
- Selective SiGe etch for GAA

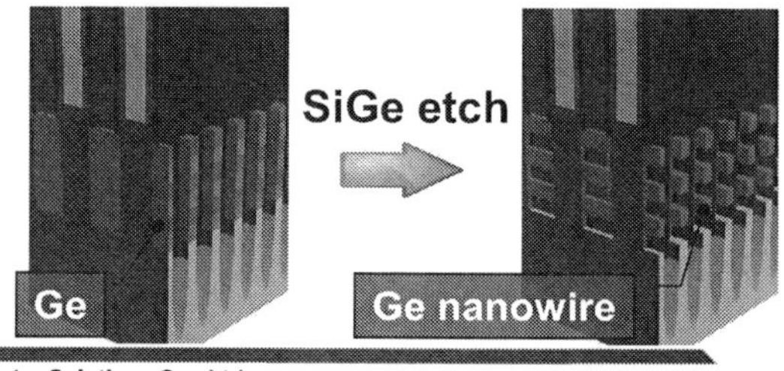

Challenges on Wet Cleaning of SiGe/Ge

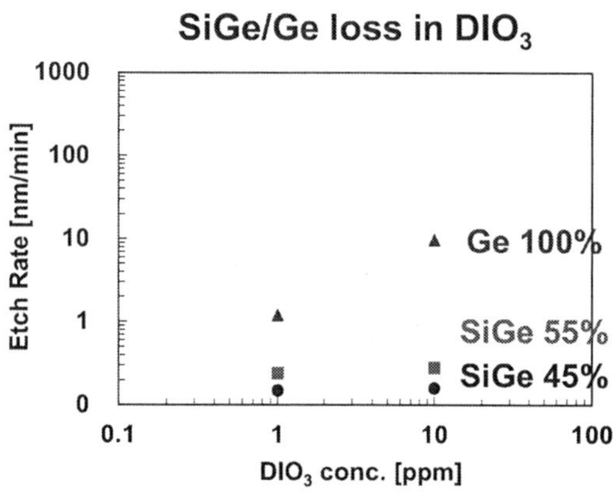

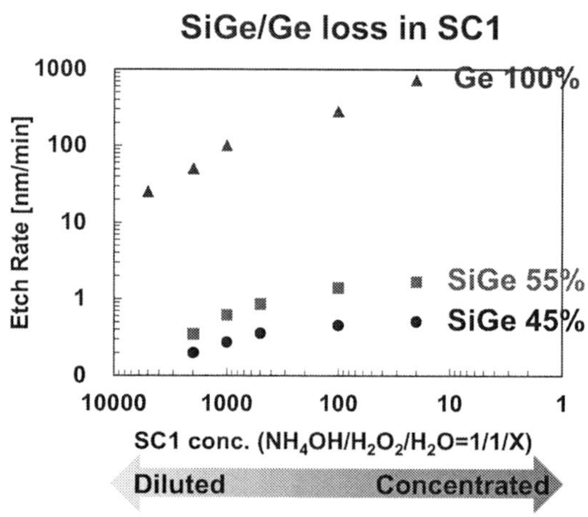

[M.Wada et al., *Solid State Phenomena* 187 (2012) 19]

- ❑ Ge is easily dissolved in oxidizing solutions, i.e. SC1, DIO_3
- ❑ Ge loss management is mandatory

- Introduction
  - Key Cleaning Steps
  - Challenges on Wet Cleaning of SiGe and Ge
- **Ge Surface Preparation**
  - PRE (Particle Removal Efficiency)
  - MRE (Metal Removal Efficiency)
- Material Removal Selective to SiGe/Ge
  - Unreacted Ni Removal
- Controlled SiGe/Ge Wet Etch
  - Controlled Ge-Fin Trimming
  - SiGe Wet Etch Selective to Ge
- Summary

PRE on Ge with Conventional Chemistries

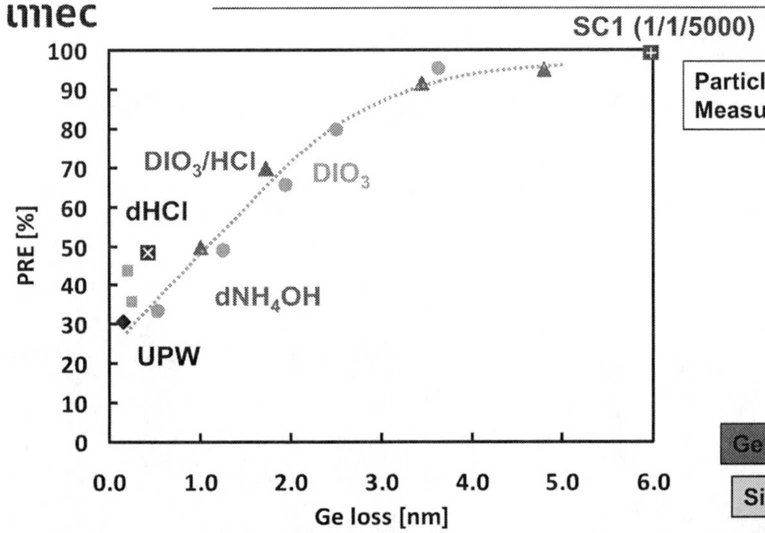

[H.Takahashi et al., *ECS Transactions* 2011 41(5), 163-170]

Particle: 30nm SiO_2-Slurry
Measurement: SP2, Haze

Applicable process:
Post gate etch clean

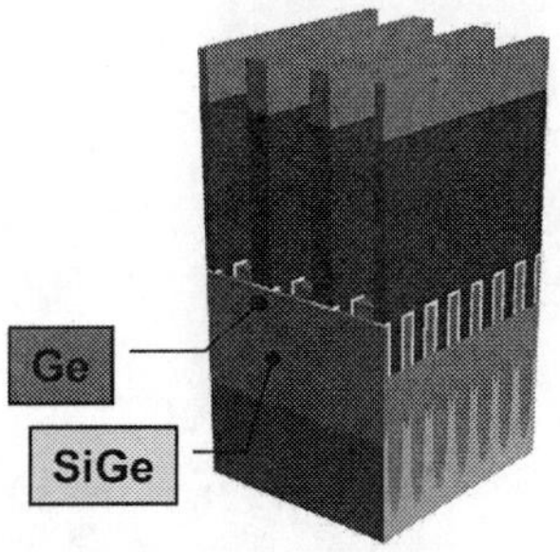

❑Surface lift-off by oxidizing chemistry is the dominant factor for sufficient particle removal
❑> 3nm Ge etch for > 90% PRE with conventional chemistries

SCREEN
imec

AOM (Ammonia/DIO$_3$ Mixture) for Ge Clean

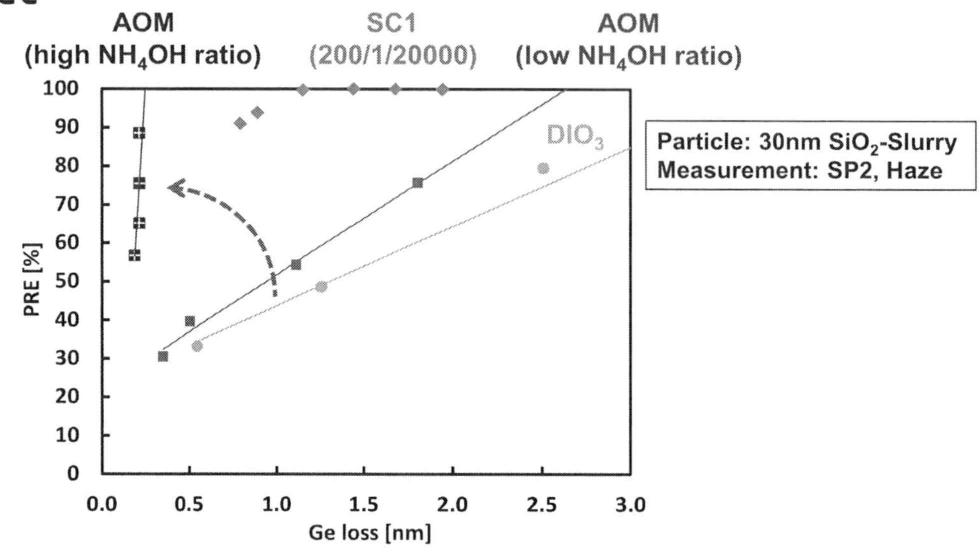

[H.Takahashi et al., *ECS Transactions* 2011 41(5), 163-170]

❑High PRE with lower Ge loss is obtained by using a clean with AOM, especially in high pH condition

SCREEN
imec

MRE on Ge Surface

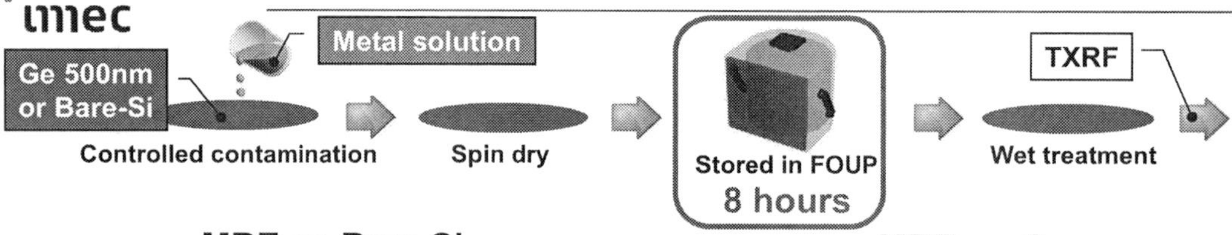

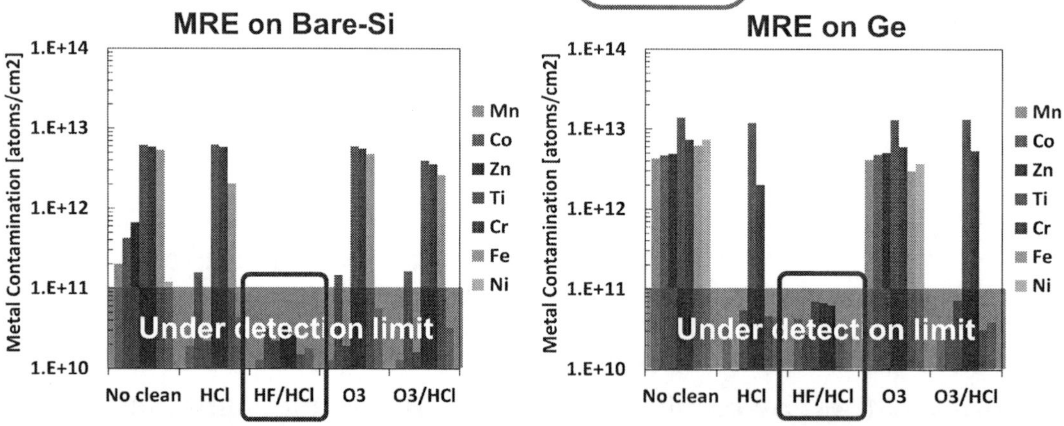

❑HF/HCl performs best: Same trend observed for Si and Ge surfaces, Native oxide removal by HF + Metal dissolution by HCl

Aging Effect on MRE on Ge Surface

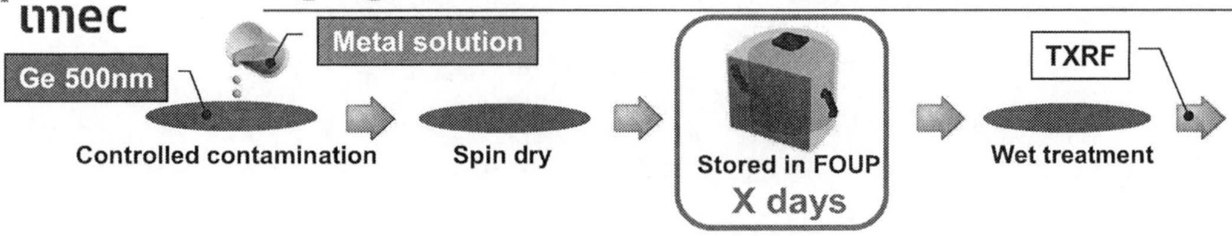

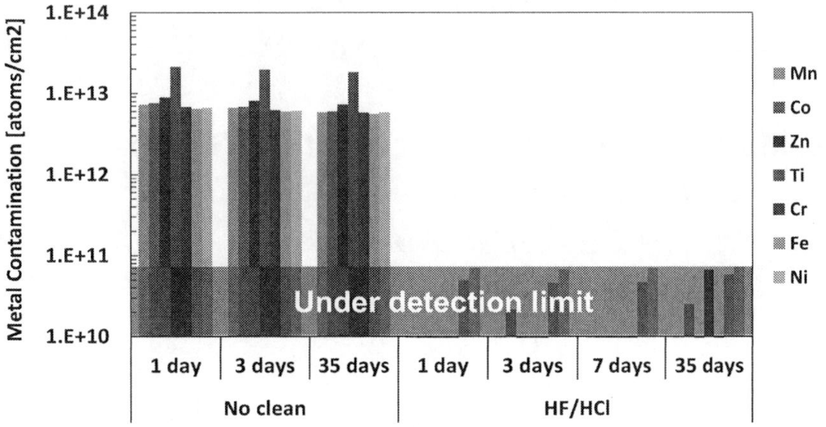

❑No aging effect on MRE with HF/HCl even after 1 month

Summary for Surface Preparations on Ge

- **Recommended cleaning combination**
  - For particles: AOM (NH_4OH/DIO_3 mixture)
  - For metals: HF/HCl mixture

SCREEN
imec

- ♦ Introduction
  - Key Cleaning Steps
  - Challenges on Wet Cleaning of SiGe and Ge
- ♦ Ge Surface Preparation
  - PRE (Particle Removal Efficiency)
  - MRE (Metal Removal Efficiency)
- ♦ **Material Removal Selective to SiGe/Ge**
  - Unreacted Ni Removal
- ♦ Controlled SiGe/Ge Wet Etch
  - SiGe Wet Etch Selective to Ge
- ♦ Summary

SCREEN
imec
SPCC2017 (Surface Preparation and Cleaning Conference), Mar 29 2017

Challenge & Approach for Galvanic Corrosion

One example: Ge corrosion during wet Ni removal using dHCl

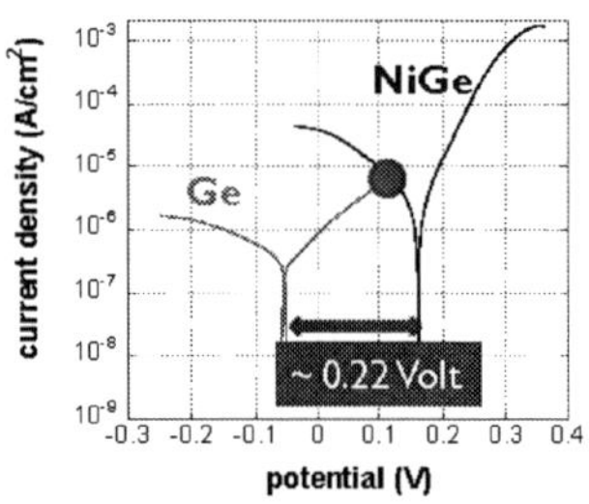

GeO dissolves in aqueous solution

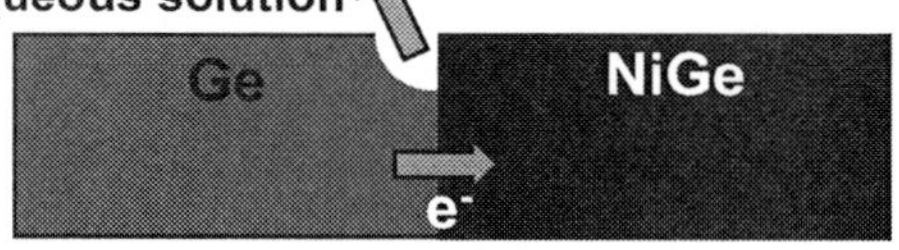

Anode:
$Ge + ½ O_2 \rightarrow GeO + e^-$

Cathode:
$H^+ + e^- \rightarrow H_2$
$O_2 + 2H_2O + 4e^- \rightarrow 4OH^-$

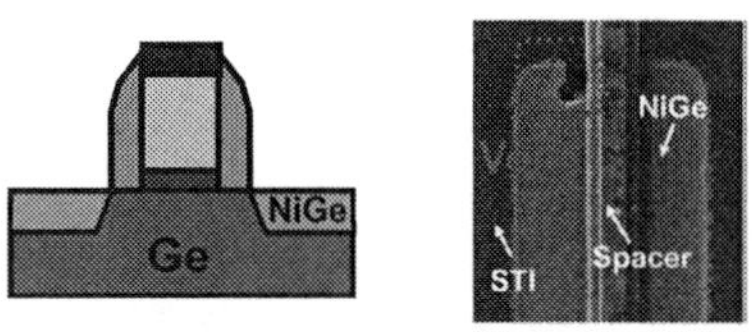

Ge planar device

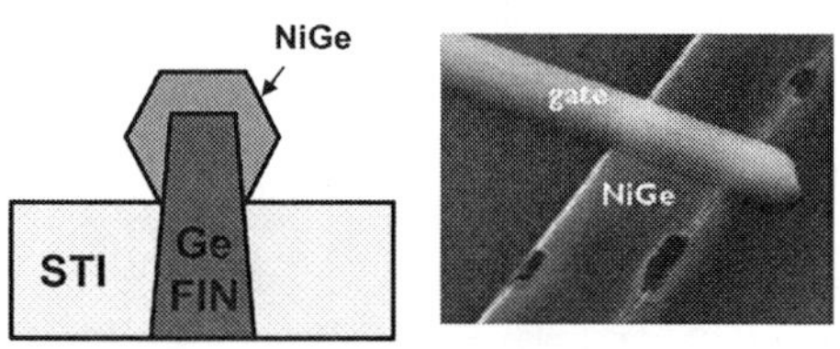

Ge-FinFET device

[F.Sebaai et al., *Solid State Phenomena* 219 (2014) 105-108]

❑Exposure of Ge & NiGe to the cleaning solution during Ge integration

SCREEN

imec

Management of Low-O_2 Processing

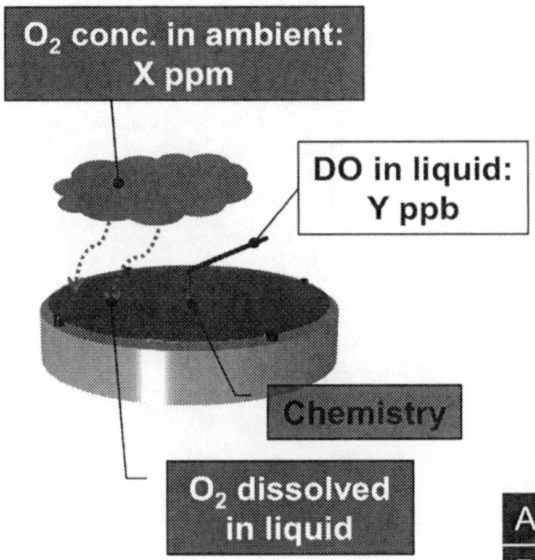

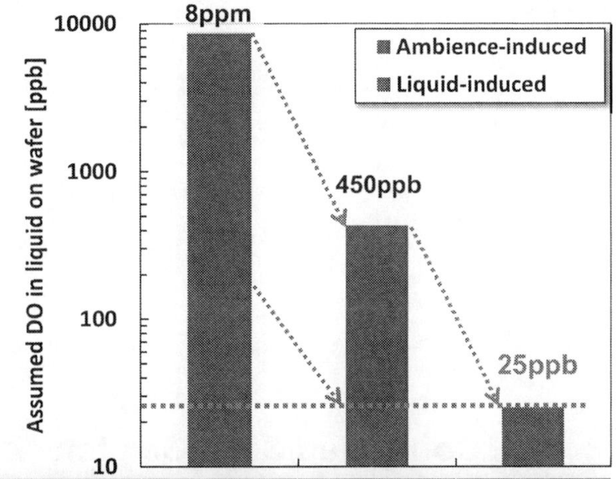

Ambient	20%	1%	10ppm
DO in liquid	170ppb	25ppb	25ppb

[Y.Yoshida et al., *Solid State Phenomena* 219 (2014) 85-88]

❑Ambient control allows to achieve low DO condition (<25ppb) in process liquid on wafer

SCREEN

imec

Selective Ni Removal on Ge-FinFET

NiGe loss with HCl

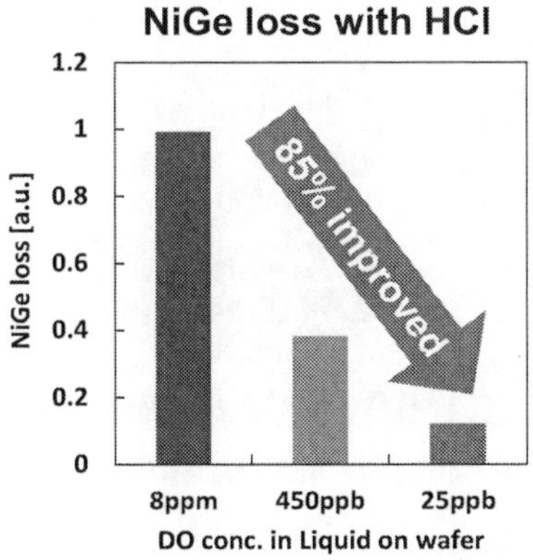

Void occurrence dependency on DO in HCl

DO	8ppm	450ppb	25ppb
SEM image	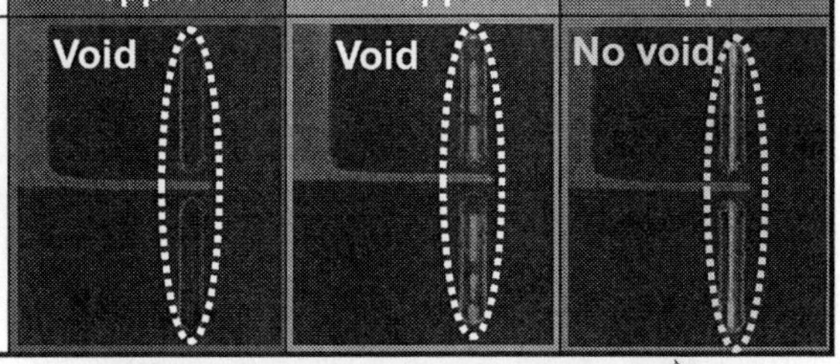		

[Y.Yoshida et al., *Solid State Phenomena* 219 (2014) 85-88]

❑Void occurrence can be suppressed by the control of the oxygen concentration in liquid & ambient

- Introduction
  - Key Cleaning Steps
  - Challenges on Wet Cleaning of SiGe and Ge
- SiGe/Ge Surface Preparation
  - PRE (Particle Removal Efficiency)
  - MRE (Metal Removal Efficiency)
- Material Removal Selective to SiGe/Ge
  - Unreacted Ni Removal
- **Controlled SiGe/Ge etch**
  - SiGe Wet Etch Selective to Ge
- Summary

SiGe Wet Etches Selective to Ge

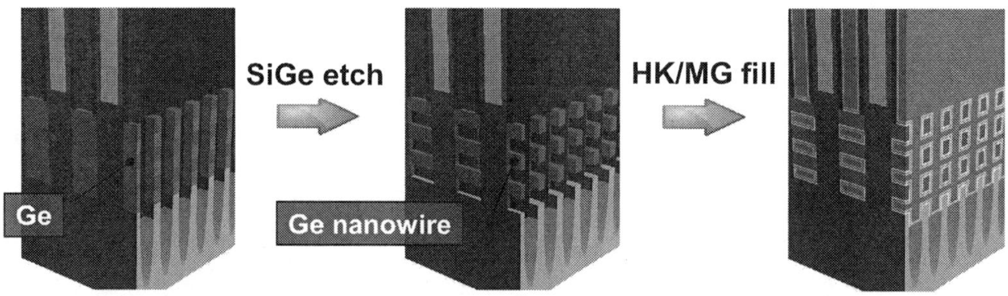

GAA (Gate All Around)

SiGe/Ge loss in SC1

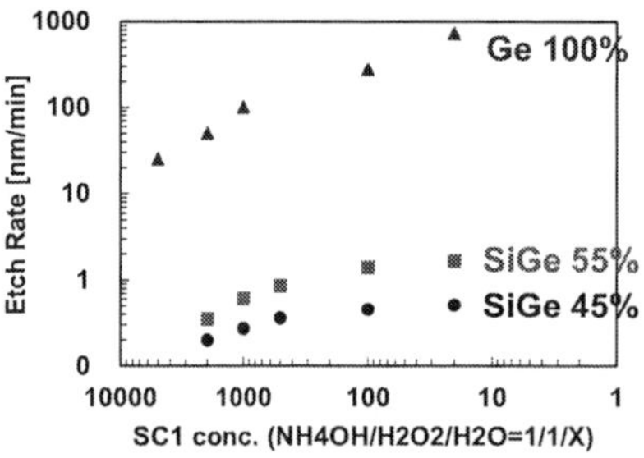

<u>Ge is too sensitive</u>

- ☐ Conventional chemistries don't work on this application

☐ **More study is needed on this kind of applications**

- Introduction
  - Key Cleaning Steps
  - Challenges on Wet Cleaning of SiGe and Ge
- SiGe/Ge Surface Preparations
  - PRE (Particle Removal Efficiency)
  - MRE (Metal Removal Efficiency)
- Material Removals Selective to SiGe/Ge
  - Unreacted Ni Removal
- Controlled SiGe/Ge Wet Etch
  - SiGe Wet Etch Selective to Ge
- **Summary**

Summary

- **SiGe/Ge Surface Preparation**
  - ☺ Particle removal on Ge surface
    - AOM is recommended for particle removal steps
      - 0.4 nm etch of Ge to achieve ~100% PRE
  - ☺ Metal removal for overall cleaning applications:
    - Sufficient removal confirmed by HF/HCl mixture cleaning
- **Material Removal Selective to SiGe/Ge**
  - ☺ Unreacted Ni removal
    - Low-O_2 is promising to full Ni remove without Ge corrosions
- **Controlled SiGe/Ge Wet Etch**
  - 😐 SiGe wet etch selective to Ge
    - Need more study

SCREEN

Fit your needs, Fit your future

SCREEN

LTM
LABORATOIRE DES TECHNOLOGIES DE LA MICROÉLECTRONIQUE

leti
ceatech

Contact:
Raynal Pierre-Edouard
Pierre-edouard.RAYNAL@cea.fr
+334 38 78 44 55
LTM CNRS/UJF CEA / LETI /
Minatec 17, avenue des Martyrs
38054 Grenoble cedex 9 – France

STUDY AND DEVELOPMENT OF ADVANCED SURFACE PREPARATIONS FOR THE INTEGRATION OF NEW MATERIALS IN MICRO/NANOELECTRONICS

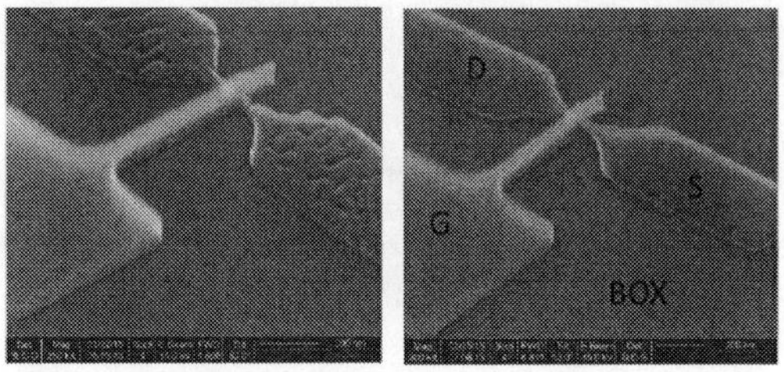

cnrs

UNIVERSITÉ Grenoble Alpes

leti

WET and Siconi™ cleaning sequences for SiGe pMOS channel

P.E. Raynal[a, b, d], V.Loup[a,d], L.Vallier[b], M. Martin[b], J. Moeyaert[b], B.Pelissier[b], P. Rodriguez[a], J.M. Hartmann[a,d], P. Besson[c]

a- Leti : Univ. Grenoble Alpes, F-38000 Grenoble, France and CEA, LETI, MINATEC Campus, F-38054 Grenoble, France
b- LTM : Univ. Grenoble Alpes, CNRS, LTM, F-38054 Grenoble, France
c- STMicroelectronics, 850 rue Jean Monnet, F-38926 Crolles Cedex, France
d- Univ. Grenoble Alpes, F-38000 Grenoble, France CEA, LETI, MINATEC Campus, F-38054 Grenoble, France

Source : Integration of Low Temperature SiGe:B Raised Sources and Drains in p-type FDSOI Field Effect Transistors. ECS Transactions, 75 (8) 51-58 (2016). The Electrochemical Society

LTM
LABORATOIRE DES TECHNOLOGIES DE LA MICROÉLECTRONIQUE

INTRODUCTION

leti
ceatech

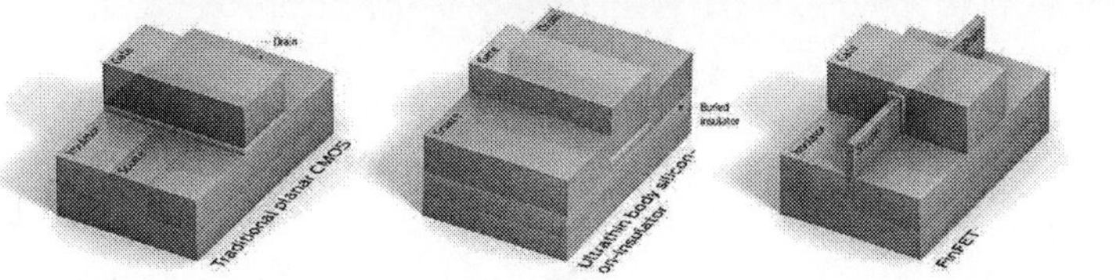

- **3D architectures** and critical **transistor dimensions** of new devices involve major challenges in terms of surfaces and **lower thermal budget** treatments.

- Integration of new materials → improve performances:
  - **SiGe** or **Ge** for pMOS transistor (hole mobility).
  - **III/V alloys materials** (InGaAS, InAs, AsGa) for nMOS transistor (electron mobility) and photonic applications (direct gap).

Material	Si	InGaAs	Ge
Electron mobility (cm²/Vs)	1,600	10,000	3,900
Hole mobility (cm²/Vs)	430	250	1,900

Existing channel material

- New materials present very reactive surfaces in ambient air
  → WET and in situ surface treatment combination

http://www.3m.com/3M/en_US/novec/industry-solutions/semiconductor

OBJECTIVE OF THE STUDY

- **SiGe source/drain epitaxial re-growth** :

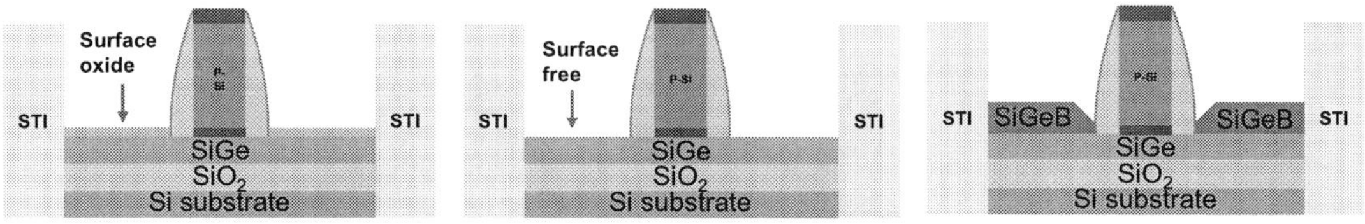

Surface preparation **WET-DRY** (**Epi-Ready**)

Epitaxy Growth

- **Remove surface oxide**
- **Low Contamination** C, O, N, Cl, F (Epi-killer)
- **Low** Roughness, defectivity, substrate consumption

CLEAN ROOM WORKING ENVIRONMENT

Substrate use :

Strained SiGe 40%layers epitaxially grown on Si bulk or SOI (wafer 300 mm)

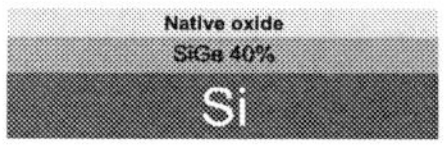

Siconi principle :

1-Etch Process (remote plasma) **2- Anneal Process (>100 °C)**

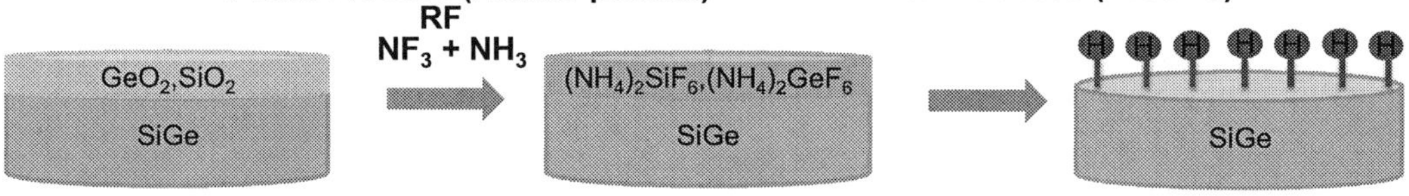

Characterization :

XPS grazing angle *thermo theta 300*, AFM, ellipsometry, FTIR-ATR and FTIR MIR

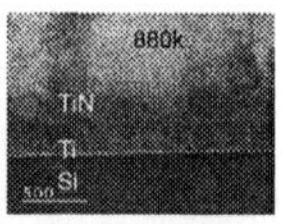

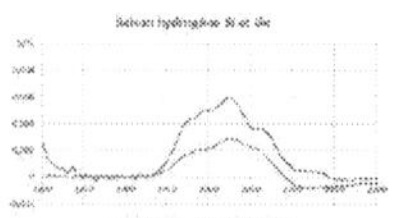

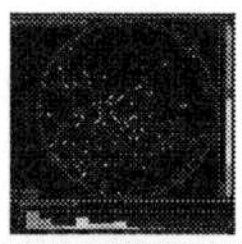

TIME EVOLUTION OF SIGE 40% SURFACE IN AIR

LTM LABORATOIRE DES TECHNOLOGIES DE LA MICROÉLECTRONIQUE — leti ceatech

Grazing XPS angular characterization :

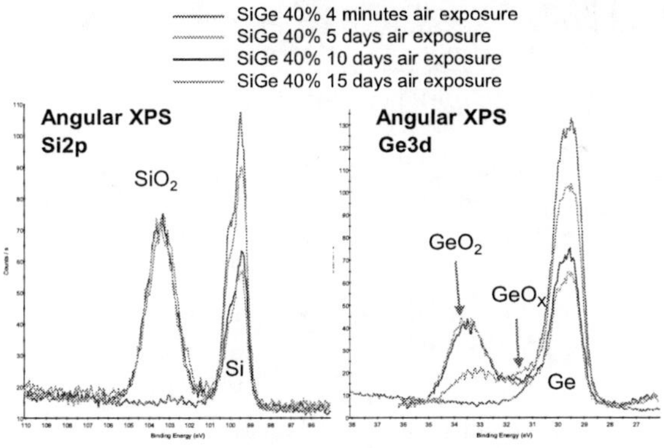

Evolution of Carbon contamination on a clean SiGe in clean room atmosphere

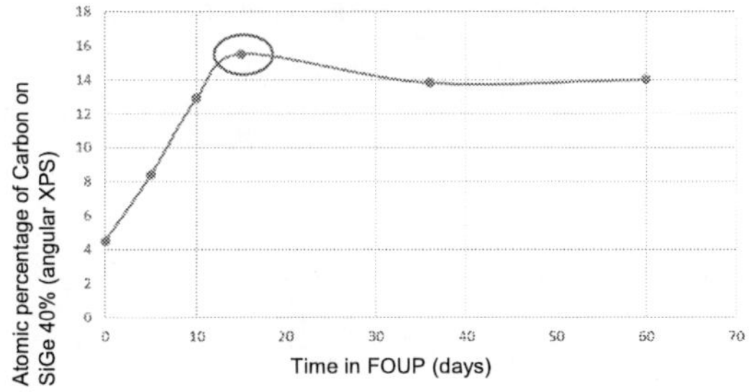

- Growth of a **mixed oxide** SiO_2, GeO_x and GeO_2 on clean room atmosphere
- **Carbon contamination increase** on surface substrate
- Maximum Carbon contamination **after 15 days**

15 days aged sample will be used for cleaning efficiency Characterization

SURFACE PREPARATION WITH HF CHEMISTRY

LTM LABORATOIRE DES TECHNOLOGIES DE LA MICROÉLECTRONIQUE — leti ceatech

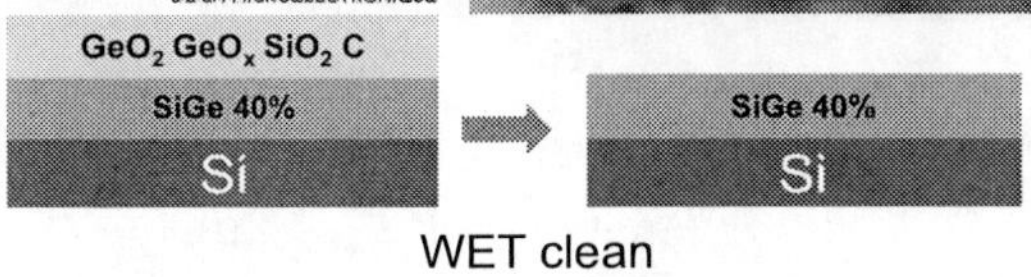

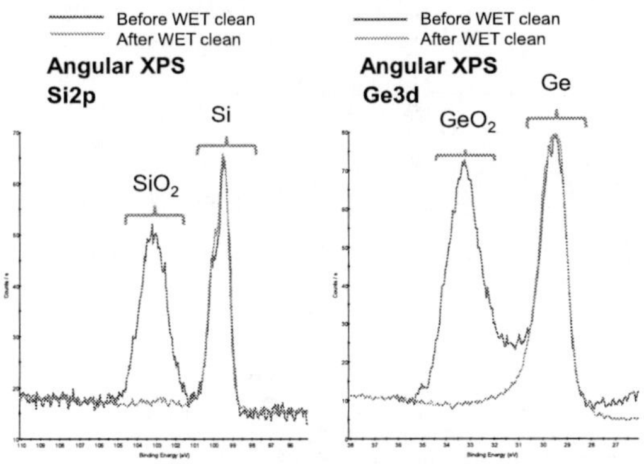

Evolution of Carbon contamination on a clean SiGe in clean room atmosphere

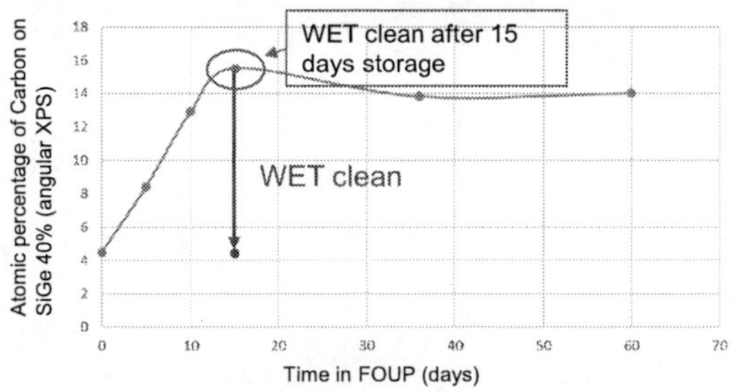

Angular XPS Quantification SiGe 40% Before and after WET clean

	Native oxide	WET clean
Air exposure		4 minutes
C % (C1s)	15,5	5
GeO_x % (Ge3d)	1,26	N.D.
GeO_2 % (Ge3d)	5,41	0
SiO_2 % (Si2p)	14,83	0

- **Roughness ≈ 0,2 nm**
- **SiGe consumption ≈ 5 Å**
- **WET cleaning :** efficiency toward oxide and C contamination removal

SURFACE PREPARATION WITH HF CHEMISTRY

LTM LABORATOIRE DES TECHNOLOGIES DE LA MICROÉLECTRONIQUE — leti ceatech

GeO_2 GeO_x SiO_2 C				
SiGe 40%	→	SiGe 40%	→	GeO_2 GeO_x SiO_2 C / SiGe 40%
Si	WET clean	Si	Air break	Si

Evolution of **IV-H bond** in atmosphere after WET clean (FTIR-MIR)

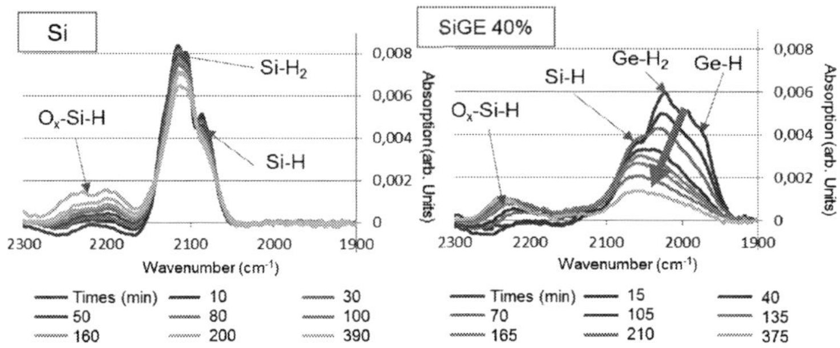

- IV-H bonds decrease after the WET clean in benefit of Si-O bonds
- **SiGe 40%Hydrogen passivation:**
  - **half of the Si-H and Ge-H loss after 2 hours for SiGe40%**
- **Air break can't be managed**

Very fast surface evolution → "in situ" clean mandatory

SURFACE PREPARATION WITH SICONI (IN-SITU)

LTM LABORATOIRE DES TECHNOLOGIES DE LA MICROÉLECTRONIQUE — leti ceatech

GeO_2 GeO_x SiO_2 C		
SiGe 40%	→	SiGe 40%
Si	Siconi clean	Si

Siconi and HF comparison :

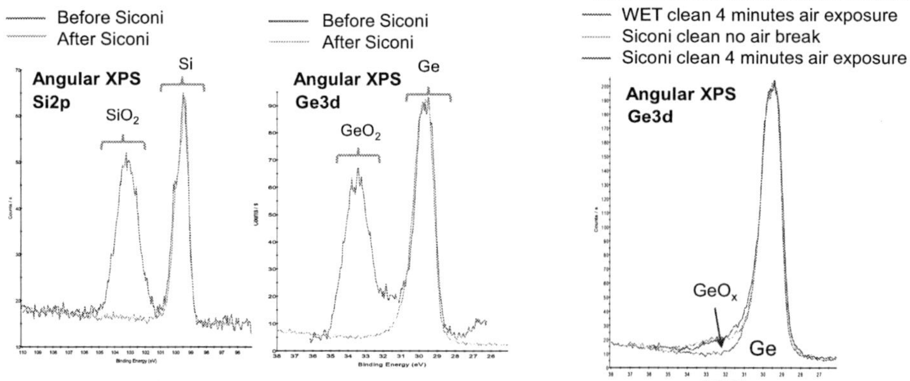

- **Roughness ≈ 0,2 nm**
- **SiGe consumption ≈ 5 Å**
- **Siconi clean : SiO_2 and GeO_2 removal efficient**
  Less efficient on GeO_x than HF

SURFACE PREPARATION WITH SICONI (IN-SITU)

Carbon contamination efficiency:

Angular XPS Quantification SiGe 40% Before and after Siconi clean

	Native oxide	Siconi
Air exposure		4 minutes
C % (C1s)	15,5	12,6
GeO_x % (Ge3d)	1,26	N.D.
GeO_2 % (Ge3d)	5,41	0
SiO_2 % (Si2p)	14,83	0

Evolution of Carbon contamination on a clean SiGe in clean room atmosphere

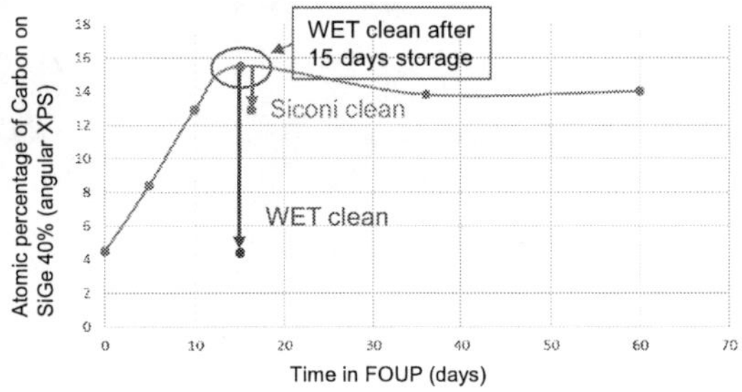

Siconi doesn't work on Carbon contamination

HF AND SICONI COMPARISON

- **WET clean (HF removal oxide):**
  - **Efficiency on** GeO_2, SiO_2, GeO_x and C contamination removal
  - **air break incompatible reactive surface**
- **Siconi treatment :**
  - Process in-situ: **no Q-time**
  - **Efficiency on** GeO_2 and SiO_2
  - **Poor efficiency** on $\mathbf{GeO_x}$
  - **Inefficient Carbon contamination**

Alternative cleaning strategy mandatory

LTM
LABORATOIRE DES TECHNOLOGIES DE LA MICROÉLECTRONIQUE

ALTERNATIVE STRATEGY PROPOSAL FOR AGED SAMPLE

leti
cea tech

GeO_2 GeO_x SiO_2 C
SiGe 40%
Si

Aged sample

SiGe 40%
Si

Goal : No oxide and carbon to be epi ready surface

SPCC Austin,Texas, USA | Pierre-Edouard Raynal | 11

LTM
LABORATOIRE DES TECHNOLOGIES DE LA MICROÉLECTRONIQUE

ALTERNATIVE STRATEGY PROPOSAL FOR AGED SAMPLE

leti
cea tech

GeO_2 GeO_x SiO_2 C
SiGe 40%
Si

Aged sample

Oxide Siconi ready
SiGe 40%
Si

- No GeO2, GeOx
- No Carbon contamination

No air break → siconi

SiGe 40%
Si

Goal : No oxide and carbon to be epi ready surface

SPCC Austin,Texas, USA | Pierre-Edouard Raynal | 11

ALTERNATIVE STRATEGY PROPOSAL FOR AGED SAMPLE

LTM LABORATOIRE DES TECHNOLOGIES DE LA MICROÉLECTRONIQUE — leti ceatech

GeO_2 GeO_x SiO_2 C
SiGe 40%
Si

Aged sample

Which wet chemistry can manage the surface composition?

Oxide Siconi ready
SiGe 40%
Si

- No GeO2, GeOx
- No Carbon contamination

No air break ➡ siconi

SiGe 40%
Si

Goal : No oxide and carbon to be epi ready surface

OXIDE COMPOSITION WITH DIFFERENT WET OXIDE

LTM LABORATOIRE DES TECHNOLOGIES DE LA MICROÉLECTRONIQUE — leti ceatech

Angular XPS Quantification SiGe 40%

	HF-HOT SPM	HF-O_3	HF- HOT SC1	HF-COLD SC1
SiO_2% (Si2p)	21,4	18,5	21,5	18
GeO_2% (Ge3d)	2,1	1,8	N.D.	N.D.
C% (C1s)	1,6	3,5	1,6	N.D.
Oxide Thickness nm	1,96	1,45	1,8	1,3

GeO_2 GeO_x SiO_2 C		SiO_2
SiGe 40%	SiGe 40%	SiGe 40%
Si	Si	Si

GeO_2 GeO_x SiO_2 C / SiGe 40% / Si ➡ HF ➡ SiGe 40% / Si ➡ SC1 based ➡ SiO_2 / SiGe 40% / Si (Germanium dissolution)

- **Roughness <0,2nm**
- GeO_2 dissolution with HOT SC1 and COLD SC1 (SiO_2 oxide growth)
- Low carbon contamination

➡ **HOT SC1 and Cold SC1 chemical oxide pre-siconi**

WET AND SICONI ASSOCIATION CLEANING

LTM LABORATOIRE DES TECHNOLOGIES DE LA MICROÉLECTRONIQUE | leti ceatech

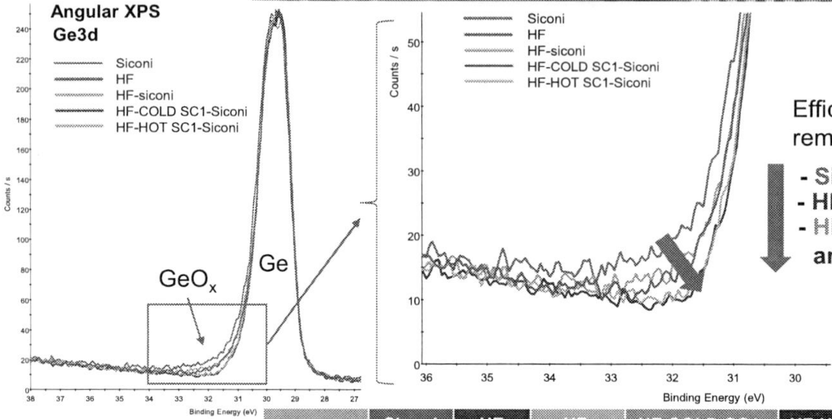

Efficiency on GeO$_x$ removal :

- Siconi only
- HF and HF-Siconi
- HF-COLD SC1-siconi and HF-Hot SC1-Siconi

Angular XPS Quantification SiGe 40%

	Siconi	HF	HF-Siconi	HF-COLD SC1-Siconi	HF-HOT SC1-Siconi
O % (O1s)	10.41	8,54	6.78	N.D.	N.D.
C % (C1s)	10.87	3.26	4.23	N.D.	N.D.

Efficiency on Oxygen and Carbon removal

- HF-COLD SC1-Siconi
- HF-HOT SC1-Siconi

→ Best Epi ready surface

HF mandatory ???

WET (WITHOUT HF) AND SICONI ASSOCIATION CLEAN

LTM LABORATOIRE DES TECHNOLOGIES DE LA MICROÉLECTRONIQUE | leti ceatech

SC1 clean only direct on aged sample

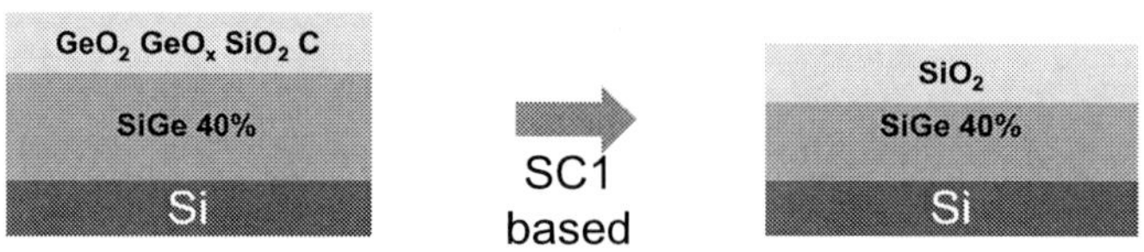

HF remove the native oxide before the SC1 | WET oxide clean without the native oxide removal

Angular XPS Quantification SiGe 40%

	NATIVE OXIDE	HF- HOT SC1	HF-COLD SC1	HOT SC1	COLD SC1
SiO_2 % (Si2p)	14,8	21,5	18	22,3	21,4
GeO_2 % (Ge3d)	5,4	N.D.	N.D.	N.D.	1,3
C % (C1s)	15,5	1,6	N.D.	N.D.	9,1

Cold SC1not efficient enough on :
- GeO_2 Dissolution
- Carbon contamination

LTM LABORATOIRE DES TECHNOLOGIES DE LA MICROÉLECTRONIQUE

WET (WITHOUT HF) AND SICONI ASSOCIATION CLEAN

leti ceatech

SC1 clean only direct on aged sample

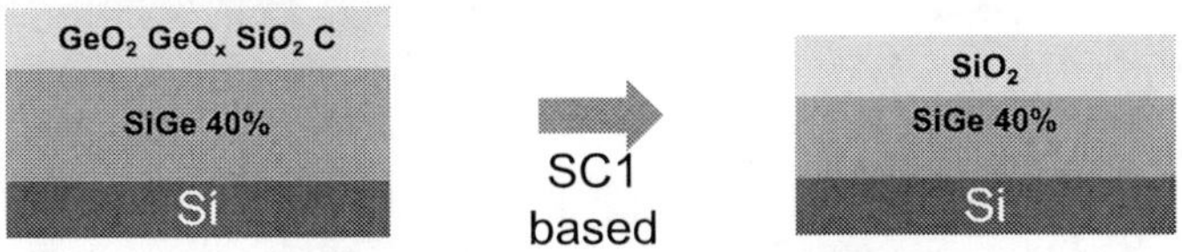

HF remove the native oxide before the SC1

WET oxide clean without the native oxide removal

Angular XPS Quantification SiGe 40%

	NATIVE OXIDE	HF- HOT SC1	HF-COLD SC1	HOT SC1	COLD SC1	COLD SC1 (Times treatment longer)
SiO_2 % (Si2p)	14,8	21,5	18	22,3	21,4	17
GeO_2 % (Ge3d)	5,4	N.D.	N.D.	N.D.	1,3	N.D
C % (C1s)	15,5	1,6	N.D.	N.D.	9,1	5,6

- SC1 based mixture alone can manage germanium oxide and carbon
- HF STEP can be removed if consumption fine tuning is mandatory

SPCC Austin,Texas, USA | Pierre-Edouard Raynal | 14

LTM LABORATOIRE DES TECHNOLOGIES DE LA MICROÉLECTRONIQUE

WET (WITHOUT HF) AND SICONI ASSOCIATION CLEAN

leti ceatech

SC1 + Siconi on aged sample

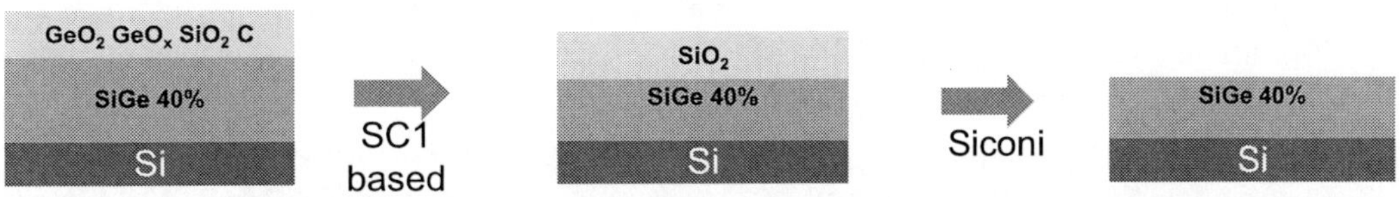

Angular XPS Quantification SiGe 40%

	HF	Siconi	HF- HOT SC1-Siconi	HF-COLD SC1-Siconi	COLD SC1-Siconi	HOT SC1-Siconi
O % (O1s)	8.2	10,4	N.D.	N.D	N.D.	N.D.
C % (C1S)	3.2	10,8	N.D.	N.D.	4.2	4.4

- SC1 based mixture + siconi permit very good epi ready surface

SPCC Austin,Texas, USA | Pierre-Edouard Raynal | 15

LTM
LABORATOIRE DES TECHNOLOGIES DE LA MICROÉLECTRONIQUE

CONCLUSION

leti
ceatech

conclusion

- **Siconi** and **HF** remove **GeO_2** and **SiO_2**
- Inefficiency of **Siconi** / **Carbon** contamination
- lower efficient of Siconi on **GeO_x**
- **SC1 based mixture** WET cleanings :
  - can **manage germanium oxide** and **carbon** (Siconi –ready)
  - **HF STEP** can be removed if **consumption fine tuning is mandatory**
- **SC1 based mixture + Siconi :**
  - permit very good **epi ready surface**

Perspectives :

- Epitaxy growth of Si on SiGe with WET-Siconi clean
  - Contamination quantification (SIMS)
  - Morphological quality (DRX, AFM…)

This work was partially supported by the LabEx Minos ANR-10-LABX-55-01

SPCC Austin,Texas, USA | Pierre-Edouard Raynal | 16

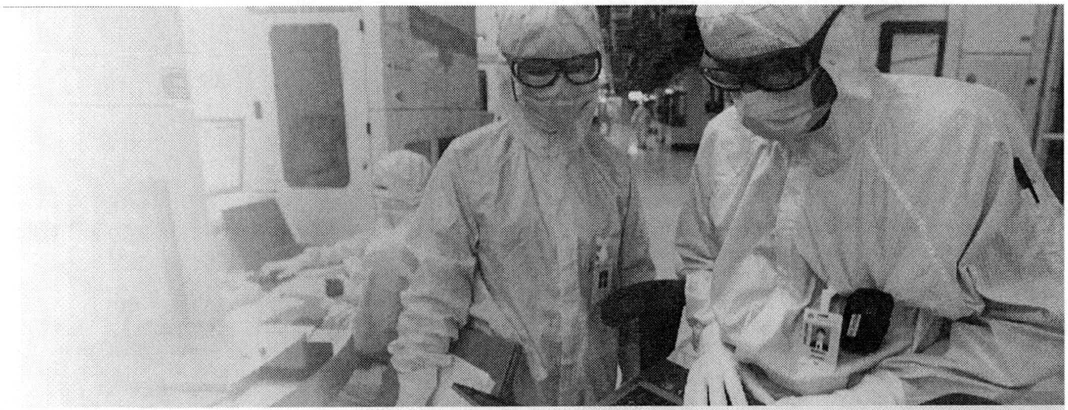

A Wet Clean Solution to Reduce Unwanted eSiGe Growth Defect in FinFET

Jian Li, Vincent Sih and Talapady Bhat
GLOBALFOUNDRIES, Malta, New York 12020, USA

SPCC, March 29, 2017

GLOBALFOUNDRIES

Outline

- Junction process flow in 3D FinFET
- Wet clean solution for unwanted growth defect in PFET (p-Channel Field-effect transistor)
- Defect reduction and line yield improvement on FinFET product wafers
- Conclusion

PFET Junction process flow

- PFET Spacer Etch
- PFET Photo Resist Ashing
- **PFET Pre eSiGe Clean**
- PFET eSiGe Growth
- Post eSiGe Clean
- PFET Defect Inspection

Gate length shrink enables performance scaling

FET is turned on its edge

- Lower supply voltage
- Reduced off-state leakage
- Faster switching speed – high drive current

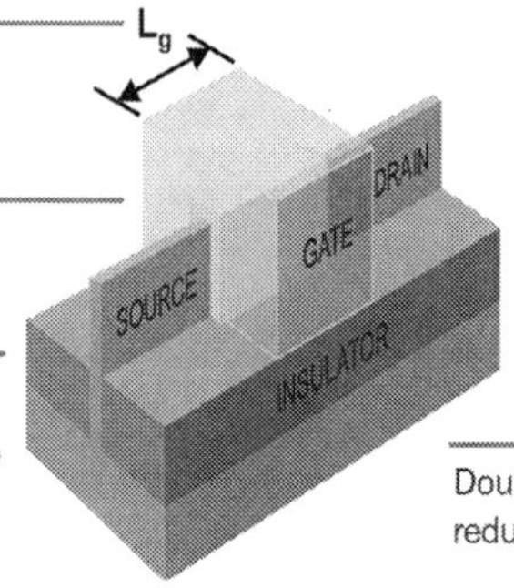

Double-gate reduces off-current

14LPP, 14nm FinFET Technology
https://www.globalfoundries.com/technology-solutions/cmos/performance/14lpp

Challenges for PFET Pre eSiGe clean

- The epitaxial growth of eSiGe is very sensitive to surface condition of substrate layer.
- Any small silicon particle (even less than 10nm) could serve as synthesis seed for eSiGe nucleation and later become a large unwanted growth defect.
- The silicon particle defect causing unwanted growth defect prior to eSiGe growth is too tiny to be caught by the state-of-the-art SP5 CFM scanning in current advanced manufacturing fab.
- The large unwanted eSiGe defect can cause degradation of device performance and raise the reliability concern.

Typical unwanted growth defect in Junction

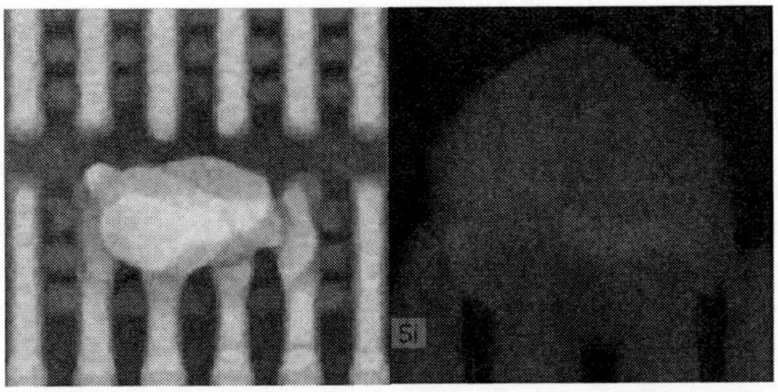

- Figures show the typical SEM images and element analysis of unwanted growth eSiGe in PFET Junction process.
- Unwanted growth defects are in the size range of 50 to 200 nm.
- Unwanted growth defects are located on top of poly gate line.

Wet clean solution for unwanted growth defect

New wet clean recipe

- Recipe designed to improve particle removal efficiency
- Recipe designed to modify surface condition
- Process parameters optimized
- Process chemistry optimized
- Process sequence optimized

Wafer defect maps on bare silicon

Pre maps-26nm bin size	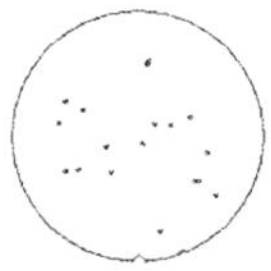	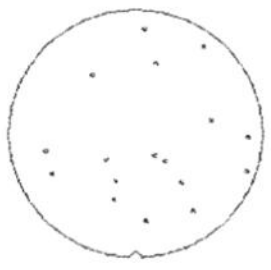	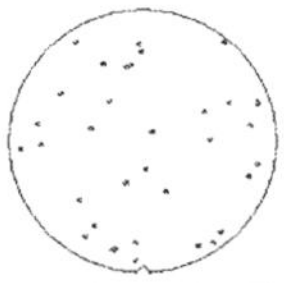	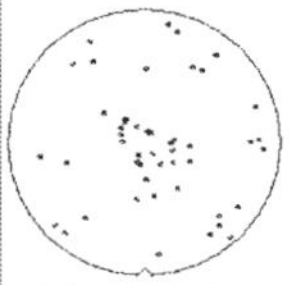
Post maps-26nm bin size	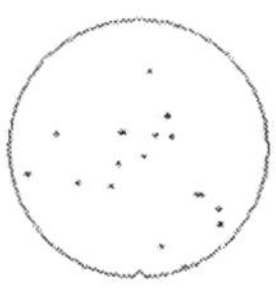	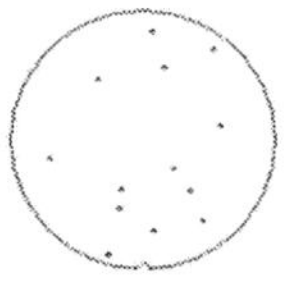	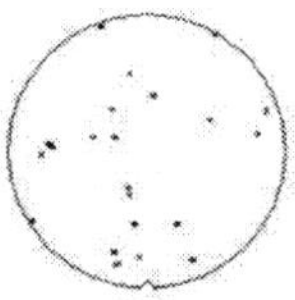	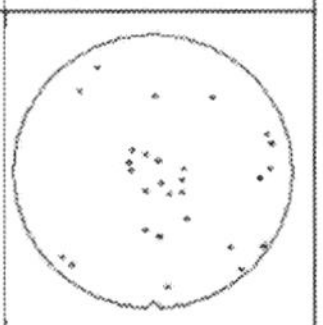
Adder number	4	1	12	4
Particle removal number	2	4	11	21

- Pre and post clean steps are based on 26 nm-size SP5 CFM scan to compare adder in different particle size categories
- The major types of particles on test wafers are Si/SiO_2 surface particles and organic residue particles

Defect reduction on FinFET product wafers

9

Normalized defect of unwanted growth

Normalized Unwanted eSiGe Defect

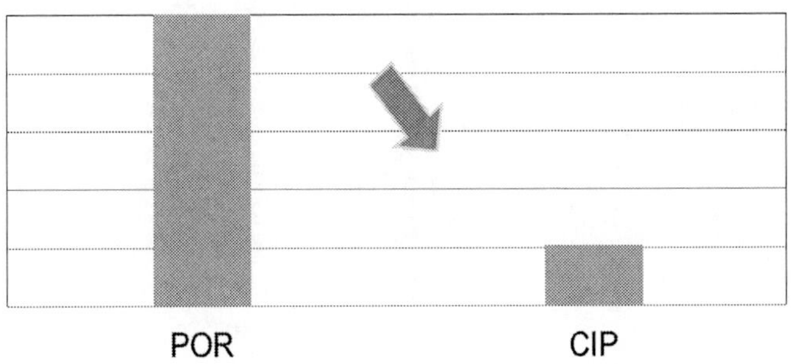

- CIP showed 79% less NDCs of unwanted growth defect comparing to POR cleaning

SPCC, March 29, 2017

GLOBALFOUNDRIES Confidential 10

Unwanted growth reduction by weeks

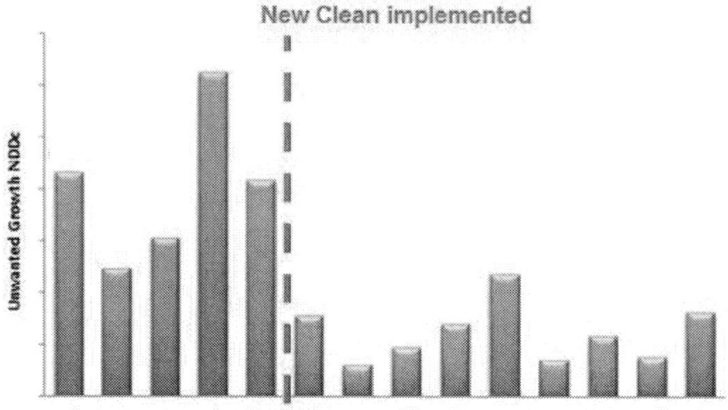

- After CIP clean implementation, unwanted growth defect was significantly reduce in weekly base. The x-axis is the time in weeks.

SPCC, March 29, 2017

VBD improvement by clean

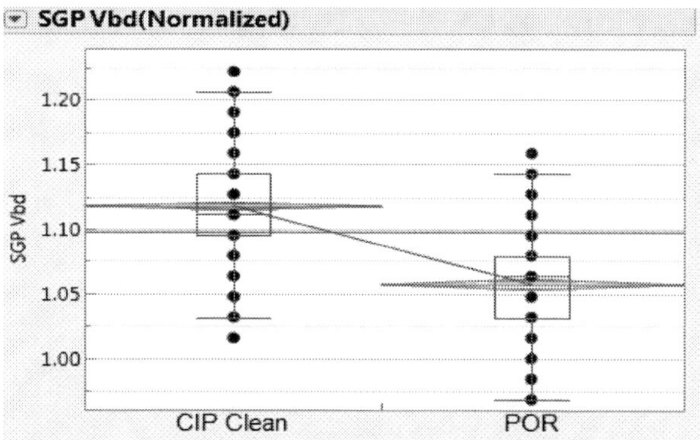

Normalized SGP Vbd comparison with POR and CIP clean process

- For normalized single-gate PFET breakdown voltage (SGP Vbd) comparison based on 224 FinFET production lots, CIP clean reduced the gate leakage (Iginv) and improved the SGP Vbd by 5%.

SPCC, March 29, 2017

Device performance comparison

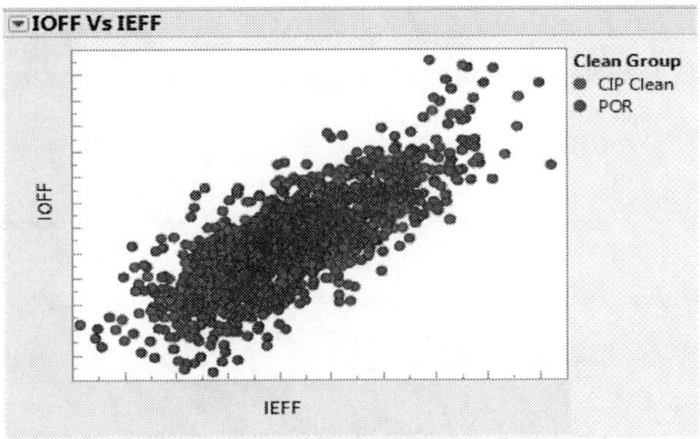

Normalized Ioff as a function of Ieff of POR and CIP clean process

- Device performance was also verified by plotting function curve of subthreshold leakage (Ioff) versus effective drain current (Ieff). CIP clean shows comparable device performance of POR clean.

Conclusion

- The new cleaning can remove tiny silicon particles more efficiently and provide friendly hydrophilic surface condition for the following eSiGe growth.
- Demonstrated in FinFET manufacturing for reducing ~79% unwanted eSiGe growth defect in PFET.
- Observed 5% line yield improvement based on Vbd data.
- The integrated wet clean solution is very promising for unwanted growth defect elimination and makes great contribution for critical rapid yield ramps in the advanced FinFET technology.

Thank you

Jian Li, Advance Module Engineering, GLOBALFOUNDRIES, Malta, New York 12020

jian.li3@globalfoundries.com

GLOBALFOUNDRIES®

Wet etching of TiN in 1-D and 2-D confined nano-spaces of FinFET transistors

Guy Vereecke, Hanne De Coster[1], Patrick Carolan, Hugo Bender, Kherim Willems, Lars-Åke Ragnarsson, Pol Van Dorpe, Naoto Horiguchi, and Frank Holsteyns

[1] UC Leuven-Limburg, Herestraat 49, 3000 Leuven, Belgium

SPCC 2017, March 28-29, Austin, TX, USA

Outline

1. Motivation
2. Study of TiN etching
   - in nanotrenches
   - in nanoholes
   - etch mechanism
3. Conclusions

RMG wet etching of WF metal in multi-Vt FinFET at N7

Nanotrench with 1-D confinement

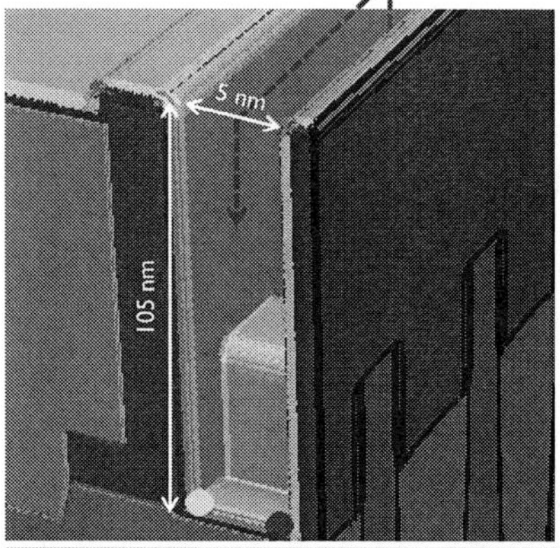

Nanohole with 2-D confinement

Feature	nm
Fin pitch	24
Fin width	4
Fin Height	50-70
Gate Pitch	42
Gate CD	14-18
Final Gate Height on Fin	25
Gate Height at RMG	105

Coventor view (not to scale)

→ Wet etching still feasible ?

imec

Overlap of electrical double layers in nanospaces

- Microchannel

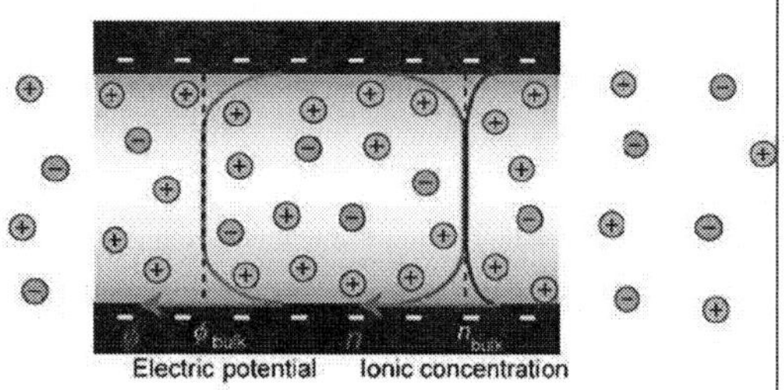

- Neutral solution in channel as in bulk
- Excess of charged ions only in diffuse part of EDL to neutralize surface charge

- Nanochannel

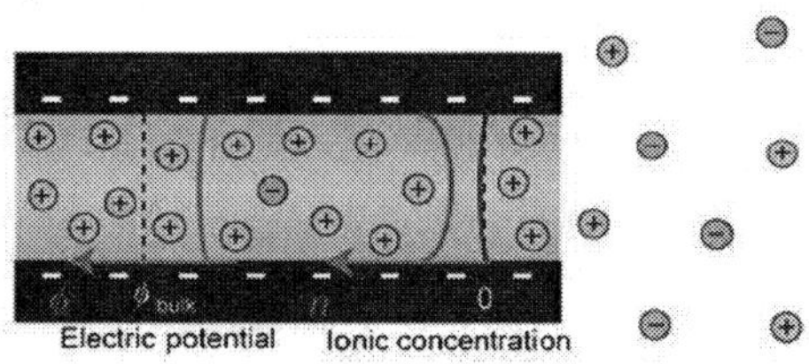

- Overlap of EDLs in channel
- Depletion of ions with same charge as surface in channel → no electroneutrality

imec

H. Daiguji, *Chem. Soc. Rev.*, 2010, **39**, 901.

Impact of EDL overlap on chemical reactions

- pH shift in 100nm-wide SiO_2 channels

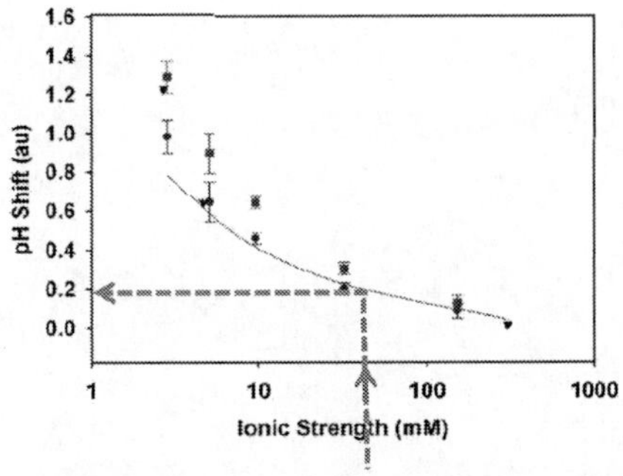

APM 1:4:20 used in etching of TiN

- Larger shifts expected in smaller FinFETs

- Etching of SiO_2 film in filled nano-slit by dHF

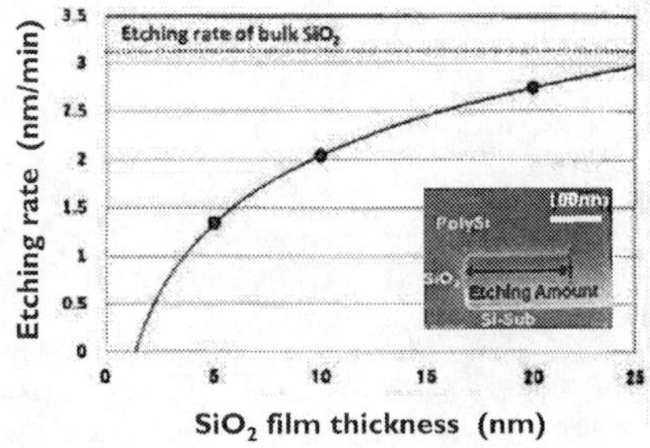

- Decreasing etch rate explained by increasing depletion of HF_2^-

imec

D.Bottenus et al., *Lab on Chip*, 2009, **9**, 219.

A. Okuyama et al., *Solid State Phenom.* 2015, **219**, 115.

TiN etching in filled nanotrenches

Test structure with filled nanotrenches

Nanotrenches covered with 1.5nm ALD HfO2 + 25nm ALD TiN

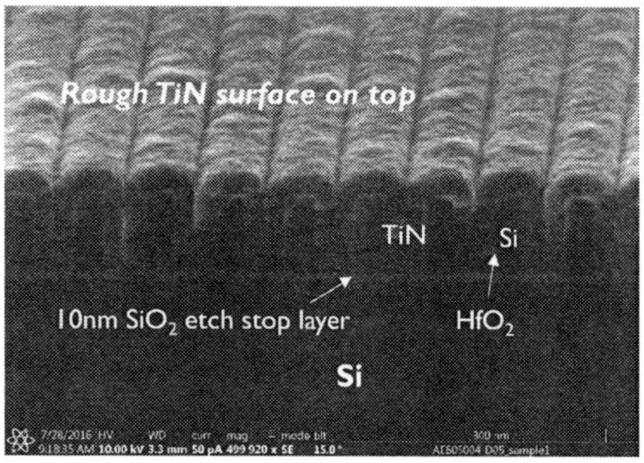

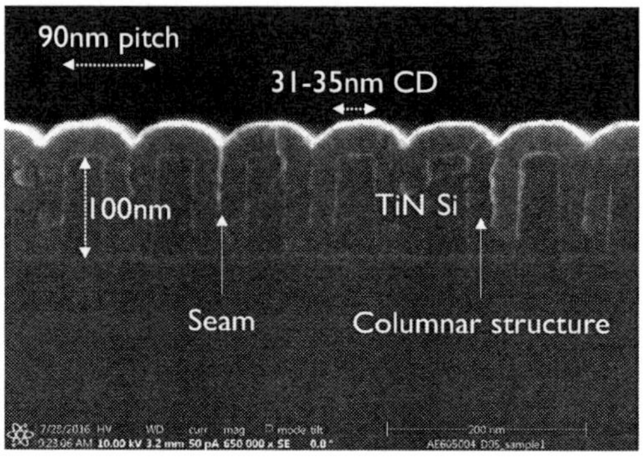

Structures after dHF clean (TiOx etched away) → ~1nm seam

imec

TiN etch in nanolines

Pre-clean: 2min HF 0.5% & etch: APM 1:4:20 at RT

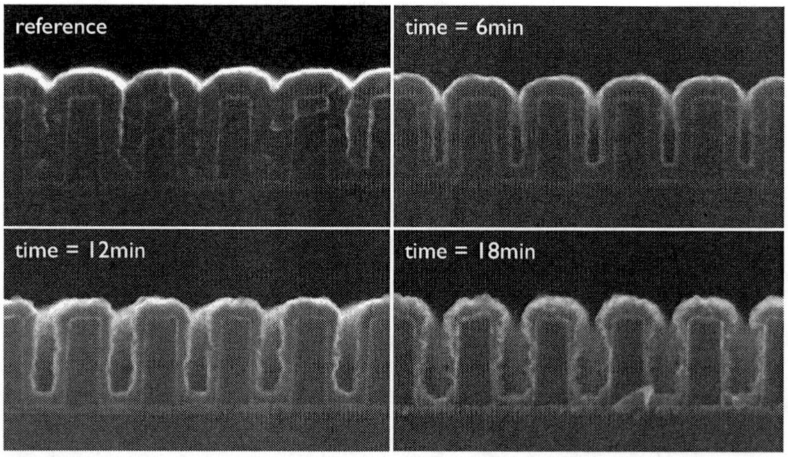

- Uniform etching along the seam → fast capillary wetting of seam
- Faster etching in the formed trench vs. top of lines
- Very rough TiN surface at end from columnar structure of TiN (non-uniform etching)

imec

TiN etch kinetics

Pre-clean: 2min HF 0.5% & etch: APM 1:4:20 at RT

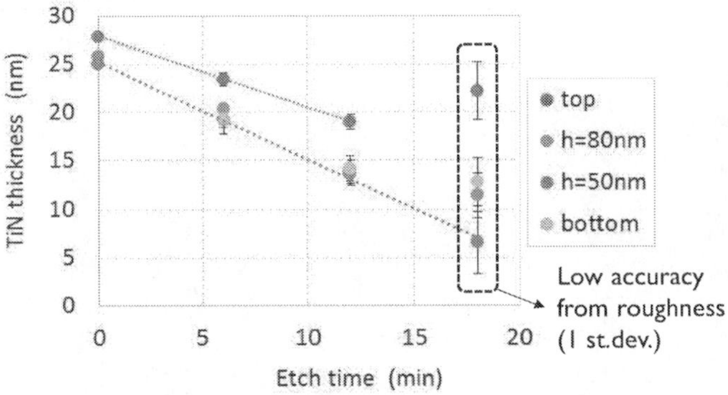

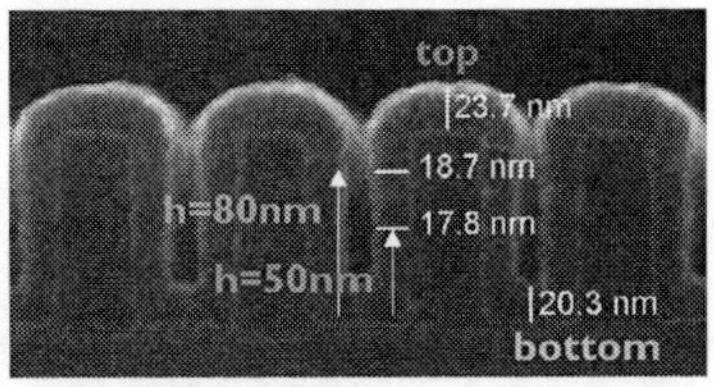

- TiN etch rate on top = 0.73±0.01nm/min
- TiN etch rate in trench (h=50nm) = 1.02±0.05nm/min

- Faster etch rate in trench vs. top of lines (38%)
- No wetting issue

imec

TiN etching in nanoholes

Nanoholes test structure

Nanoholes covered with 1.5nm ALD HfO2 + 5nm ALD TiN

- FIB-TEM inspection: Uniform TiN deposition from top to bottom

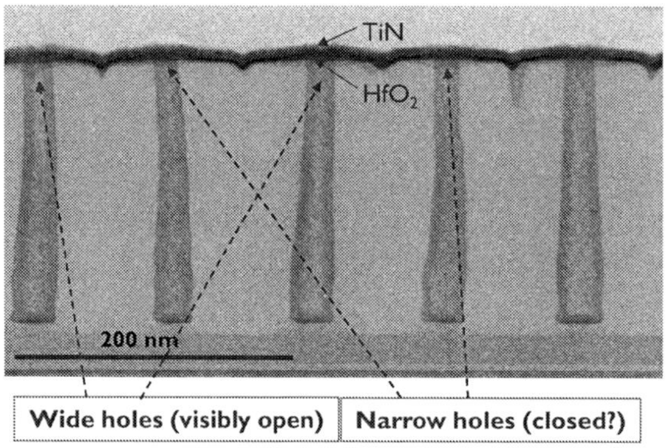

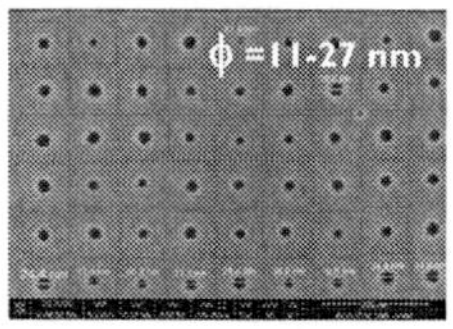

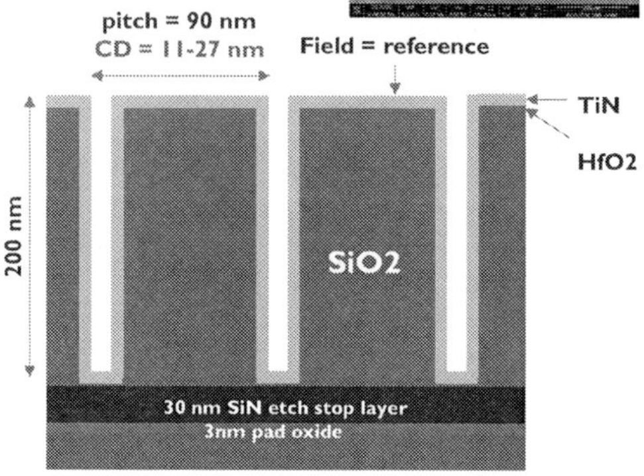

imec

11

FIB-TEM nanoholes

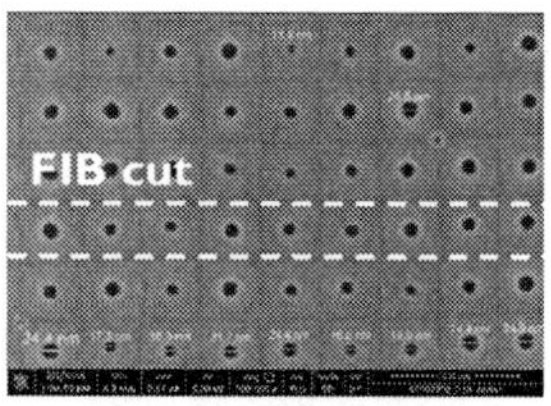

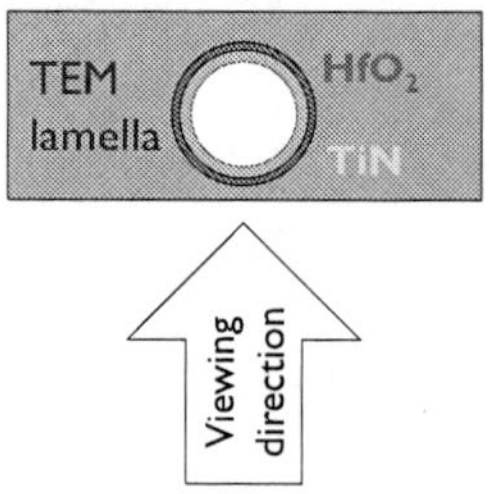

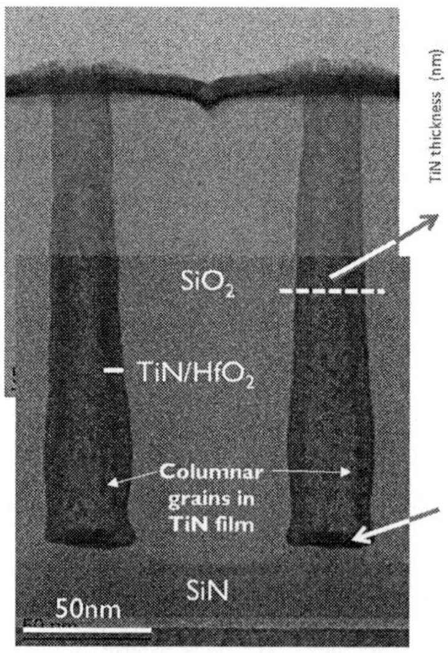

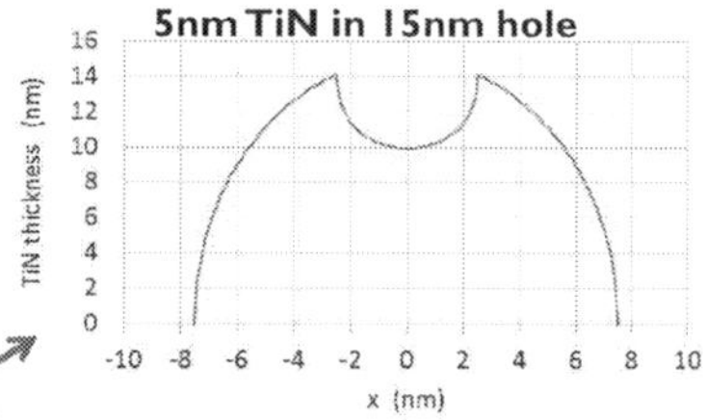

- In projection the layers on front/backside of the hole are seen as well, decreasing the contrast
- At the bottom the HfO_2 & TiN are seen in projection over the full diameter of the holes → higher contrast

imec

12

TiN etching at RT

Similar results with 9min etch

2min HF 0.5% + 3min APM 1:4:20 at RT

narrow wide

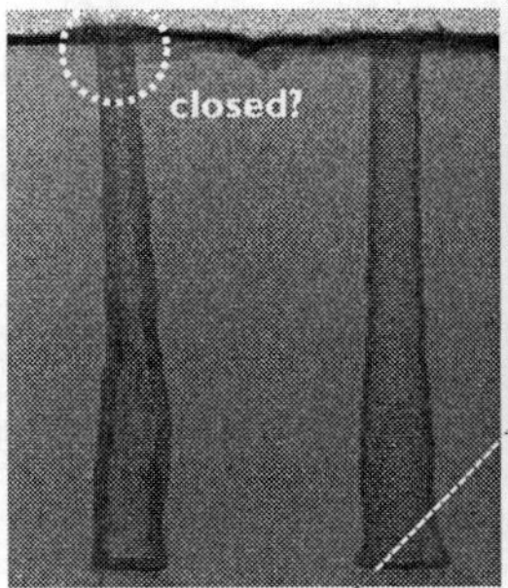

EDS maps

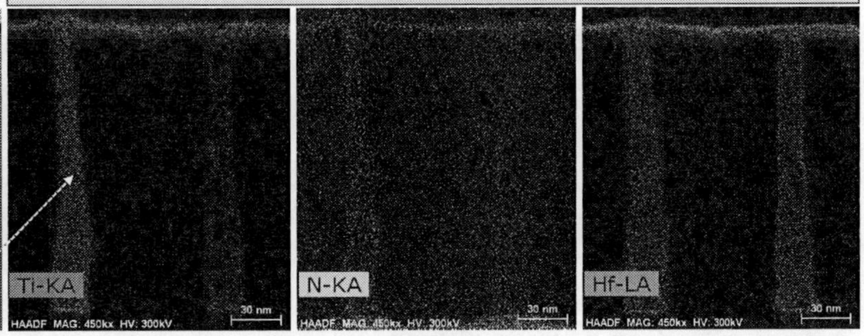

A little of Ti & N left in wide hole

No etching of HfO_2

No etching in narrow (closed?) hole

- Wide holes: >> 50% TiN removed
- Etch rate in wide hole (>0.9nm/min) > planar film (0.73nm/min)

imec

TiN etching at T = 65°C

2min HF 0.5% + 3min APM 1:4:20 at 65°C

etch rate of planar TiN = 26nm/min at 65°C

Nanoholes with 1.5nm HfO_2 + 10nm TiN

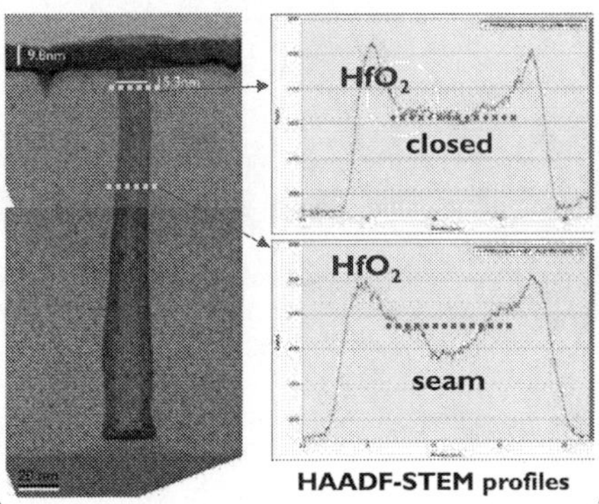

HAADF-STEM profiles

0.5min etch: partial removal

narrow wide

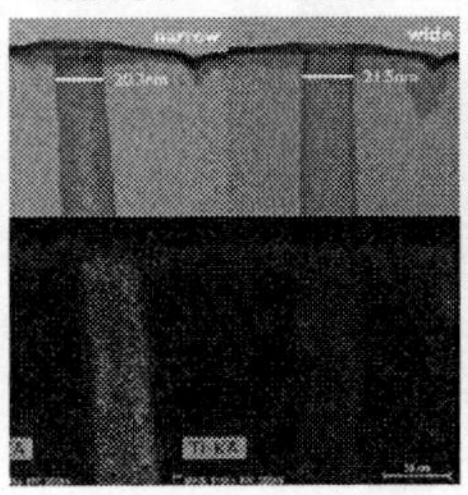

- No etching in narrow hole
- Ti mostly gone in wide hole & in the field

1.5min etch: TiN completely removed

Narrow hole

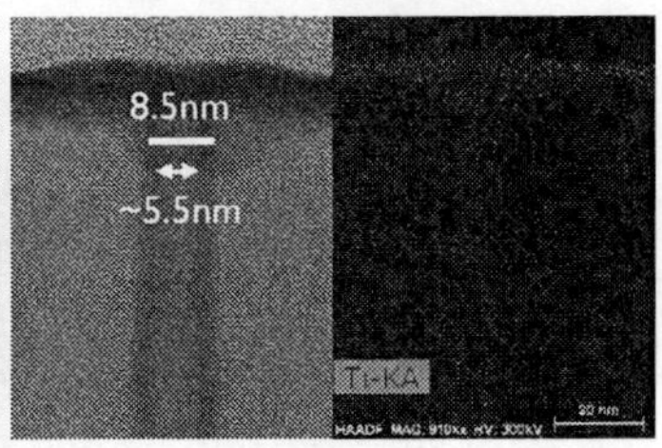

- Ti mostly removed in the field
- Ti not clearly detected above noise level in the holes

imec

Etching mechanism

Etch mechanisms proposed in literature

Rate determining step vary depending on authors (marked with *)

1. Oxidation of TiN to TiO_x
   * S. O'Brien et al., *Proc. UCPSS,* 1996, 205.
   * A. Philipossian et al., *Proc. UCPSS,* 1994, 275
   1. by $HO^{\cdot}$ radicals generated in-situ by Ti-catalyzed H_2O_2 decomposition
   2. by HOO^- anions from H_2O_2 ionization
2. Dissolution of TiO_x by complexation
   1. Complexation by HOO^- anion
      in analogy with an established titration method for Ti
      * S.Verhaverbeke et al., *Mat. Res. Soc. Symp. Proc.* 1997, **477**, 447.
   2. Complexation by NH_4^+ cation
      in analogy with the cleaning of metallic impurities by APM
      * this work

imec

16

Numerical simulations: Concentration of ions in TiN nanoholes

Solution of Poisson-Boltzmann and Poisson-Nernst-Planck equations for APM 1:4:20 at RT

B.Lu et al., *Phys. Rev. E* 2012, **86**, 011921. Implementation in Comsol.

Relative concentration map of [NH4$^+$] ≈ [H$^+$]

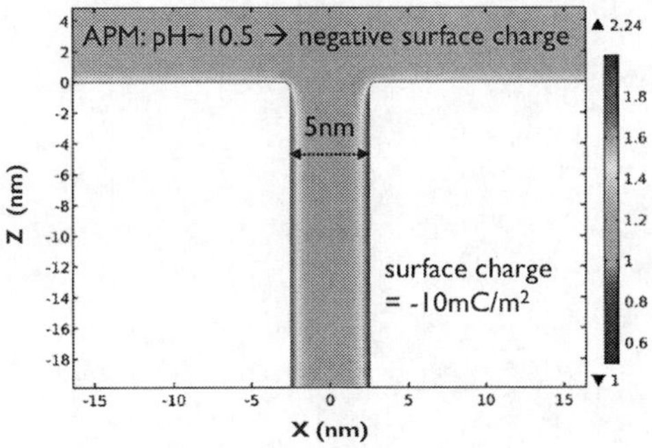

Relative concentration profiles

surface charge = -20mC/m²

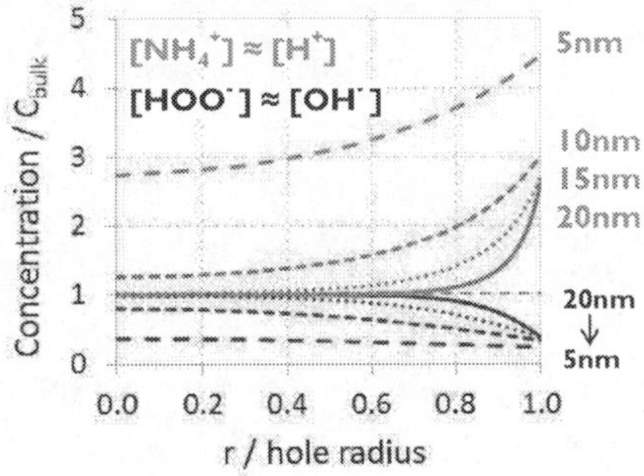

- Negative surface charge → cations enrichment and anions depletion
- Loss of electroneutrality in confined solution for CD < 10-15nm

Rate determining step in nanospaces

Considering impact of EDL overlap on kinetics

1. Oxidation of TiN to TiO_x
   1. by HO$^{\cdot}$ radicals generated in-situ by Ti-catalyzed H_2O_2 decomposition
   2. by HOO$^-$ anions from H_2O_2 ionization
2. Dissolution of TiO_x by complexation
   1. Complexation by HOO$^-$ anion
      in analogy with an established titration method for Ti
   2. Complexation by NH_4^+ cation
      in analogy with the cleaning of metallic impurities by APM

- Faster etch rate in nanolines vs. planar films suggests that complexation by NH4$^+$ is dominant

- Constant etch rate in nanolines a.f.o. time, i.e. at increasing opening, implies that reactions with HOO$^-$ would also play a role

imec

Conclusions

Conclusions

- No issue found for APM etching of TiN in advanced FinFET's RMG module, at the contrary
  - No wetting issue in narrow openings (~1nm seam after dHF)
  - Faster etch rate in narrow trenches vs. top of structures (~40%)
  - Faster etch rate in open nanoholes vs. top of structures (only qualitative)
  - Closed nanoholes with 10nm TiN cleared by APM@65°C in 0.5-1.5min
- Learnings on etching mechanism
  - Faster etch rate in nanospaces suggest kinetics dominated by TiOx complexation-dissolution by $NH4^{+}$
  - No dependency of etch rate on opening size in nanotrenches
    → kinetics cannot be described by a single ionic reaction

imec

imec
embracing a better life
imec
PUBLIC

Hilltop Economics

Shifting Global Economic Forces Shaping the Semiconductor Outlook

Surface Preparation and Cleaning Conference (SPCC)
March 28, 2017

Duncan Meldrum
Chief Economist, Hilltop Economics, LLC

duncan.meldrum@hilltopeconomics.com
610-709-7951

**A global outlook based on Consensus Economics, Inc.
CONSENSUS FORECASTS®*

Hilltop Economics

Outline

- Global outlook
- Semi MSI outlook (ESF March 2017)
- Risks in the economy
- A look back at SPCC 2016 forecast

Hilltop Economics

Global Economic Outlook Summary: 2017-18

- Modest growth, faster than 2016
- Improving investment climate
- Slightly higher inflation, diverging interest rates
- Currency turmoil, stronger U.S. dollar
- Much higher economic policy uncertainty

Duncan Meldrum
SPCC March 2017

3

Hilltop Economics

World Real GDP Growth Slowly Accelerates

Annual Trend: 95-07: +3.2% 11-16: +2.5%
2016: +2.2% 2017: +2.7% 2018: +2.9%

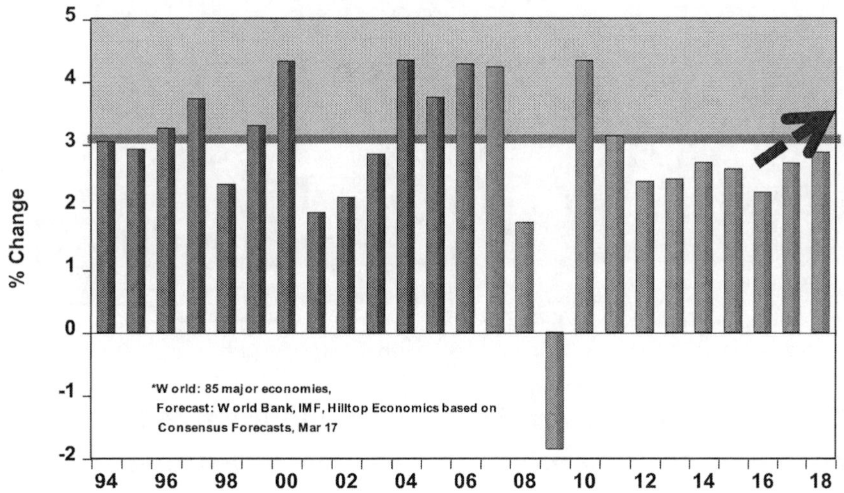

Duncan Meldrum
SPCC March 2017

4

Hilltop Economics

Key U.S. & Other Policy Initiatives
Uncertainty Factor & Most Likely Impact on Growth

- **Monetary Policy: Low Uncertainty (neutral for growth for now)**
  - Interest rates rising
  - Dollar strengthening
  - Regulations in response to financial crisis in place and operational
- **Fiscal (Tax & Spend) Policy: High Uncertainty**
  - Tax reform (important for growth)
  - Budget shift to defense (mixed for growth)
  - Regulatory reform (important for growth)
  - Healthcare reform (now on indefinite hold)
- **International Policy: High Uncertainty (no longer financial crisis-driven)**
  - GLOBAL increase in anti-globalization sentiment negative for growth
  - Immigration (negative for growth)
  - Pull-back from trade agreements: Negative to uncertain impact on growth
  - Brexit impact on EU
  - China policy shift to reforms with a lower growth policy target (& upcoming Communist Party Congress)

Duncan Meldrum
SPCC March 2017

5

Hilltop Economics

Economic Policy Uncertainty

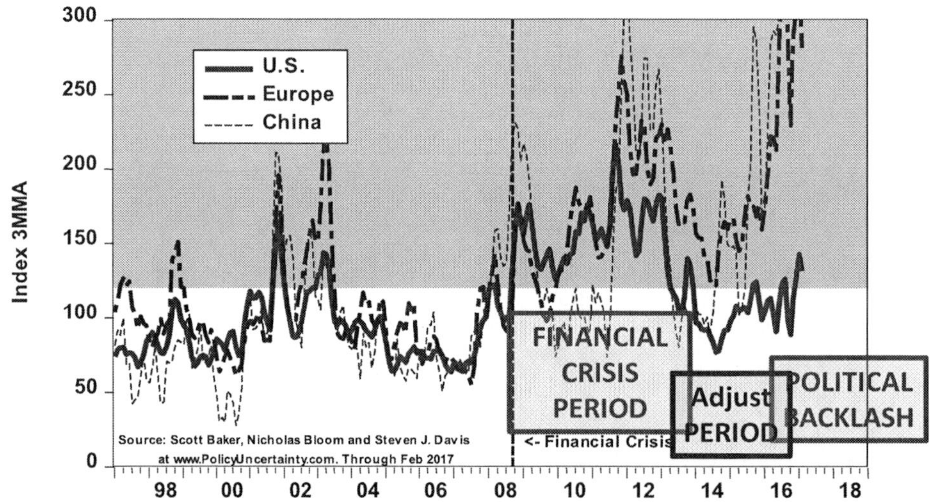

The grey area: Uncertainty is high enough to alter economic decision making (especially investment)

Duncan Meldrum
SPCC March 2017

6

Hilltop Economics

Steady Global Consumer Spending Growth

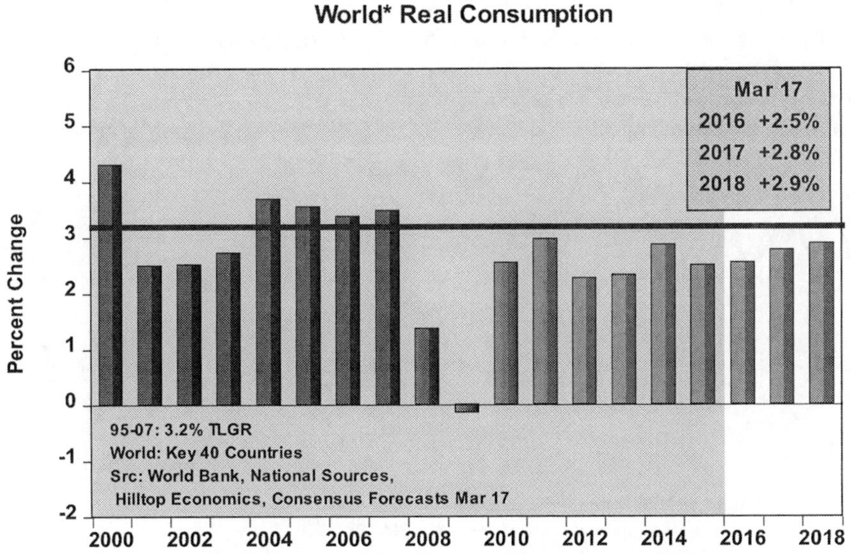

Consumer Spending Roughly 2/3 of GDP Demand

Duncan Meldrum
SPCC March 2017

7

Hilltop Economics

Global Investment Climbing Out of a Weak Cycle

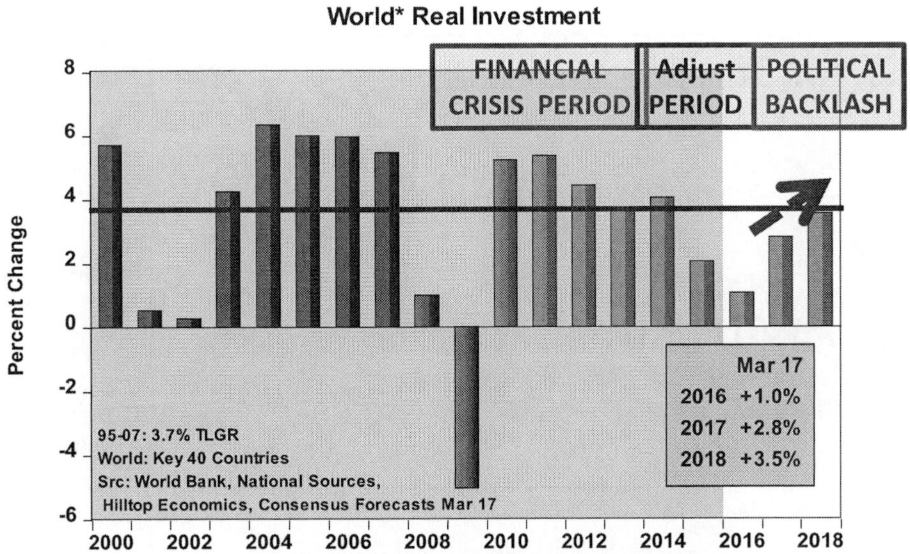

Investment Roughly 1/4 of GDP Demand

Duncan Meldrum
SPCC March 2017

8

Hilltop Economics

Spending on Technology Goods

An small but growing share of consumer spending. Until 2014, a decreasing share of investment in the U.S. Trends likely repeated globally.

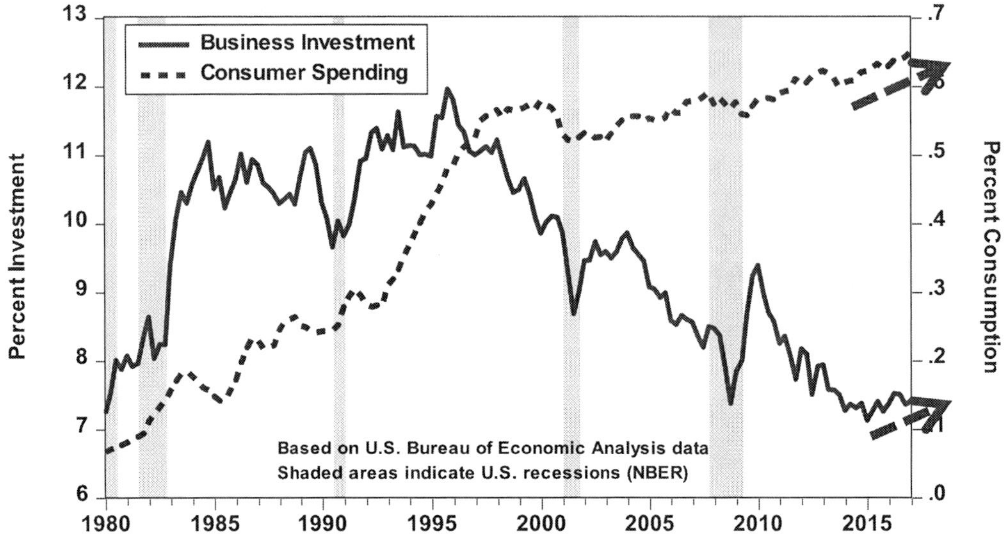

Duncan Meldrum
SPCC March 2017

9

Hilltop Economics

Real Investment Essential for Semiconductors

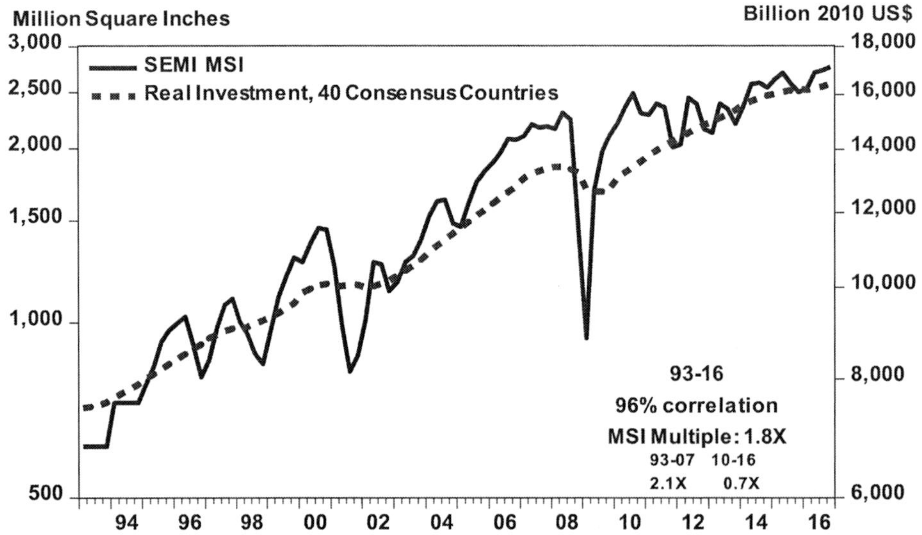

Final spending on fixed investment – plant, equipment & structures - is even more highly related to MSI over the past 23 years than total real GDP

Duncan Meldrum
SPCC March 2017

10

Hilltop Economics

Semiconductor Equipment Investment: Ahead of the Rest of the World

North America + Japan Semiconductor Equipment Billed

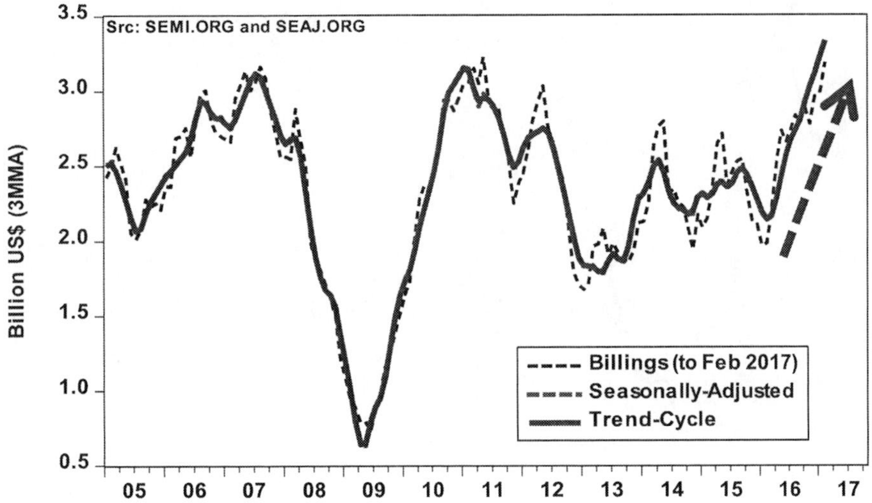

Duncan Meldrum
SPCC March 2017

11

Hilltop Economics

CPI Inflation Higher, But Not a Problem

CPI % Change	2015	2016	2017	2018
China	1.4	2.0	2.5	2.3
Japan	0.8	(0.1)	0.7	1.0
South Korea	0.7	1.0	1.8	1.8
Taiwan	(0.3)	1.4	1.5	1.4
Germany	0.2	0.5	1.8	1.7
France	0.0	0.2	1.5	1.3
Italy	0.1	(0.1)	1.3	1.3
UK	0.0	0.7	2.7	2.7
US	0.1	1.3	2.5	2.3

Hilltop Economics, Based on Consensus Forecasts, March 17

Sign of stronger demand + tighter supply
Businesses have a chance to get price increases

Duncan Meldrum
SPCC March 2017

12

Hilltop Economics

Improving Economy: Rising Interest Rates

A sign that we are finally moving back to "normal"

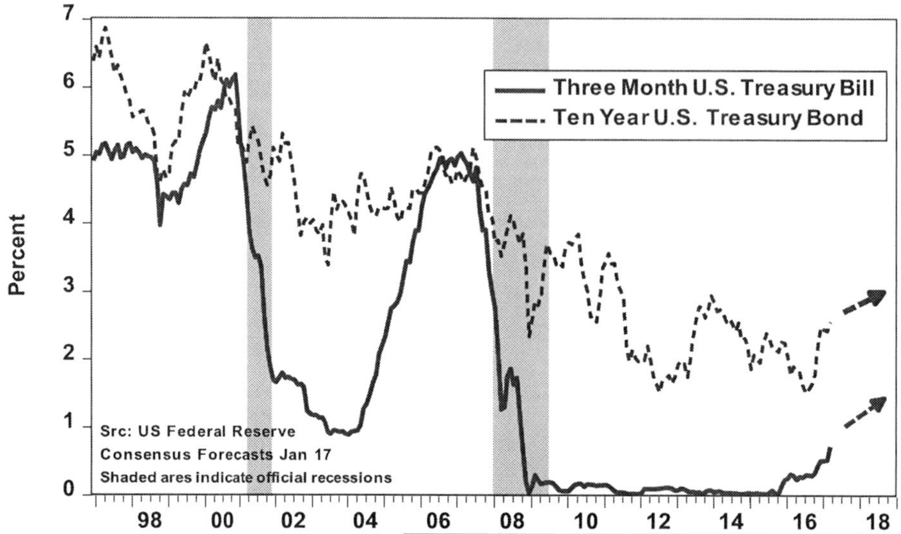

Sign of optimisim: stronger demand + tighter supply
Not too high to dampen investment (through 18)

Duncan Meldrum
SPCC March 2017

13

Hilltop Economics

Dollar Remains Strong

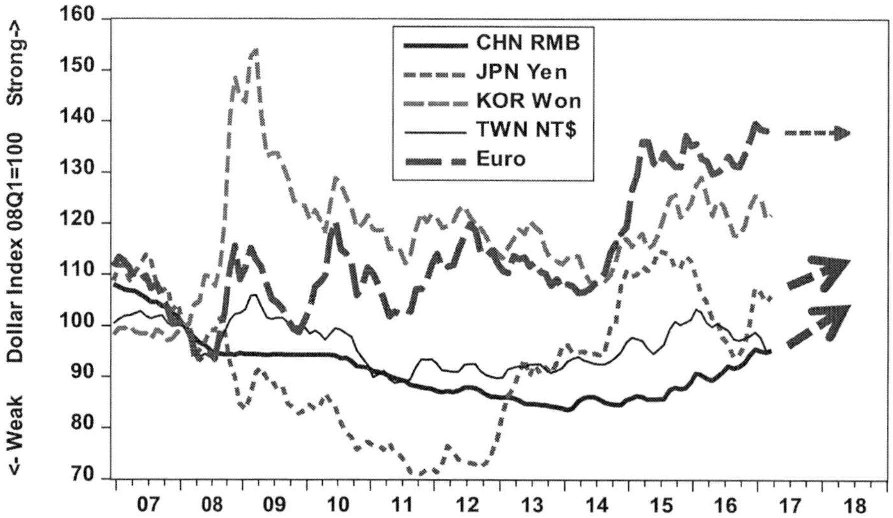

Duncan Meldrum
SPCC March 2017

14

Hilltop Economics

Semiconductor MSI Outlook

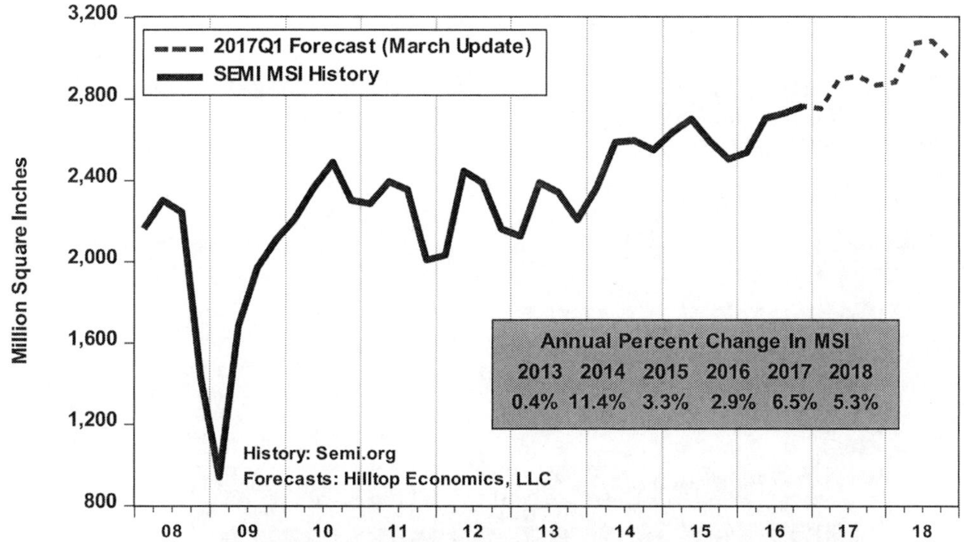

Despite uncertainty (or maybe because of it), the demand-driven outlook for the semiconductor industry is strong

Duncan Meldrum
SPCC March 2017

15

Hilltop Economics

MSI Becoming Less Seasonal

A plus for suppliers as semiconductors ubiquitous in final demand

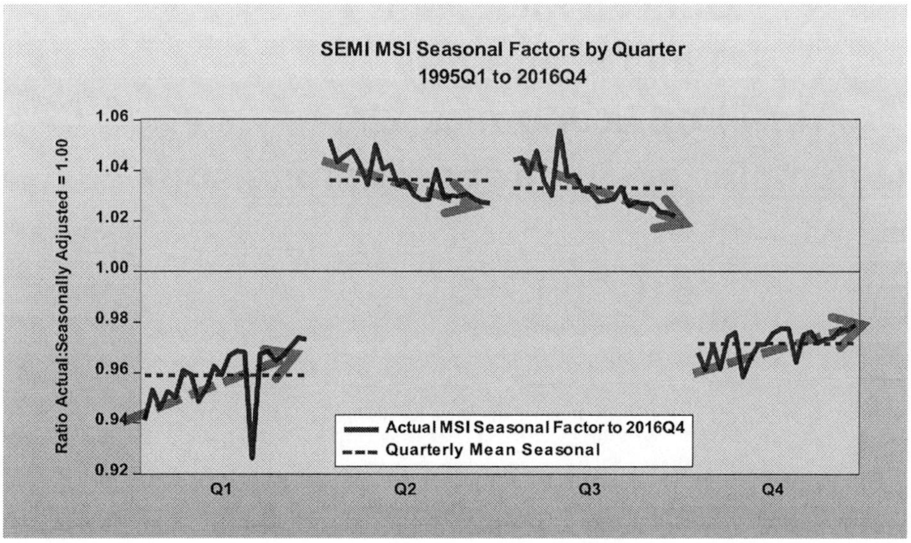

Duncan Meldrum
SPCC March 2017

Src: Hilltop Economics using US Census Bureau X-13 Seasonal Adjustment Program

16

Hilltop Economics

Semiconductor MSI Outlook Summary
March 2017 vs March 2016

MSI Forecast	2014	2015	2016	2017	2018
MSI – March 17 Forecast	10,097	10,434	10,738	11,432	12,036
% Change	*11.4%*	*3.3%*	**2.9%**	**6.5%**	**5.3%**
Global Real GDP % Change	*2.7%*	*2.6%*	**2.2%**	**2.7%**	**2.9%**
MSI - March 16 Forecast	10,097	10,434	*10,465*	*11,264*	*11,744*
% Change	*11.4%*	*3.3%*	*0.3%*	*7.6%*	*4.3%*
Global Real GDP % Change	*2.6%*	*2.4%*	*2.4%*	*2.8%*	*3.0%*

This year versus last year:
MORE SEMICONDUCTORS on SLIGHTLY LOWER ECONOMIC GROWTH

HIGHER RISK BANDS AROUND THE FORECAST

Duncan Meldrum
SPCC April 19-20 2016

Source: Linx_Consulting Econometric Forecasting Service (ESF) & Hilltop Economcs, March 2016 & March 2017

17

Hilltop Economics

Duncan Meldrum
Chief Economist
Hilltop Economics, LLC
duncan.meldrum@hilltopeconomics.com

610.709.7951

Duncan Meldrum
SPCC April 19-20 2016

18

Hilltop Economics

This Year Versus Last Year

This Year

- ***MSI: Up a little in 2016, rises above trend in 2017, eases to around trend in 2018***
- **Global final demand growth marginally better in 2017 & 2018**
  - **Real GDP in 2017 better than 2016, up even more in 2018**
  - **Business investment stalled in 2016, recovers in 2017**
  - **Consumer spending rises remains a little below 3% trend**
  - **PC unit sales fell modestly in 2016; increase slightly in 2017**
  - **Mobile phone sales growth slowed, now single digits at best**
  - **No major new markets**

Last Year

- *MSI: Flat in 16 on poor investment outlook, snaps back in 2017, falls to around trend in 2018*
- Global final demand grows marginally in 2016, better in 2017 & 2018
  - Real GDP should increase in 2016, reach "trend" by 2017
  - Business investment stalls in 2016, begins to recover in 2017
  - Consumer spending rises back toward trend of 3%
  - PC unit sales down sharply in 2015; down modestly in 2016; increase slightly in 2017
  - Mobile phone sales growth slows
  - No major new markets

Duncan Meldrum
SPCC April 19-20 2016

19

Hilltop Economics

PC Market Stabilizes With Investment Uptick (Less important for semiconductors)

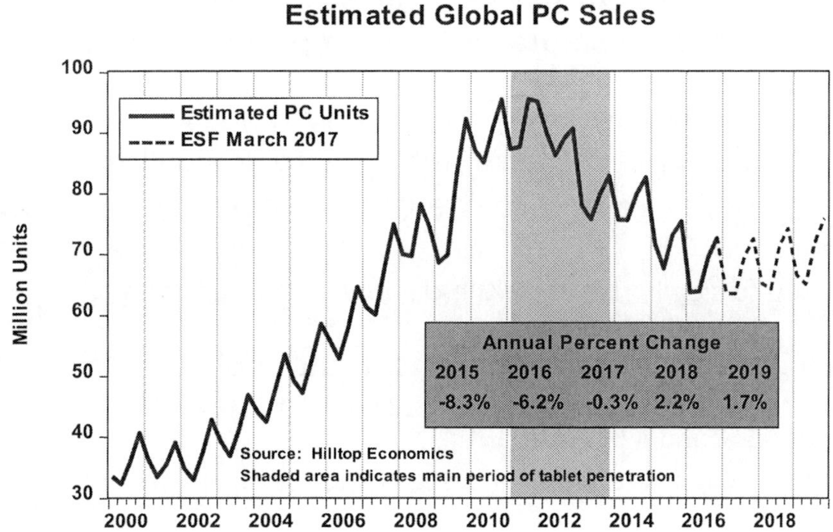

Duncan Meldrum
SPCC March 2017

20

Hilltop Economics

Seasonal Factors Drive PC Quarterly Change

Moving AWAY from 100%, Indicating RISING Seasonality of PCs

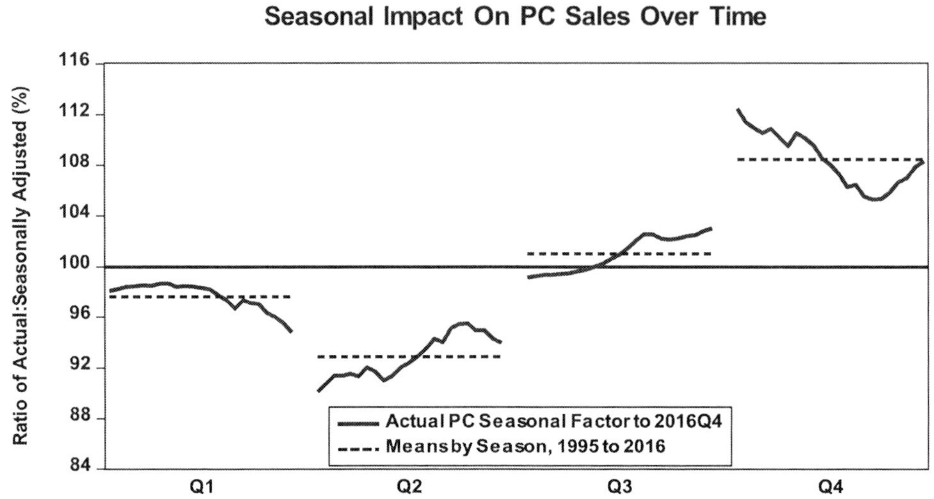

Duncan Meldrum
SPCC March 2017

Src: Hilltop Economics using US Census Bureau X-13 Seasonal Adjustment Program

21

Hilltop Economics

Things to Watch: Risks

- Yield Curve (expected to stay near current levels)
- Federal Reserve interest rate increases impact on financial markets
  - (forecasters don't expect more than two in 2016)
- Magnitude of oil price increase (expected to gradually rise)
- Value of the Dollar (expect to strengthen modestly)
- Inflation
  - (expected to slowly rise to ~2% in US, lower in other parts of the developed world)
- Jobs keep growing and unemployment rates hold
  - (forecasters expect unemployment rate in US to stay in the range of 4.5% and 4.9% in 2016)
- Forecasts getting lowered in a systematic fashion through the year

SPCC
April 2016

22

Hilltop Economics

The Best Single Recession Predictor: Treasury Yield Curve Inverts (Short Rates > Long Rates)

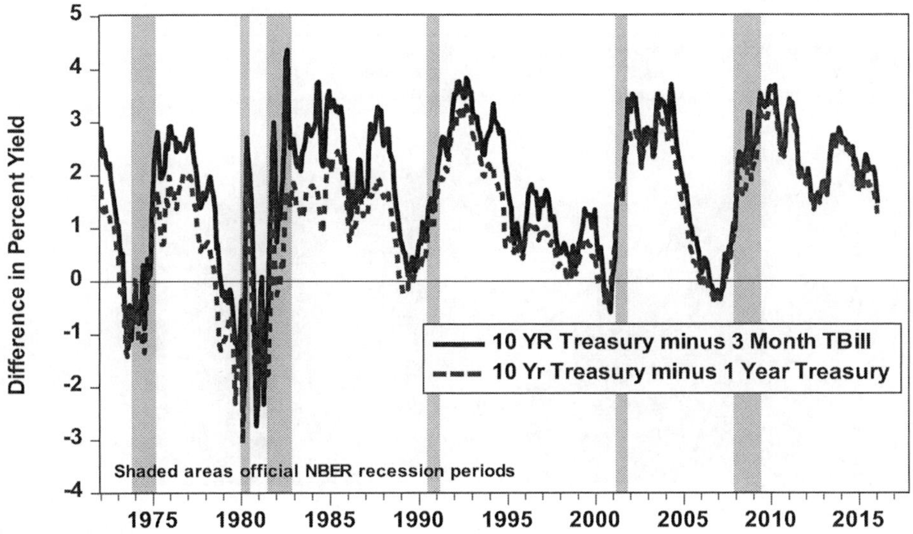

SPCC
April 2016

23

Hilltop Economics

The Best Single Recession Predictor: Treasury Yield Curve Inverts (Short Rates > Long Rates)

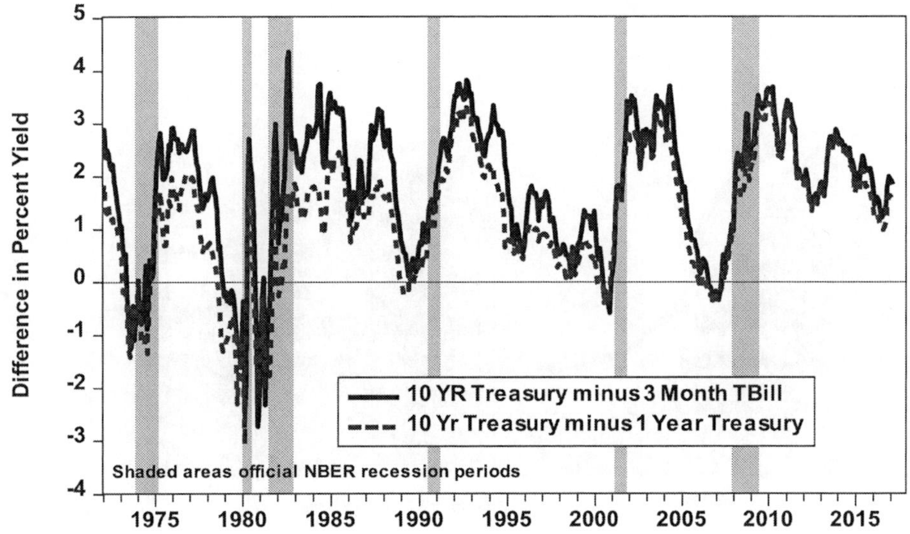

Duncan Meldrum
SPCC March 2017

24

Hilltop Economics

Forecasters continue the trend: "next year" gets closer to "normal" but 2017 already below "normal" 3+% trend

Steady reduction in '16 and '17 forecasts

Global Consensus-based Real GDP Forecasts

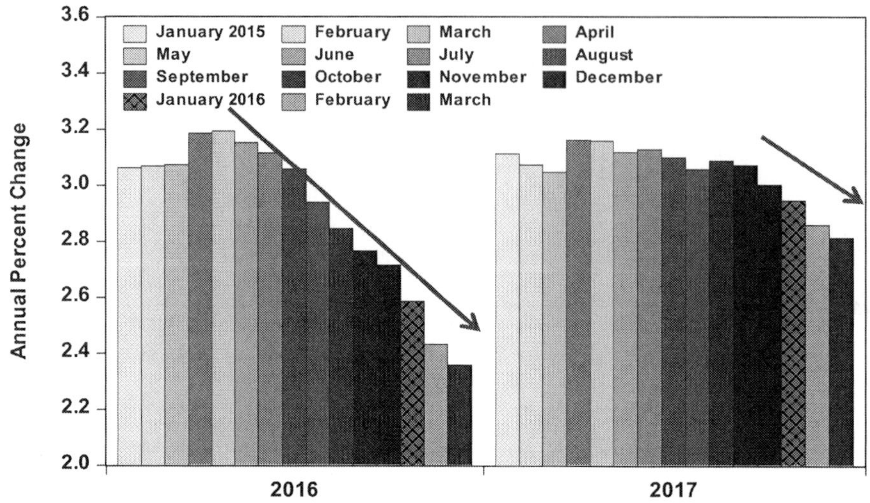

Forecast for Year By Month Forecast Made

SPCC April 2016

25

Hilltop Economics

Forecasters began to settle on growth outlook in mid-2016

Steady reduction in '16 and '17 forecasts

Global Consensus Real GDP Forecasts

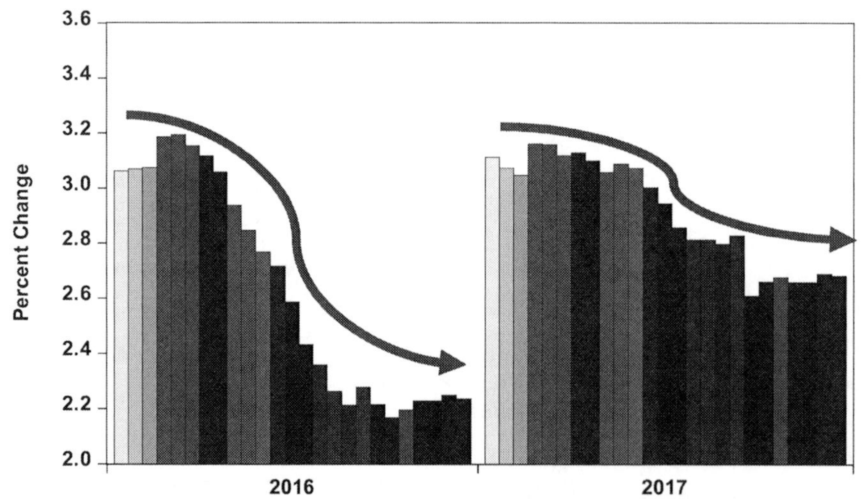

Forecasts from Jan 15 to Jan 17 (left to right)

Duncan Meldrum
SPCC March 2017

26

Hilltop Economics

Oil Prices Rise a Little

Modest demand growth plus plenty of supply limit increases

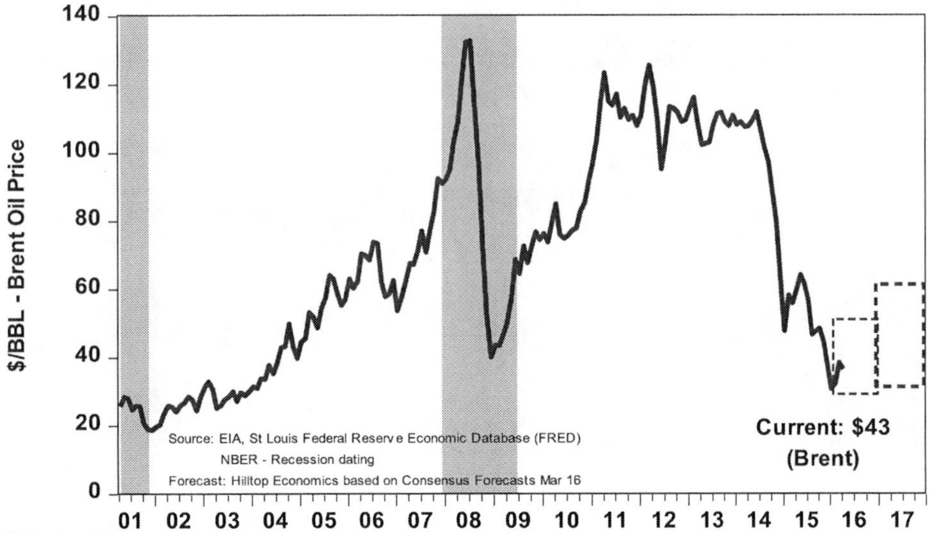

SPCC
April 2016

27

Hilltop Economics

Oil Prices Rise a Little

Modest demand growth plus plenty of supply limit increases

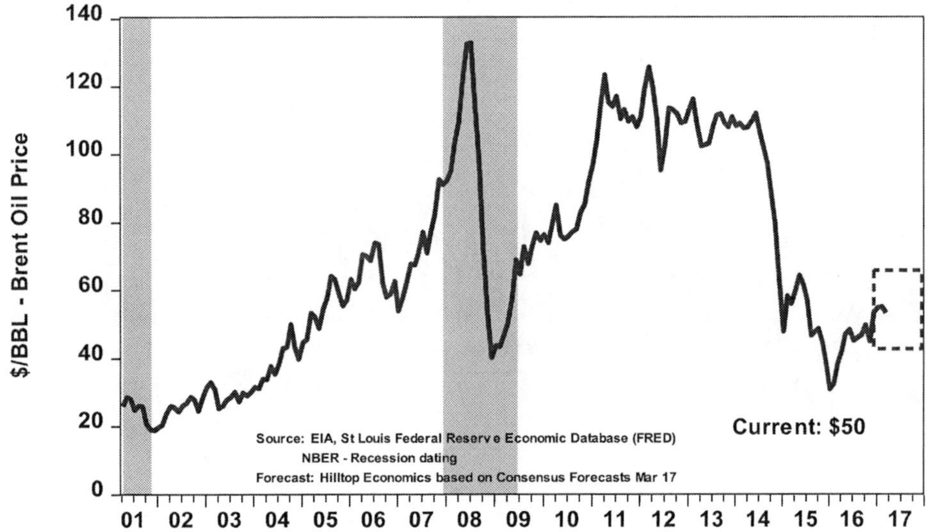

Duncan Meldrum
SPCC March 2017

28

Hilltop Economics

Consumers Globally Benefit From Low Energy Prices

(U.S. consumers alone gained almost 2% of disposable income since the end of 2013: almost $250 BN)

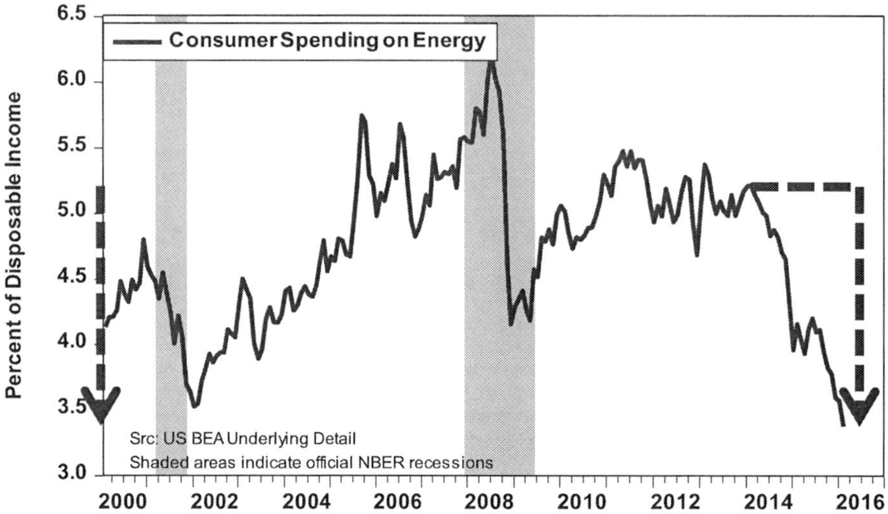

SPCC
April 2016

29

Hilltop Economics

Consumers Globally Benefit From Low Energy Prices

(U.S. consumers alone gained almost 2% of disposable income since the end of 2013: almost $250 BN)

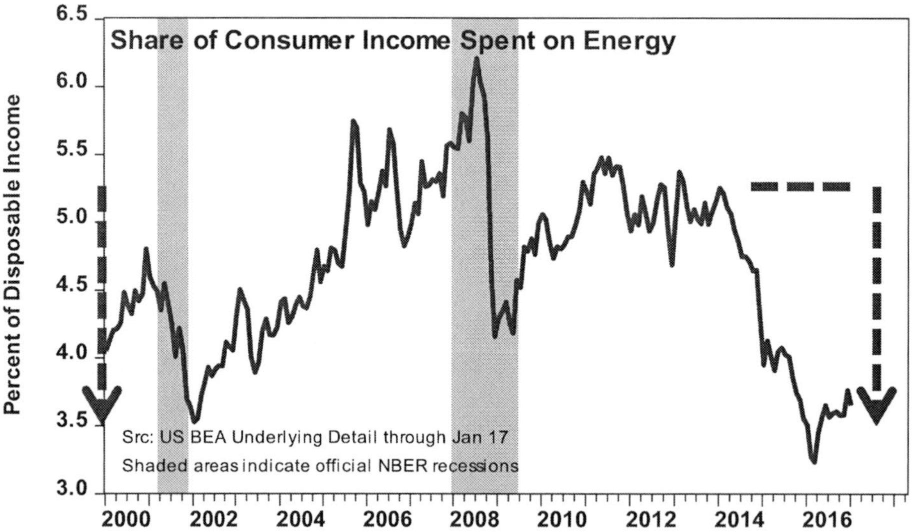

Duncan Meldrum
SPCC March 2017

30

Hilltop Economics

U.S. Household Debt Drops as Government Rises

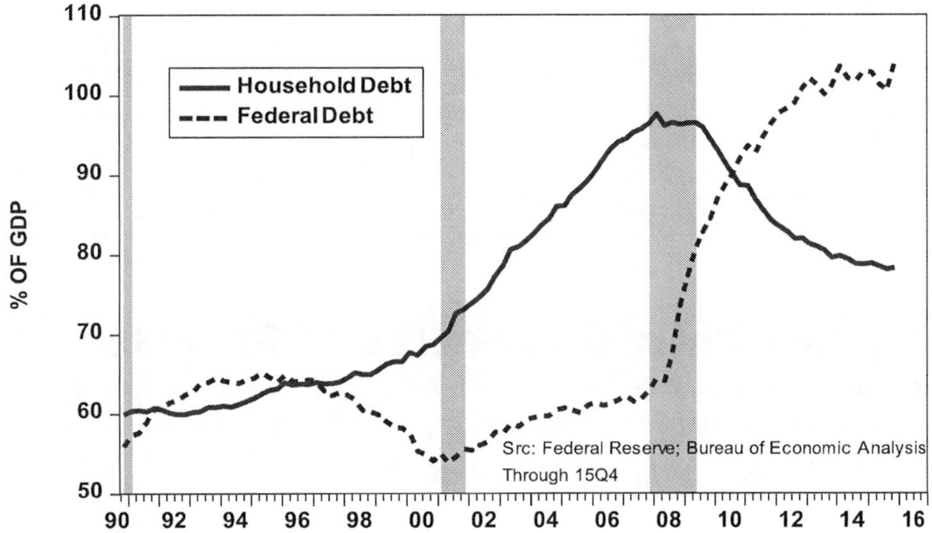

SPCC
April 2016

31

Hilltop Economics

U.S. Household Debt Drops as Government Rises

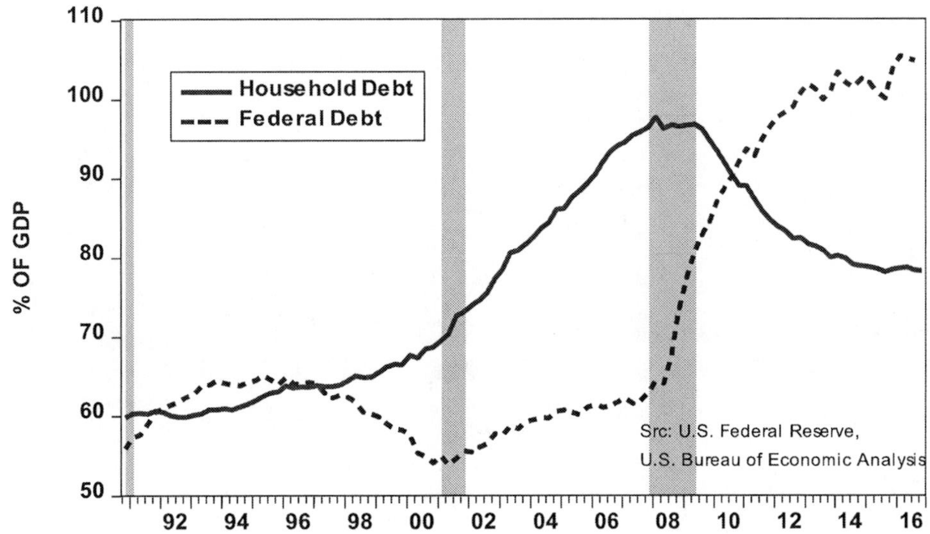

Duncan Meldrum
SPCC March 2017

32

Hilltop Economics

Recent Dollar Retreat Expected to Reverse

U.S. Interest rates are rising, key trading partners (Europe, China, Japan) are keeping rates steady or cutting: capital will flow to U.S., pushing dollar stronger

Dollar Strong VS Yen and Euro

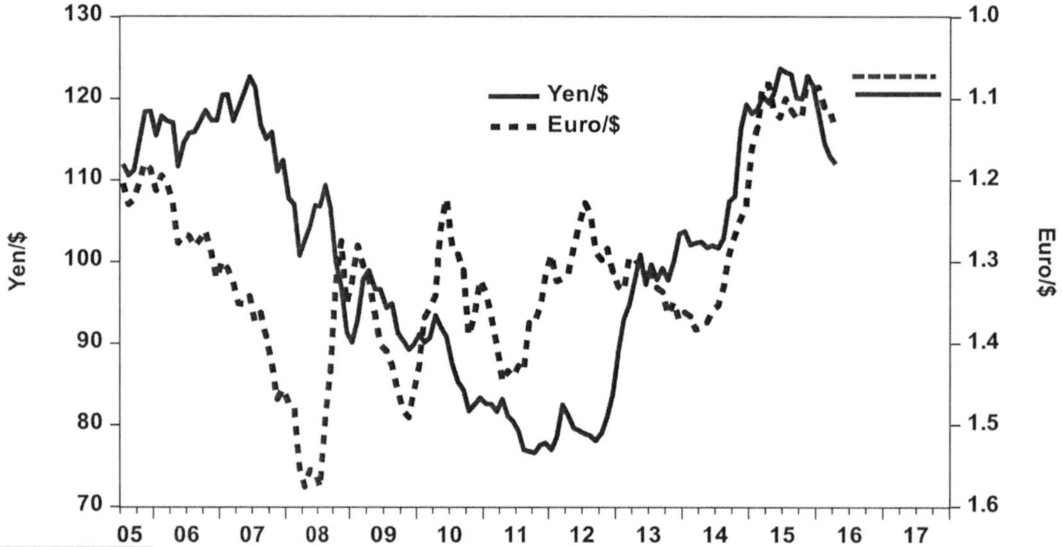

SPCC
April 2016

33

ACS APPLIED MATERIALS & INTERFACES

Research Article

www.acsami.org

Thermal Atomic Layer Etching of SiO_2 by a "Conversion-Etch" Mechanism Using Sequential Reactions of Trimethylaluminum and Hydrogen Fluoride

Jaime W. DuMont,[†] Amy E. Marquardt,[†] Austin M. Cano,[†] and Steven M. George[*,†,‡]

[†]Department of Chemistry and Biochemistry, and [‡]Department of Mechanical Engineering, University of Colorado at Boulder, Boulder, Colorado 80309, United States

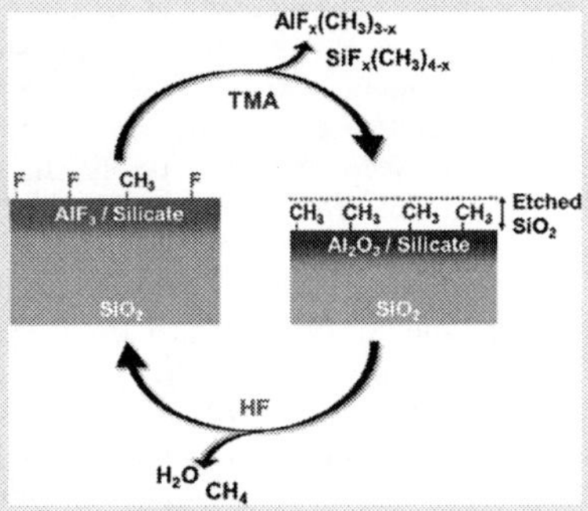

ABSTRACT: The thermal atomic layer etching (ALE) of SiO_2 was performed using sequential reactions of trimethylaluminum (TMA) and hydrogen fluoride (HF) at 300 °C. Ex situ X-ray reflectivity (XRR) measurements revealed that the etch rate during SiO_2 ALE was dependent on reactant pressure. SiO_2 etch rates of 0.027, 0.15, 0.20, and 0.31 Å/cycle were observed at static reactant pressures of 0.1, 0.5, 1.0, and 4.0 Torr, respectively. Ex situ spectroscopic ellipsometry (SE) measurements were in agreement with these etch rates versus reactant pressure. In situ Fourier transform infrared (FTIR) spectroscopy investigations also observed SiO_2 etching that was dependent on the static reactant pressures. The FTIR studies showed that the TMA and HF reactions displayed self-limiting behavior at the various reactant pressures. In addition, the FTIR spectra revealed that an Al_2O_3/aluminosilicate intermediate was present after the TMA exposures. The Al_2O_3/aluminosilicate intermediate is consistent with a "conversion-etch" mechanism where SiO_2 is converted by TMA to Al_2O_3, aluminosilicates, and reduced silicon species following a family of reactions represented by $3SiO_2 + 4Al(CH_3)_3 \rightarrow 2Al_2O_3 + 3Si(CH_3)_4$. Ex situ X-ray photoelectron spectroscopy (XPS) studies confirmed the reduction of silicon species after TMA exposures. Following the conversion reactions, HF can fluorinate the Al_2O_3 and aluminosilicates to species such as AlF_3 and SiO_xF_y. Subsequently, TMA can remove the AlF_3 and SiO_xF_y species by ligand-exchange transmetalation reactions and then convert additional SiO_2 to Al_2O_3. The pressure-dependent conversion reaction of SiO_2 to Al_2O_3 and aluminosilicates by TMA is critical for thermal SiO_2 ALE. The "conversion-etch" mechanism may also provide pathways for additional materials to be etched using thermal ALE.

KEYWORDS: *atomic layer etching, SiO_2, trimethylaluminum, hydrogen fluoride, Al_2O_3, aluminosilicate*

I. INTRODUCTION

Silicon dioxide (SiO_2) is the most abundant mineral in the Earth's crust. SiO_2 is also an important semiconductor material. SiO_2 has many uses, such as for a gate oxide, interlayer dielectric or spacer, in semiconductor devices. SiO_2 etching is needed for multiple steps during semiconductor manufacturing.[1] The progressive miniaturization of semiconductor devices to sub-10 nm dimensions has imposed more demanding requirements on etching. As device structures become increasingly smaller and more complex, high precision etching is necessary at the sub-1 nm level using atomic layer etching (ALE) methods.[2]

There are many approaches for SiO_2 etching using HF/H_2O solutions, HF/H_2O vapor mixtures, and halogen plasmas. Etching SiO_2 in aqueous HF solutions is one of the oldest and useful SiO_2 etching methods.[3] The overall chemical reaction during SiO_2 etching in HF solutions is $SiO_2 + 6HF \rightarrow H_2SiF_6 + 2H_2O$.[4,5] The mechanism of SiO_2 etching in HF solutions involves the progressive fluorine replacement of hydroxyl groups on Si–OH silanol surface species using HF or HF_2^- to form Si–F surface species.[4−6] H_2O or HF can also break the Si–O–Si siloxane linkage in SiO_2 to produce additional Si–OH silanol and Si–F surface species.

SiO_2 can also be etched by HF/H_2O in the gas phase. The gas phase approach is an extension of the solution phase approach and occurs when H_2O and HF are present at pressures sufficient to form a condensed HF/H_2O layer on the SiO_2 surface.[7,8] H_2O is essential for HF etching of SiO_2.[9] Vapor phase SiO_2 etch rates are in agreement with the etch rates expected from the composition of the condensed HF/H_2O layer present at the given HF and H_2O partial pressures.[7] The chemistry of vapor phase SiO_2 etching is the same as observed in aqueous HF solutions. In addition, H_2SiF_6 is removed by the reaction $H_2SiF_6 \rightarrow SiF_4 + 2HF$.[7]

Fluorocarbon-based plasma etching can also be employed for SiO_2 etching.[10−12] F atoms in the plasma have been identified as the active etching reactant. SiF_4 has been detected as the main Si-containing reaction product. Fluorocarbon plasmas are also used in newly developed plasma ALE methods based on an Ar plasma producing low energy Ar^+ ions together with a periodic fluorocarbon-based plasma exposure.[13,14] The fluorocarbon-based plasma deposits a fluorocarbon layer on the

Received: January 24, 2017
Accepted: February 27, 2017
Published: February 27, 2017

DOI: 10.1021/acsami.7b01259
ACS Appl. Mater. Interfaces 2017, 9, 10296−10307

SiO_2 surface. The Ar^+ ion bombardment then desorbs the fluorocarbon layer and also removes a controlled amount of the SiO_2 layer.

Thermal ALE processes have been introduced recently for Al_2O_3, HfO_2, ZrO_2, AlF_3, and AlN ALE based on sequential fluorination and ligand-exchange reactions.[15−22] In these thermal reactions, the metal oxide is first fluorinated to produce the metal fluoride.[17,19,22] The metal fluoride is then removed by a metal precursor in a ligand-exchange transmetalation reaction.[17,19,22] In this ligand-exchange reaction, the metal precursor accepts fluorine from the metal fluoride and donates one of its ligands to the metal fluoride. The resulting reaction product formed from the metal fluoride can then desorb from the surface if this metal species is stable and volatile.[20,22]

In this paper, a new gas-phase thermal SiO_2 ALE process is reported based on sequential exposures of $Al(CH_3)_3$ (trimethylaluminum (TMA)) and HF. SiO_2 film thicknesses were measured versus number of ALE cycles using ex situ spectroscopic ellipsometry to obtain the SiO_2 etching rates. These studies determined the linearity of the SiO_2 etching and the dependence of the SiO_2 etching on the TMA and HF reactant pressures. Ex situ X-ray photoelectron spectroscopy (XPS) measurements were also employed to determine the silicon oxidation state after the TMA and HF reactions. In addition, in situ Fourier transform infrared (FTIR) spectroscopy measurements were used to measure the loss of absorbance from Si−O stretching vibrations and to monitor the surfaces species present after each TMA and HF exposure at various pressures.

Recent studies of ZnO ALE using TMA and HF as the reactants have revealed a new "conversion-etch" reaction mechanism where a ZnO surface layer is converted to an Al_2O_3 surface layer and volatile $Zn(CH_3)_2$ reaction products during the TMA exposure.[23] In this investigation, the FTIR and XPS studies determined that there was a similar conversion of SiO_2 to Al_2O_3 by TMA. The observation of Al_2O_3, aluminosilicate species, and silicon in reduced oxidation states established that the "conversion-etch" mechanism also occurs for SiO_2 ALE. In addition, the reactant pressures were critical for SiO_2 ALE and larger etch rates occurred at higher reactant pressures. The generality of the "conversion-etch" mechanism could expand the number of materials that can be etched using thermal ALE methods.

II. EXPERIMENTAL SECTION

1. Static Reactor, Silicon Wafers, and Ex Situ Film Analysis. ALE reactions were performed in a hot wall ALD reactor.[24] A proportional-integral-derivative (PID) temperature controller (2604, Eurotherm) maintained the reaction temperature at 300 °C. A capacitance manometer (Baratron 121A, MKS) monitored the pressure in the reactor. The starting SiO_2 substrate was a thermal SiO_2 layer with a thickness of 40 nm on Si(100) wafers with a diameter of 100 mm (University Wafer). The silicon wafers were cut into samples with dimensions of 2.5 cm × 2.5 cm before the SiO_2 ALE.

The SiO_2 ALE reactions were accomplished using sequential static exposures of TMA (97%, Sigma-Aldrich) and HF derived from HF−pyridine (70 wt % HF, Sigma-Aldrich). HF−pyridine was transferred to a gold-plated stainless steel bubbler using a dry N_2-filled glovebag. The TMA and HF−pyridine precursors were held at room temperature. Each reactant exposure consisted of a static dose of 20 s followed by a purge of 120 s with 550 sccm of ultrahigh purity (UHP) N_2. The total N_2 gas flow of 550 sccm produced a base pressure of ~2 Torr in the reactor during the purge steps. The N_2, precursor, and reaction byproducts were pumped by a mechanical pump (Pascal 2010SD, Alcatel Adixen).

The ex situ X-ray reflectivity (XRR) scans were measured by a high resolution X-ray diffractometer (Bede D1, Jordan Valley Semiconductors) using Cu $K\alpha$ (λ = 1.540 Å) radiation. The filament voltage and current in the X-ray tube were 40 kV and 35 mA, respectively. A 5 arcsec step size and a 10 s acquisition time were used for recording all XRR scans with a range of 300−4500 arcsec. The analysis software (Bede REFS, Jordan Valley Semiconductors) determined film thickness, film density, and surface roughness by fitting the XRR scans.

Ellipsometry was performed using a spectroscopic ellipsometer (J. A. Woollam M-2000D). This ellipsometer has a spectral range of 240−1700 nm and utilizes an incidence angle of 75°. A Cauchy model was used to determine the film thickness. This modeling was performed using analysis software (CompleteEASE, J. A. Woollam). SiO_2 film thicknesses were measured at the same position on each silicon sample before and after etching.

Thermal SiO_2 layers with a thickness of 40 nm on Si(100) wafers (University Wafer) were used for XPS analysis to determine the surface species after TMA-only and TMA/HF exposures. These SiO_2 samples were exposed to 8 Torr of TMA for 5 min or 8 Torr of TMA for 5 min followed by 4 Torr of HF for 10 min. After the static reactant exposures, the samples were immediately transferred to a PHI 5600 X-ray photoelectron spectroscopy (XPS) spectrometer to analyze the films. A SiO_2 sample without TMA or HF exposures was also analyzed with XPS.

A monochromatic Al $K\alpha$ X-ray source (1486.6 eV) was used to collect survey scans with a pass energy of 93.9 eV and step size of 0.400 eV. An XPS software package (Auger Scan, RBD Instruments) was used to collect the XPS spectra. Casa XPS software (Casa XPS, Casa Software) was used for peak analysis of the C 1s, O 1s, Si 2p, Al 2p, and F 1s XPS signals. All peaks were calibrated to the C 1s peak centered at 284.8 eV.

The oxidation state of Si was determined through fitting the Si 2p peak profile using Gaussian−Lorentzian line shapes after a Shirley background subtraction. XPS analysis of the SiO_2 control sample was used to determine the Si 2p binding energy position and full width at half-maximum (fwhm) of the 4+ oxidation state peak. The binding energy (104.3 eV) and fwhm (2.1 eV) of the 4+ oxidation state were imposed in the fittings of the TMA-only and TMA/HF exposed samples. The binding energy positions for the 3+, 2+, 1+, and 0 oxidation states were defined as −1, −2, −3, and −4 eV with respect to the 4+ peak.[25,26] The fwhm values of the 3+, 2+, 1+, and 0 oxidation states were set to 1.2 ± 0.1, 1.1 ± 0.1, 1.0 ± 0.1, and 1.0 ± 0.1 eV, respectively.[25−27]

2. In Situ Fourier Transform Infrared (FTIR) Spectroscopy Studies. The surface chemistry during sequential exposures of TMA (97%, Sigma-Aldrich) and HF derived from HF−pyridine (70 wt % HF, Sigma-Aldrich) on SiO_2 at 300 °C was studied using in situ FTIR spectroscopy. The in situ FTIR studies were performed in a reactor equipped with an FTIR spectrometer that has been described previously.[28] The FTIR experiments utilized high surface area Si nanoparticles (>98%, US Research Nanomaterials) with an average diameter of 20−30 nm. The large surface area provided by the nanopowder sample improved the signal-to-noise ratio for infrared absorption. The Si nanoparticles were mechanically pressed into a tungsten grid support to facilitate the transmission FTIR measurements.[29] The tungsten grid was 2 × 3 cm^2, 50 μm thick, with 100 grid lines per inch.

The tungsten grid could be resistively heated using a dc power supply (6268B, 20 V/20 A, Hewlett-Packard). The voltage output of the power supply was controlled by a PID temperature controller (Love Controls 16B, Dwyer Instruments). A type K thermocouple was attached to the bottom of the tungsten grid with epoxy (Ceramabond 571, Aremco) that attached and electrically isolated the thermocouple. To create a uniform SiO_2 surface, the Si nanoparticles were annealed at 650 °C for 1 h under ~25 Torr of H_2O vapor.

Figure 1a shows the absolute FTIR spectra of the SiO_2 coating on the Si nanoparticles before and after the H_2O oxidation process.

DOI: 10.1021/acsami.7b01259
ACS Appl. Mater. Interfaces 2017, 9, 10296−10307

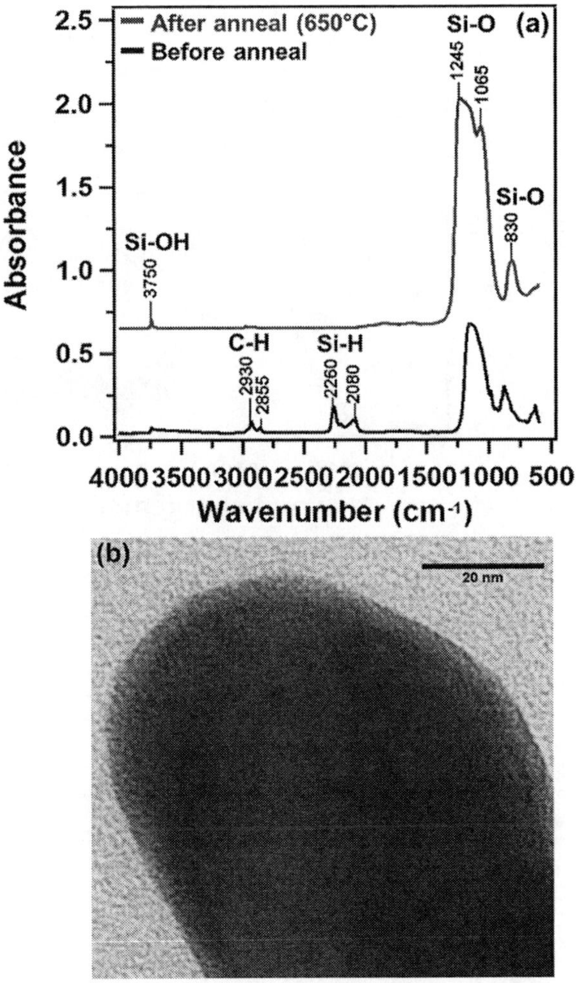

Figure 1. (a) Absolute FTIR spectra of silicon particles with a SiO_2 coating before and after oxidation with ~25 Torr H_2O at 650 °C for 1 h. (b) TEM image of a silicon nanoparticle with a ~3.5 nm thick SiO_2 coating after the H_2O thermal oxidation.

Distinctive absorption features for the unannealed sample are observed in four frequency regions: 835–1250 cm^{-1}, 2080–2260 cm^{-1}, 2855–2930 cm^{-1}, and 3250–3750 cm^{-1}. The strong and broad absorption bands between 835 and 1250 cm^{-1} are consistent with the TO/LO modes that arise from Si–O symmetric and asymmetric stretching vibrations.[30] The absorption bands observed between 2080 and 2260 cm^{-1} are attributed to SiH_x and SiO_xH_{3-x} species.[31−33] Absorption bands between 2855 and 2930 cm^{-1} are attributed to the C–H stretching vibrations of adventitious hydrocarbons. The broad absorption feature observed between 3250 and 3750 cm^{-1} is consistent with isolated and hydrogen bonded Si–OH species. These results are in agreement with a partially oxidized hydrogen-terminated silicon surface.

After thermal oxidation with H_2O, evidence of silicon oxidation is observed in each distinctive absorption band. Between 835 and 1250 cm^{-1}, large absorbance gains are observed that are consistent with the addition of bulk Si–O–Si stretching vibrations. These results are consistent with previous studies that have demonstrated the formation of a thermal oxide layer on silicon surfaces exposed to water vapor at elevated temperatures.[34,35] Furthermore, a distinct blue shift from 1045 to 1065 cm^{-1} and from 1160 to 1245 cm^{-1} is observed for the TO/LO modes after thermal oxidation. This behavior has been observed previously and is related to the stoichiometry of the SiO_x film. As more oxygen is incorporated into the film, a more electronegative environment leads to shorter Si–O bonds lengths and higher stretching frequencies.[30]

The disappearance of the vibrational features between 2080 and 2260 cm^{-1} is attributed to Si–H removal via H_2 evolution. H_2 desorption has been shown to occur on hydrogen-covered Si surfaces at 325–625 °C.[36,37] Thermal Si–H removal results in Si surface dangling bond sites that are susceptible to oxidation.[32,38,39] Absorption features between 2855 and 2930 cm^{-1} are also lost due to the volatilization of the carbon contaminants in the oxidizing environment. Lastly, there is an absorbance gain observed at 3750 cm^{-1} that is attributed to the formation of isolated Si–OH species. Previous studies have also observed the formation of isolated Si–OH species during oxidation of hydrogen-passivated silicon surfaces.[32]

This H_2O thermal oxidation procedure consistently removed any carbon-related vibrational modes and resulted in a SiO_2 coating on the Si nanoparticles with absorbances attributed only to the O–H hydroxyl stretching vibrations and bulk SiO_2 vibrational features. Transmission electron microscopy (TEM) analysis was utilized to confirm the presence of the SiO_2 coating on the Si nanoparticles. Figure 1b shows the TEM image of an annealed silicon nanoparticle with a SiO_2 coating with a thickness of ~3.5 nm. The SiO_2 coating is very uniform and conformal on the Si nanoparticle.

III. RESULTS AND DISCUSSION

1. 1. X-ray Reflectivity Measurements of SiO_2 Etching. Figure 2 displays XRR measurements of the change in the SiO_2 film thickness versus ALE cycles at 300 °C for various static reactant pressures. The reaction sequence was defined by reactant exposures with a duration of 20 s and purges with a duration of 120 s. This reaction sequence is designated 20–120–20–120. The reactant pressures for the TMA and HF exposures were the same. The initial SiO_2 thermal oxide film on the silicon samples had a thickness between 390 and 397 Å.

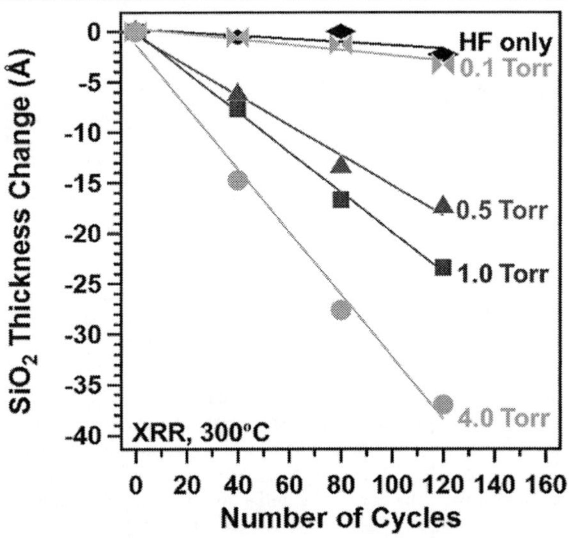

Figure 2. SiO_2 film thickness change versus number of SiO_2 ALE cycles determined by X-ray reflectivity measurements at TMA/HF pressures of 0.1, 0.5, 1.0, and 4.0 Torr. HF only experiments were also performed using an HF pressure of 1.0 Torr.

Figure 2 shows that the SiO_2 film thickness change versus number of ALE cycles is linear at each reactant pressure. The

SiO_2 etch rates were 0.027, 0.15, 0.20, and 0.31 Å/cycle for static reactant pressures of 0.1, 0.5, 1.0, and 4.0 Torr, respectively. Previous results that utilized lower reactant pressures of ~0.08 Torr in a viscous flow mode showed no SiO_2 etching by TMA and HF at 300 °C.[18] The larger SiO_2 etch rates observed at higher precursor pressures suggest a pressure-dependent reaction mechanism for SiO_2 ALE.

Control experiments were also performed using only HF exposures of 1.0 Torr. The results in Figure 2 indicate that HF exposures by themselves resulted in negligible etching of the SiO_2 film at 300 °C. These results are consistent with earlier results that concluded that HF alone cannot etch SiO_2.[7,9] H_2O vapor needs to be present together with HF for SiO_2 etching.

2. Spectroscopic Ellipsometry Measurements of SiO_2 Etching. Figure 3 displays SE measurements of the change in

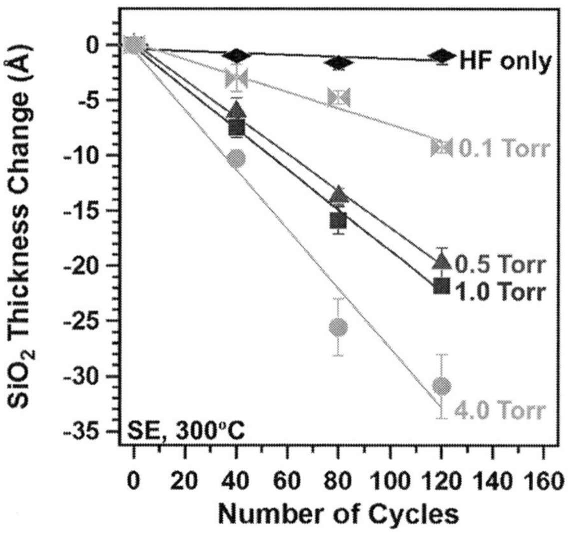

Figure 3. SiO_2 film thickness change versus number of SiO_2 ALE cycles determined by spectroscopic ellipsometry measurements at TMA/HF pressures of 0.1, 0.5, 1.0, and 4.0 Torr. HF only experiments were also performed using an HF pressure of 1.0 Torr.

the SiO_2 film thickness versus ALE cycles at 300 °C for various static reactant pressures. The reaction sequence was defined by reactant exposures with a duration of 20 s and purges with a duration of 120 s. The reactant pressures for the TMA and HF exposures were the same. The initial SiO_2 thermal oxide film on the silicon samples had a thickness between 390 and 397 Å.

Figure 3 shows that the SiO_2 film thickness change versus number of ALE cycles is linear at each reactant pressure. The etch rates were 0.07, 0.17, 0.18, and 0.27 Å/cycle for static reactant pressures of 0.1, 0.5, 1.0, and 4.0 Torr, respectively. The etch rates from the SE measurements are in good agreement with the etch rates from the XRR measurements. Control experiments were also performed using only HF exposures of 1.0 Torr. The results in Figure 3 demonstrate that only HF exposures lead to negligible SiO_2 etching.

3. FTIR Spectroscopy Measurements of SiO_2 Etching. Figure 4 displays the absolute FTIR spectra from 700 to 1500 cm^{-1} for Si particles covered with a SiO_2 coating during static TMA and HF exposures at 0.5, 1, and 4 Torr at 300 °C. Each reactant exposure consisted of a static dose at the specified pressure for 30 s followed by a 240 s purge. The most

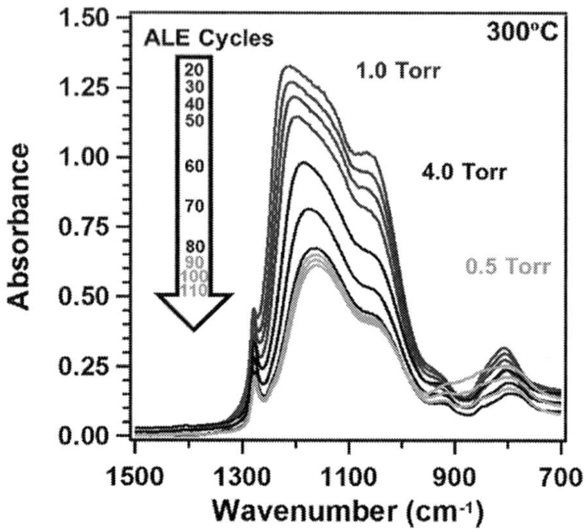

Figure 4. Absolute FTIR spectra showing the absorbance loss from Si−O stretching vibrations between 700 and 1500 cm^{-1} versus SiO_2 ALE cycles at TMA/HF pressures of 0.5, 1.0, and 4.0 Torr at 300 °C.

prominent vibrational features are attributed to the transverse optical (TO) and longitudinal optical (LO) modes arising from asymmetric Si−O stretching vibrations of SiO_4 tetrahedra at 1065 and 1230 cm^{-1}, respectively.[30] The less intense peak at 820 cm^{-1} is attributed to the corresponding TO and LO modes arising from symmetric Si−O stretching vibrations.[30] The absorbance features from the Si−O stretching vibrations decrease progressively versus the number of ALE cycles. An absorbance feature also appears at 1275 cm^{-1} after the first TMA exposure that is attributed to a deformation mode of Si−CH_3 surface species.[40] This result is consistent with previous studies that have observed the formation of Si−CH_3 species during reactions of TMA with SiO_2 surfaces.[41−43]

The FTIR spectra in Figure 4 were recorded after the TMA exposures. The first 50 cycles consisted of static reactant exposures at 1 Torr and showed a progressive loss of the absorbance for the Si−O stretching vibrations versus the number of ALE cycles. The nanoparticles were subsequently exposed to 30 cycles of static reactant exposures at 4 Torr and then 30 cycles of static reactant exposures at 0.5 Torr. The absorbance change per cycle of the Si−O stretching vibrations increased significantly at 4 Torr. This more rapid absorbance change indicates a larger SiO_2 etch rate at the higher reactant pressure. The absorbance change per cycle then decreased during the subsequent 0.5 Torr static exposures.

The absolute integrated absorbances of the asymmetric Si−O stretching vibrations between 900 and 1400 cm^{-1} versus ALE cycle number are plotted in Figure 5. At each pressure, the absorbance decreases linearly versus sequential TMA and HF exposures. ΔAbs/cycle values of −1.1, −1.9, and −3.6 are obtained for reactant pressures of 0.5, 1, and 4 Torr, respectively. These results are in agreement with the ex situ XRR and SE measurements that revealed larger SiO_2 etch rates at higher reactant pressures.

The etch rates and absorbance losses per cycle versus reactant pressure obtained by XRR, SE, and FTIR measurements are plotted in Figure 6. The relative etch rates determined by XRR and SE measurements at various reactant

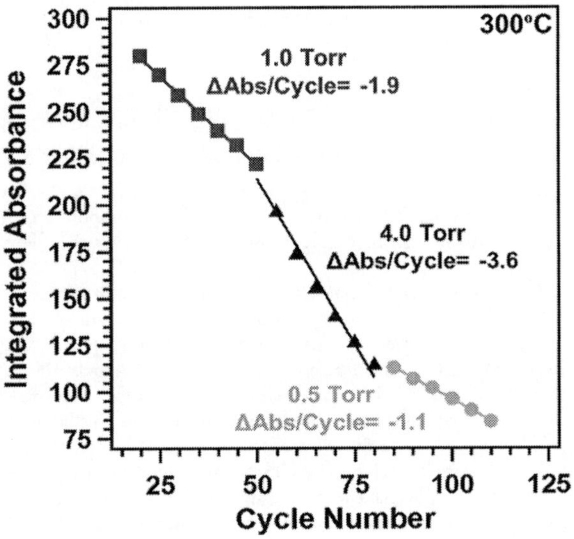

Figure 5. Absolute integrated absorbance of the Si–O stretching vibrations between 900 and 1400 cm^{-1} versus ALE cycle number at TMA/HF pressures of 0.5, 1.0, and 4.0 Torr. The integrated absorbance change per cycle was −1.1, −1.9, and −3.6 at pressures of 0.5, 1.0, and 4.0 Torr, respectively.

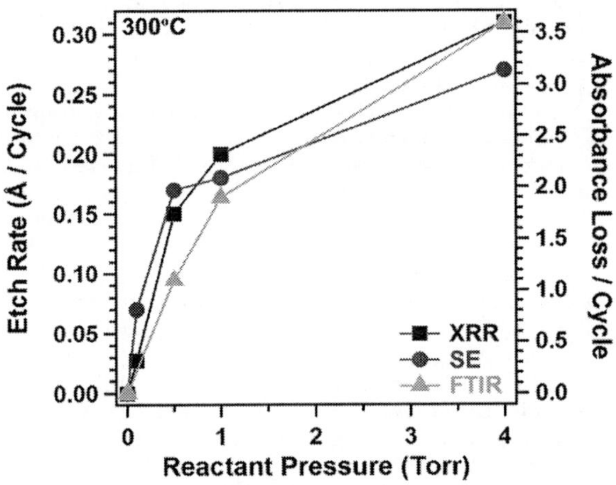

Figure 6. Etch rates from the XRR measurements in Figure 2 and the SE measurements in Figure 3 compared with the absorbance loss versus cycle from Figure 5 versus reactant pressure.

pressures are in good agreement with the absorbance loss per cycle determined using FTIR spectroscopy. All of the measurement techniques showed that the etch rate increases up to 1 Torr and then begins to level off at higher pressures. The leveling off of the etch rate at higher pressures suggests a diffusion-limited process similar to the oxidation of silicon substrates.[34] The surface layers formed during either the TMA or HF reactions act as diffusion barriers. These surface layers progressively limit the transport of reactants as the thickness of the surface layer increases.

4. FTIR Studies of Surface Chemistry and Thin Film Changes. The FTIR spectra can examine the surface chemistry and thin film changes during SiO_2 ALE. Figure 7 displays the FTIR difference spectra between 500 and 1700 cm^{-1} for the first TMA and HF exposures on SiO_2 at 300 °C. Each reactant exposure consisted of a ~1 Torr static dose for 30 s followed by a purge for 240 s. Figure 7a shows the difference spectrum defined by the spectrum recorded after the first TMA exposure referenced to the spectrum of the initial SiO_2 coating on the Si particles (TMA − SiO_2). Negative absorbance features at 975 and 1250 cm^{-1} are observed that are consistent with the loss of Si–OH and Si–O species, respectively.[30,44] These absorbance losses are accompanied by absorbance gains at 670, 750, 845, 1212, and 1275 cm^{-1}.

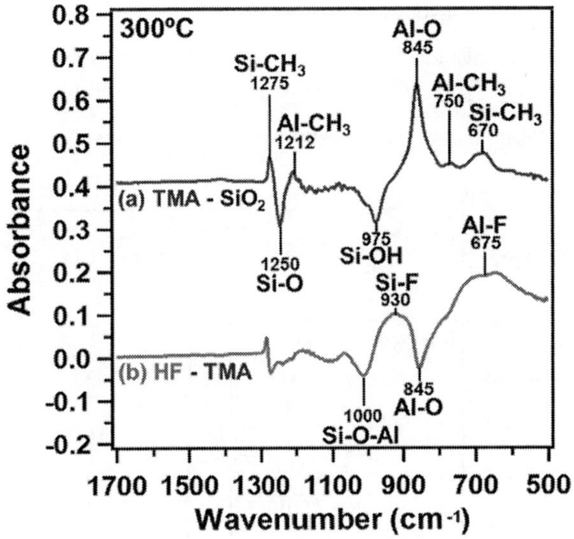

Figure 7. Difference infrared absorbance spectra from 500–1700 cm^{-1} during the first TMA and HF exposures on SiO_2 at 300 °C. (a) Spectrum after the first TMA exposure referenced to the spectrum for the initial SiO_2 layer on the silicon particles. (b) Spectrum after the first HF exposure referenced to the spectrum after the first TMA exposure.

The absorbance gains at 670 and 1275 cm^{-1} in Figure 7a are attributed to the addition of Si–CH_3 species.[41,45] The absorbance gains at 750 and 1212 cm^{-1} are consistent with the addition of Al–CH_3 species. The absorbance gain at 845 cm^{-1} is assigned to the formation of Al–O species.[46−48] The resulting surface could contain $AlO_{3-x}(CH_3)_x$ species where x = 1 or 2 and $SiO_{4-y}(CH_3)_y$ species where y = 1, 2, or 3. These FITR measurements are in agreement with previous studies that have reported that TMA chemisorbs on silica surfaces primarily through reaction with hydroxyl groups and siloxane bridges.[41,43,49] TMA reactions with hydroxyl groups result in the formation of siloxy-methylaluminum surface species and methane. During dissociation of TMA on siloxane bridges, methyl groups are transferred from aluminum to silicon.

Figure 7b displays the infrared difference spectrum after the subsequent HF exposure. The Si–O–Al and Al–O vibrational features that were added during the previous TMA exposure are lost after the subsequent HF exposure. These absorbance losses are accompanied by absorbance gains at 675 and 930 cm^{-1} that are consistent with Al–F and Si–F species, respectively.[50,51] These results suggest that the Al–O species and Si–O–Al species formed during the TMA exposures are converted to Al–F and Si–F species during HF exposures, respectively. The reaction of HF with Al_2O_3 to produce AlF_3 has been observed in previous ALE studies.[17−19] This reaction occurs as

$$Al_2O_3 + 6HF \rightarrow 2AlF_3 + 3H_2O \quad (1)$$

Thermochemical calculations indicate that this reaction is spontaneous at 300 °C with a Gibbs free energy value of $\Delta G = -41.1$ kcal.[52]

Figure 8 shows the FTIR difference spectra between 500 and 1700 cm^{-1} for two consecutive TMA and HF exposures

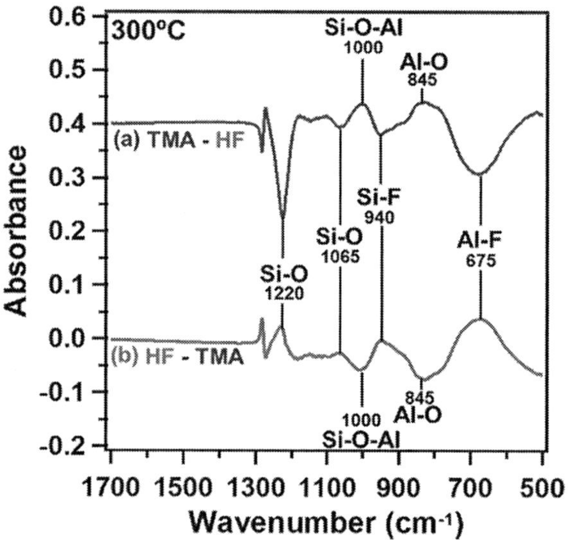

Figure 8. Difference infrared absorbance spectra from 500 to 1700 cm^{-1} during the sixth ALE cycle at 300 °C. (a) Spectrum after the sixth TMA exposure referenced to the spectrum after the fifth HF exposure. (b) Spectrum after the sixth HF exposure referenced to the spectrum after the sixth TMA exposure.

recorded during the sixth ALE cycle at 300 °C. These difference spectra are referenced using the spectra after the previous reactant exposure and are displaced for clarity in presentation. Each reactant exposure consisted of a ~1 Torr static dose for 30 s followed by a purge for 240 s. Figure 8a displays the difference spectrum after the TMA exposure referenced to the spectrum collected after the previous HF exposure (TMA − HF).

After TMA exposures, a broad absorbance loss is observed at 675 cm^{-1} that is consistent with the removal of Al−F stretching vibrations.[51] Etching of AlF_3 has been observed between 250 and 300 °C using TMA.[16,18] The etching of AlF_3 is consistent with the reaction

$$AlF_3 + 2Al(CH_3)_3 \rightarrow 3AlF(CH_3)_2 \quad (2)$$

In this ligand-exchange reaction, TMA accepts fluorine from the AlF_3 surface layer to form volatile $AlF(CH_3)_2$ species. $AlF(CH_3)_2$ is stable and has been shown to have a vapor pressure of 80 Torr at 100 °C.[53] Methyl groups from TMA are also transferred to the AlF_3 surface layer to form additional volatile $AlF(CH_3)_2$ reaction products. This ligand-exchange reaction removes Al−F species from the surface.

Negative absorbance features are also observed in Figure 8a at 1065 and 1220 cm^{-1} that are consistent with the loss of Si−O asymmetric stretching vibrations.[30] These absorbance losses are accompanied by absorbance gains at 845 and 1000 cm^{-1} that are attributed to the addition of Al−O and Al−O−Si stretching vibrations, respectively.[46−48,54,55] These absorbance changes are consistent with the reaction of TMA with SiO_2 to form Al_2O_3 and aluminosilicate surface species. The reaction of TMA with SiO_2 to form Al_2O_3 could occur according to

$$4Al(CH_3)_3 + 3SiO_2 \rightarrow 2Al_2O_3 + 3Si(CH_3)_4 \quad (3)$$

This reaction is predicted to be spontaneous with a Gibbs free energy change of $\Delta G = -235$ kcal/mol at 300 °C.[52] The reaction between TMA and SiO_2 has also been computed to be thermodynamically favorable using density functional theory methods.[56]

Infrared analysis, temperature-programmed desorption, and X-ray photoelectron spectroscopy have also been used to analyze the Si−methyl and Al−methyl surface species that result from the chemisorption of TMA on SiO_2. Above 175 °C, the aluminum alkyl surface complex has been shown to decompose, yielding more methyl groups bonded to silicon. At 325 °C, Si−CH_3 species were observed to react further to form volatile tetramethylsilane ($Si(CH_3)_4$) as suggested in eq 3.[41]

Previous studies have also investigated the reaction of Al with SiO_2.[57−59] Aluminum films deposited on thermally grown SiO_2 surfaces have been shown to reduce SiO_2 and yield Al_2O_3 and Si according to

$$4Al + 3SiO_2 \rightarrow 2Al_2O_3 + 3Si \quad (4)$$

This reaction is thermodynamically favorable with a Gibbs free energy value of $\Delta G = -68$ kcal/mol at 300 °C.[52] There are similarities between the reactions of TMA and Al with SiO_2. During thermal ALE, TMA may act like metallic aluminum to produce Al_2O_3 as expected for thermite reactions.[60]

Figure 8b shows the absorbance changes after a subsequent HF exposure. An absorbance gain at 675 cm^{-1} is accompanied by absorbance losses at 845 and 1000 cm^{-1}. These results are similar to those observed in Figure 7b and are consistent with HF exposures acting to fluorinate the Al_2O_3/aluminosilicate layer formed during TMA exposures. This fluorinated layer can then be removed during subsequent TMA exposures. TMA can also act to further displace silicon at the surface to form a new Al_2O_3/aluminosilicate layer that can then be etched during the subsequent ALE cycle.

5. FTIR Studies of Self-Limiting Behavior and Pressure Dependence. The self-limiting behavior of the TMA and HF reactions at a given reactant pressure was determined by monitoring the absorbance changes for the Si−O stretching vibrations between 1050 and 1225 cm^{-1} during TMA exposures and the Al−F stretching vibrations between 550 and 750 cm^{-1} during HF exposures. Figure 9 examines the self-limiting behavior of the SiO_2 ALE reactions with TMA and HF at 300 °C. Each data point was collected after a ~0.5 Torr static dose for 5 s.

Figure 9a displays the normalized integrated absorbance change for the Si−O stretching vibration between 1050 and 1225 cm^{-1} after TMA exposures. Self-limiting behavior is observed after a TMA exposure of 15 Torr s. The surface Al_2O_3/aluminosilicate layer acts as a diffusion barrier that limits Al diffusion through the layer to reach unreacted SiO_2. Figure 9b shows the normalized integrated absorbance change for the Al−F stretching vibrations between 550 and 750 cm^{-1} after HF exposures. Self-limiting behavior is observed after a HF exposure of 10 Torr s. The self-limiting behavior is consistent with the fluorination of the Al−O species that were added during previous TMA exposures. After all of the Al−O species are converted to Al−F species, no further reaction occurs.

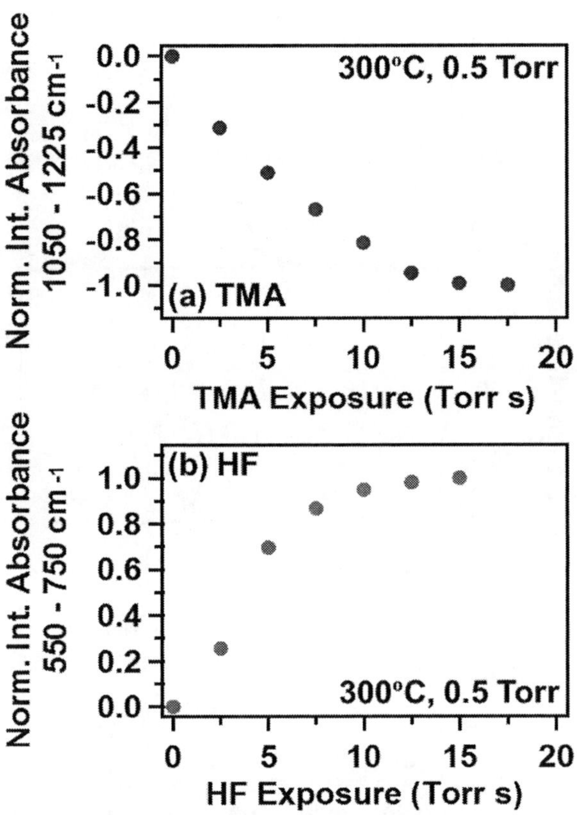

Figure 9. (a) Normalized integrated absorbance of the Si—O stretching vibration between 1050 and 1225 cm^{-1} during the TMA exposure. (b) Normalized integrated absorbance of the Al—F stretching vibration between 550 and 750 cm^{-1} during the HF exposure.

FTIR studies also observed the absorbance changes at different reactant pressures. Figure 10 displays the FTIR difference spectra between 450 and 1450 cm^{-1} for increasing TMA pressures at 300 °C. The difference spectra are referenced to the previous HF exposure. This previous exposure consisted of a ~1 Torr static HF dose for 30 s. Figure 10a was obtained after a self-passivating reaction with a TMA static exposure of 0.5 Torr. The resulting vibrational features are similar to those presented in Figure 8. The TMA exposure of 0.5 Torr shows the loss of Al—F species at 675 cm^{-1} and Si—O—Si species at 1220 cm^{-1}. Absorbance gains are also observed at 845 and 1000 cm^{-1} that are consistent with the addition of Al—O and Si—O—Al species.

Figure 10b and Figure 10c show the resulting difference spectra when the TMA reactant pressure is increased to 1.0 and 4.0 Torr, respectively. These difference spectra were recorded after self-passivating reactions with TMA static exposures of 1.0 and 4.0 Torr, respectively. Little change is observed in the Al—F vibrational region. This behavior suggests that Al—F species are removed from the surface before Al reacts with the underlying SiO_2 layer. Further absorbance loss at 1220 cm^{-1} and absorbance gains at 845 and 1000 cm^{-1} are consistent with additional conversion of Si—O species to Al—O species at the higher TMA pressures.

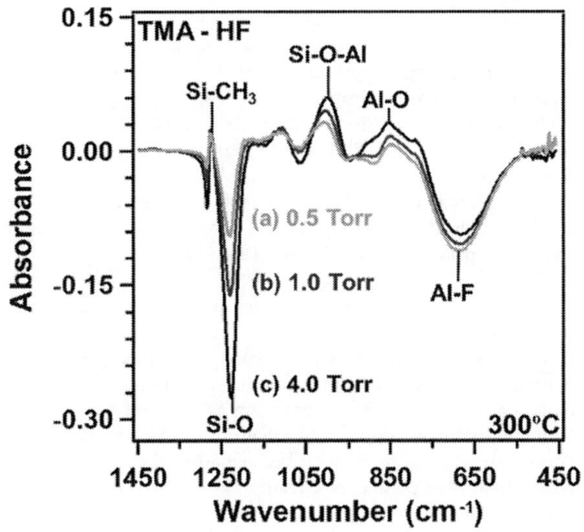

Figure 10. Difference infrared absorption spectra between 450 and 1450 cm^{-1} for increasing TMA pressures of (a) 0.5, (b) 1.0, and (c) 4.0 Torr at 300 °C. Difference spectra are referenced to the spectrum after the previous HF exposure at 1 Torr for 30 s.

Figure 11 shows the FTIR difference spectra between 450–1450 cm^{-1} for increasing HF pressures at 300 °C. The difference spectra are referenced to the previous TMA exposure. This previous exposure consisted of a ~1 Torr static TMA dose for 30 s. Figure 11a was obtained after a self-passivating reaction with a HF static exposure of 0.5 Torr. The HF exposure of 0.5 Torr causes absorbance losses at 845 and 1000 cm^{-1} and absorbance gains at 675 and 955 cm^{-1}. These absorbance changes are consistent with the conversion of Al—O and Si—O—Al species to Al—F and Si—F species, respectively.

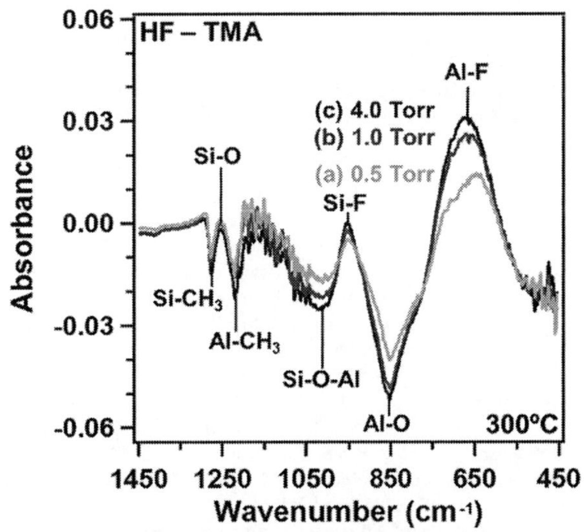

Figure 11. Difference infrared absorption spectra between 450 and 1450 cm^{-1} for increasing HF pressures of (a) 0.5. (b) 1.0. and (c) 4.0 Torr at 300 °C. Difference spectra are referenced to the spectrum after the previous TMA exposure at 1 Torr for 30 s.

Figure 11b was collected after a self-passivating HF reaction when the HF pressure was increased to 1.0 Torr. Figure 11c was recorded after a self-passivating reaction when the HF pressure was increased further to 4.0 Torr. These spectra show additional Al−O and Si−O−Al to Al−F and Si−F conversion in each absorbance region at the higher HF pressures. The absorbance changes observed after the HF exposure at 4.0 Torr are much less than the absorbance changes observed after the HF exposures at 0.5 and 1.0 Torr. These results are consistent with a self-passivating fluorination reaction.

The self-passivating Al_2O_3/aluminosilicate and fluoride surface layers are similar to the self-passivating oxide layers that form on metal[61,62] or silicon[34,63,64] surfaces and protect the metal or silicon surfaces from further oxidation. These self-passivating reactions can be understood in terms of diffusion-limited growth. The surface layers act as diffusion barriers that limit the conversion and fluorination reactions. Diffusion-limited growth has been described earlier in the classic Deal−Grove treatment of silicon oxidation.[34]

In the Deal−Grove approach,[34] the flux of reactants, F, is dependent on the concentration gradient across the SiO_2 surface layer according to

$$F = D(C_0 - C_s)/x \qquad (5)$$

where D is the diffusion coefficient, C_0 is the concentration of reactants on the SiO_2 surface, C_s is the concentration of reactants at the SiO_2/Si interface, and x is the thickness of the SiO_2 surface layer. Solution of the diffusion equations reveals that the thickness of the SiO_2 surface layer will increase parabolically versus time and display self-passivating behavior at longer times.[34]

The larger amounts of conversion and fluorination at higher pressures can also be understood in terms of diffusion-limited growth. The pressure dependence of the thickness is determined by C_0, the concentration of reactants on the surface. C_0 is defined by the reactant pressure, P, and the Henry's law constant, H, according to $C_0 = HP$. Higher pressures lead to higher surface reactant concentrations. These higher concentrations produce larger concentration gradients across the surface layer and thicker surface layers.

6. XPS Studies of Silicon Oxidation State after TMA and HF Reactions with SiO_2. The reaction of TMA with SiO_2 given by eq 3 reduces silicon because CH_3 is much less electronegative than oxygen. The CH_3 group has an electronegativity of χ = 2.525.[65] In contrast, oxygen has an electronegativity of χ = 3.5. The reaction of TMA with SiO_2 will lead to more electron density on silicon in the $SiO_{4-y}(CH_3)_y$ suboxide products.

The reduction of SiO_2 and formation of Al_2O_3/aluminosilicates after TMA exposure were observed by ex situ XPS analysis. Figure 12 displays the Si 2p XPS spectra of (a) bare SiO_2, (b) SiO_2 after an 8 Torr TMA exposure for 5 min at 300 °C, and (c) SiO_2 after an 8 Torr TMA exposure for 5 min followed by a 4 Torr HF exposure for 10 min at 300 °C. The higher reactant pressures and exposure times were chosen to ensure that changes in the surface species were detectable by XPS analysis.

The Si 2p peaks in Figure 12 were deconvoluted into silicon oxidation states of 4+, 3+, 2+, 1+, and 0. The fits were constrained by the fwhm and binding energies of the individual Si oxidation states as described in section II.1. The binding energies of the silicon oxidation states of 4+, 3+, 2+, 1+, and 0 are shown by the dashed lines in Figure 12.[25−27] Figure 12a depicts the SiO_2 control sample with a broad peak at 104.3 eV. This peak position is consistent with literature values for SiO_2 and silicon atoms in the Si^{4+} oxidation state.[66]

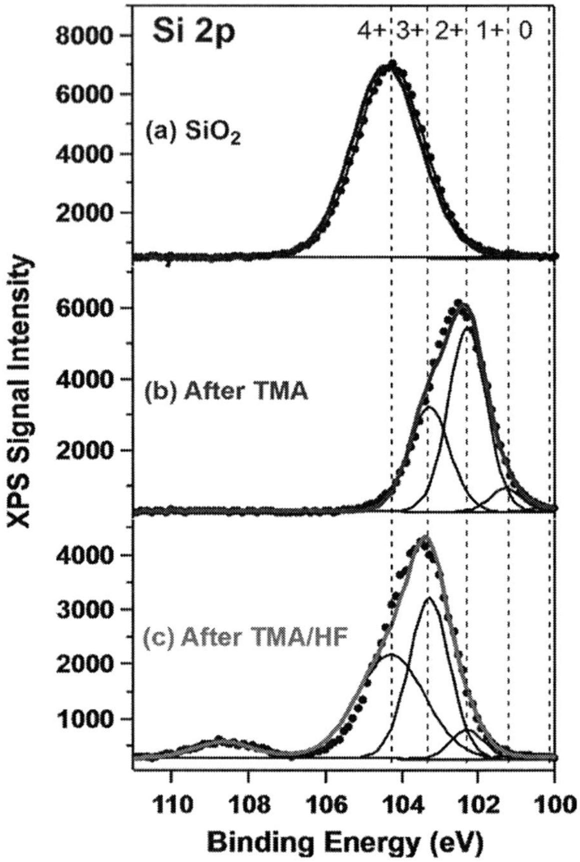

Figure 12. XPS results for the Si 2p peak from (a) initial SiO_2 film, (b) SiO_2 film after TMA exposure, and (c) SiO_2 film after TMA and HF exposures. The dashed lines correspond to the 4+, 3+, 2+, 1+, and 0 oxidation states for the Si 2p peak.

Figure 12b shows the results after the TMA exposure. The Si 2p peak shifts to lower binding energies centered around 102.5 eV that is composed of 3+, 2+, and 1+ oxidation states. The disappearance of the 4+ oxidation state indicates the TMA exposure reduces the Si^{4+} in SiO_2 into Si suboxides. As suggested by the reaction in eq 3, these silicon suboxides may be $SiO_3(CH_3)$, $SiO_2(CH_3)_2$, and $SiO(CH_3)_3$ surface species.

Figure 12c displays the XPS results for the Si 2p peak after sequential TMA and HF exposures. The HF exposure shifts the Si 2p peak back to higher energies and can be deconvoluted into silicon oxidation states of 4+, 3+, and 2+. These higher silicon oxidation states are consistent with HF reacting with $SiCH_3$ surface species according to $SiCH_3 + HF \rightarrow SiF + CH_4$. This reaction could form $SiFO_3$, $SiFO_2(CH_3)$, and $SiFO(CH_3)_2$ surface species with silicon oxidation states of 4+, 3+, and 2+, respectively. These higher oxidation states are expected because fluorine has a high electronegativity of χ = 4.0. In addition, Figure 12c also displays a higher energy peak at 108.8 eV. This higher energy peak can be attributed to a shake-up satellite peak.[67] Shake-up satellite peaks can occur when the emitted photoelectron excites an ion. These shake-up peaks from Si 2p

DOI: 10.1021/acsami.7b01259
ACS Appl. Mater. Interfaces 2017, 9, 10296−10307

can be observed when Si is present in a siloxane polymer structure.[67]

The reduction of SiO_2 and formation of Al_2O_3/aluminosilicates after TMA exposure were also monitored by XPS analysis of the Si 2p XPS spectra after a 4 Torr TMA exposure on SiO_2 for 20 s at 300 °C. The Si 2p XPS spectra displayed a shift from the 4+ oxidation state for SiO_2 to a distribution of 4+, 3+, 2+, and 1+ oxidation states after the TMA exposure. The peak of the spectrum occurred at the 3+ oxidation state compared with a peak at the 2+ oxidation state for the TMA exposure at 8 Torr shown in Figure 12b. The Si 2p XPS spectrum after a 4 Torr TMA exposure on SiO_2 for 20 s followed by a 4 Torr HF exposure for 20 s at 300 °C displayed a shift back to higher oxidation states similar to the results shown in Figure 12c.

The Al 2p XPS peak was also consistent with the formation of Al_2O_3 at the surface after TMA exposure. Figure 13a and

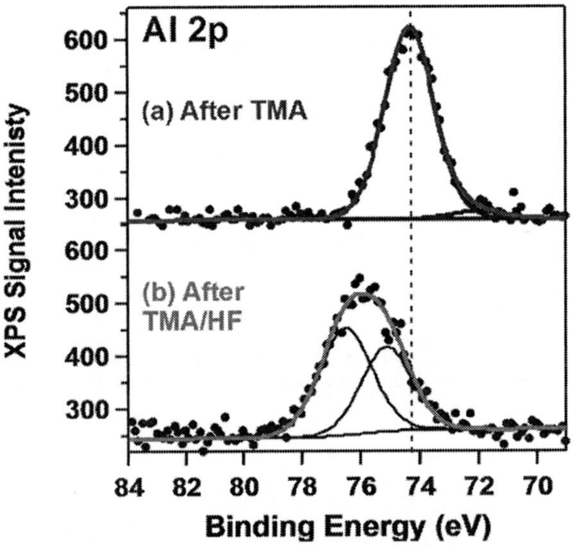

Figure 13. XPS results for the Al 2p peak from (a) SiO_2 film after TMA exposure and (b) SiO_2 film after TMA and HF exposures. These XPS spectra were obtained from the same samples as shown in Figure 12. The dashed line corresponds to the binding energy for Al in Al_2O_3.

Figure 13b show the Al 2p XPS spectra after the TMA exposure and after sequential TMA and HF exposures, respectively. These XPS spectra were obtained from the same samples as shown in Figure 12. Figure 13a displays a peak at 74.3 eV after the TMA exposure that corresponds to the formation of Al_2O_3 or aluminosilicates.[66,68] There is also a smaller shoulder on this peak at ~72 eV that corresponds to more metallic Al. This peak may result from some TMA decomposition.[69]

Figure 13b shows that the Al 2p XPS peak shifts to higher binding energies after sequential TMA and HF exposures. The broader peak at ~76 eV is consistent with the fluorination of Al_2O_3.[68] This peak may be composed of AlF_3 at 77 eV and oxyaluminum fluoride species at lower energies.[66,68] Oxygen is less electronegative and will not shift the Al 2p binding energy as much as fluorine.

7. Mechanism for SiO_2 ALE and Future Possibilities. A simplified schematic for the proposed SiO_2 ALE chemistry during the sequential TMA and HF exposures on SiO_2 is shown in Figure 14. After the first TMA and HF exposures, subsequent TMA exposures react with the AlF_3 species on the

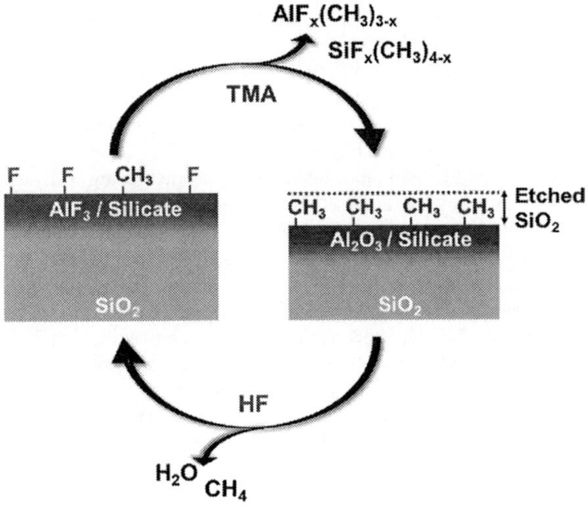

Figure 14. Schematic of the proposed reaction mechanism for SiO_2 ALE showing the sequential TMA and HF reactions.

SiO_2 substrate to form volatile $AlF(CH_3)_2$ etch products. After removal of the AlF_3 species, TMA reacts with the underlying SiO_2 to form Al_2O_3 and aluminosilicates with the possible release of $Si(CH_3)_4$ or $Si(CH_3)_xF_{4-x}$, as volatile reaction products. Mass spectrometry studies are required to confirm the volatile reaction products. In addition, suboxides such as $AlO_{3-x}(CH_3)_x$ where $x = 1$ or 2 and $SiO_{4-y}(CH_3)_y$ where $y = 1$, 2, or 3 may be present in the surface layer.

The subsequent HF exposures converts the Al_2O_3 and aluminosilicates to AlF_3 and releases H_2O as a volatile reaction product. The $AlO_{3-x}(CH_3)_x$ and $SiO_{4-y}(CH_3)_y$ suboxides may also lead to the formation of CH_4 during the HF exposure. Moreover, silicon species such as $SiFO_3$, $SiFO_2(CH_3)$, and $SiFO(CH_3)_2$ may be produced in the near surface layer. The shake-up satellite XPS peak from Si 2p also suggests that the Si may be in polymer structures similar to siloxane.

The basics of this mechanism are similar to the recently observed "conversion-etch" mechanism for ZnO ALE.[23] For ZnO ALE, TMA exposures converted ZnO to Al_2O_3. HF exposures then fluorinated the Al_2O_3 to AlF_3. Subsequent TMA exposures removed the AlF_3 and converted more ZnO to Al_2O_3.[23] ZnO ALE by the "conversion-etch" mechanism occurred at lower reactant pressures of 60−80 mTorr. In contrast to the results for SiO_2 ALE, higher reactant pressures of >0.1 Torr were not required for ZnO ALE.

The lower pressures for ZnO ALE suggest that "conversion-etch" may be more favorable for ZnO ALE than SiO_2 ALE. However, thermochemical calculations do not support this viewpoint. The free energy change for the reaction $3ZnO + 2Al(CH_3)_3 \rightarrow Al_2O_3 + 3Zn(CH_3)_2$ is $\Delta G = -168$ kcal at 300 °C.[52] In comparison, the free energy change for the reaction $1.5SiO_2 + 2Al(CH_3)_3 \rightarrow Al_2O_3 + 1.5Si(CH_3)_4$ is $\Delta G = -197$ kcal at 300 °C.[52] There must be other kinetic factors that favor ZnO ALE by the "conversion-etch" mechanism.

The "conversion-etch" mechanism may be general and may be realized when the metal oxide formed by the conversion reaction is more stable than the initial metal oxide. Guidelines

DOI: 10.1021/acsami.7b01259
ACS Appl. Mater. Interfaces 2017, 9, 10296−10307

for determining the relative stability of metal oxides can be obtained from thermochemical calculations or Ellingham diagrams.[70] Some materials may not have viable pathways for thermal ALE because the ligand-exchange reactions may not produce stable and volatile reaction products.[20] By converting the initial material to a different material, the "conversion-etch" mechanism may enlarge the number of materials that can be etched using thermal ALE. For example, TMA is anticipated to convert other metal oxides, such as SnO_2 and In_2O_3, to Al_2O_3.[23] Thermal Al_2O_3 ALE has been demonstrated using HF and either $Sn(acac)_2$ or TMA as the reactants.[17−19] Likewise, TMA is expected to convert various nitrides, such as Si_3N_4, to AlN.[23] Thermal AlN ALE has been demonstrated using HF and $Sn(acac)_2$ as the reactants.[21]

Thermal SiO_2 ALE also suggests a possible procedure for thermal Si ALE. The silicon surface can first be oxidized to form a self-passivating oxide layer on the silicon surface.[63,64] The resulting SiO_2 layer can then be etched by the "conversion-etch" mechanism using sequential TMA and HF exposures. By repetition of the silicon oxidation and sequential TMA and HF exposures, silicon should be etched with atomic layer precision. Similar procedures may also be defined for the thermal ALE of metals using sequential oxidation, fluorination, and ligand-exchange reactions.

IV. CONCLUSIONS

Sequential reactions of trimethylaluminum (TMA) and hydrogen fluoride (HF) were employed for the thermal atomic layer etching (ALE) of SiO_2 at 300 °C. The etch rates during SiO_2 ALE were dependent on reactant pressures. SiO_2 etch rates of 0.027, 0.15, 0.20, and 0.31 Å/cycle were observed at static reactant pressures of 0.1, 0.5, 1.0, and 4.0 Torr, respectively, by ex situ X-ray reflectivity measurements. A similar increase in SiO_2 etch rate with reactant pressure was confirmed by ex situ spectroscopic ellipsometry measurements. The pressure dependence of the etch rate for SiO_2 ALE was also observed by in situ FTIR studies. In addition, the FTIR measurements monitored the pressure-dependent conversion of the SiO_2 surface to an Al_2O_3/aluminosilicate layer. Ex situ XPS measurements also monitored the reduction of the silicon oxidation state after the TMA reaction.

The Al_2O_3/aluminosilicate intermediate is attributed to the conversion of SiO_2 to Al_2O_3 and reduced silicon species by the TMA exposure according to reactions like $3SiO_2 + 4Al(CH_3)_3 \rightarrow 2Al_2O_3 + 3Si(CH_3)_4$. The HF exposure can then fluorinate Al_2O_3 to AlF_3 and form silicon fluoride species after this conversion reaction. Subsequently, the next TMA exposure can remove AlF_3 by a ligand-exchange transmetalation reaction and then convert additional SiO_2 to Al_2O_3 and reduced silicon species. The sequential TMA and HF reactions lead to SiO_2 ALE because the individual TMA and HF reactions display self-limiting behavior at the various reactant pressures.

The pressure-dependent conversion reaction of the SiO_2 surface to an Al_2O_3/aluminosilicate layer by TMA is critical for thermal SiO_2 ALE. The "conversion-etch" mechanism for SiO_2 ALE is observed at reactant pressures of >0.1 Torr. Other conversion reactions will be able to transform initial materials that cannot be etched to different materials that may have possible etching pathways. The "conversion-etch" mechanism will increase the number of materials that can be etched by thermal ALE methods. Oxidation and conversion reactions may also lead to procedures for thermal Si ALE.

■ AUTHOR INFORMATION

Corresponding Author

*E-mail: steven.george@colorado.edu. Phone: 303-492-3398. Fax: 303-492-5894.

ORCID

Steven M. George: 0000-0003-0253-9184

Notes

The authors declare no competing financial interest.

■ ACKNOWLEDGMENTS

This research was funded by the National Science Foundation by Grants CHE-1306131 and CHE-1609554 and the Defense Advanced Projects Agency (DARPA) by Grant W911NF-13-1-0041. The authors thank Huaxing Sun for the XPS measurements. The authors also thank Joel Clancey for assistance with the TEM analysis of the silicon nanoparticles.

■ REFERENCES

(1) Nojiri, K. *Dry Etching Technology for Semiconductors*; Springer: Heidelberg, Germany, 2012.

(2) Carver, C. T.; Plombon, J. J.; Romero, P. E.; Suri, S.; Tronic, T. A.; Turkot, R. B. Atomic Layer Etching: An Industry Perspective. *ECS J. Solid State Sci. Technol.* **2015**, *4*, N5005−N5009.

(3) Blumberg, A. A.; Stavrinou, S. C. Tabulated Functions for Heterogeneous Reaction Rates - The Attack of Vitreous Silica by Hydrofluoric Acid. *J. Phys. Chem.* **1960**, *64*, 1438−1442.

(4) Monk, D. J.; Soane, D. S.; Howe, R. T. A Review of the Chemical Reaction Mechanisma and Kinetics for Hydrofluoric Acid Etching of Silicon Dioxide for Surface Micromachining Applications. *Thin Solid Films* **1993**, *232*, 1−12.

(5) Spierings, G. A. C. M. Wet Chemical Etching of SIlicate-Glasses in Hydrofluoric Acid Based Solutions. *J. Mater. Sci.* **1993**, *28*, 6261−6273.

(6) Knotter, D. M. Etching Mechanism of Vitreous Silicon Dioxide in HF-Based Solutions. *J. Am. Chem. Soc.* **2000**, *122*, 4345−4351.

(7) Helms, C. R.; Deal, B. E. Mechanisms of the HF/H_2O Vapor Phase Etching of SiO_2. *J. Vac. Sci. Technol., A* **1992**, *10*, 806−811.

(8) Vanderheide, P. A. M.; Baan Hofman, M. J.; Ronde, H. J. Etching of Thin SiO_2 Layers Using Wet HF Gas. *J. Vac. Sci. Technol., A* **1989**, *7*, 1719−1723.

(9) Kang, J. K.; Musgrave, C. B. The Mechanism of HF/H_2O Chemical Etching of SiO_2. *J. Chem. Phys.* **2002**, *116*, 275−280.

(10) Mogab, C. J.; Adams, A. C.; Flamm, D. L. Plasma Etching of Si and SiO_2 - Effect of Oxygen Additions to CF_4 Plasmas. *J. Appl. Phys.* **1978**, *49*, 3796−3803.

(11) Rueger, N. R.; Beulens, J. J.; Schaepkens, M.; Doemling, M. F.; Mirza, J. M.; Standaert, T.; Oehrlein, G. S. Role of Steady State Fluorocarbon Films in the Etching of Silicon Dioxide using CHF_3 in an Inductively Coupled Plasma Reactor. *J. Vac. Sci. Technol., A* **1997**, *15*, 1881−1889.

(12) Zhang, D.; Kushner, M. J. Investigations of Surface Reactions during C_2F_6 Plasma Etching of SiO_2 with Equipment and Feature Scale Models. *J. Vac. Sci. Technol., A* **2001**, *19*, 524−538.

(13) Metzler, D.; Bruce, R. L.; Engelmann, S.; Joseph, E. A.; Oehrlein, G. S. Fluorocarbon Assisted Atomic Layer Etching of SiO_2 using Cyclic Ar/C_4F_8 Plasma. *J. Vac. Sci. Technol., A* **2014**, *32*, 020603.

(14) Oehrlein, G. S.; Metzler, D.; Li, C. Atomic Layer Etching at the Tipping Point: An Overview. *ECS J. Solid State Sci. Technol.* **2015**, *4*, N5041−N5053.

(15) Lee, Y.; DuMont, J. W.; George, S. M. Atomic Layer Etching of HfO_2 Using Sequential, Self-Limiting Thermal Reactions with $Sn(acac)_2$ and HF. *ECS J. Solid State Sci. Technol.* **2015**, *4*, N5013−N5022.

(16) Lee, Y.; DuMont, J. W.; George, S. M. Atomic Layer Etching of AlF_3 Using Sequential, Self-Limiting Thermal Reactions with Sn-

(acac)$_2$ and Hydrogen Fluoride. *J. Phys. Chem. C* **2015**, *119*, 25385–25393.

(17) Lee, Y.; DuMont, J. W.; George, S. M. Mechanism of Thermal Al_2O_3 Atomic Layer Etching Using Sequential Reactions with Sn(acac)$_2$ and HF. *Chem. Mater.* **2015**, *27*, 3648–3657.

(18) Lee, Y.; DuMont, J. W.; George, S. M. Trimethylaluminum as the Metal Precursor for the Atomic Layer Etching of Al_2O_3 Using Sequential, Self-Limiting Thermal Reactions. *Chem. Mater.* **2016**, *28*, 2994–3003.

(19) Lee, Y.; George, S. M. Atomic Layer Etching of Al_2O_3 Using Sequential, Self-Limiting Thermal Reactions with Sn(acac)$_2$ and HF. *ACS Nano* **2015**, *9*, 2061–2070.

(20) Lee, Y.; Huffman, C.; George, S. M. Selectivity in Thermal Atomic Layer Etching Using Sequential, Self-Limiting Fluorination and Ligand-Exchange Reactions. *Chem. Mater.* **2016**, *28*, 7657–7665.

(21) Johnson, N. R.; Sun, H.; Sharma, K.; George, S. M. Thermal Atomic Layer Etching of Crystalline Aluminum Nitride Using Sequential, Self-Limiting Hydrogen Fluoride and Sn(acac)$_2$ Reactions and Enhancement by H_2 and Ar Plasmas. *J. Vac. Sci. Technol., A* **2016**, *34*, 050603.

(22) George, S. M.; Lee, Y. Prospects for Thermal Atomic Layer Etching Using Sequential, Self-Limiting Fluorination and Ligand-Exchange Reactions. *ACS Nano* **2016**, *10*, 4889–4894.

(23) Zywotko, D. R.; George, S. M. Thermal Atomic Layer Etching of ZnO by a "Conversion-Etch" Mechanism Using Sequential Exposures of Hydrogen Fluoride and Trimethylaluminum. *Chem. Mater.* **2017**, *29*, 1183–1191.

(24) Elam, J. W.; Groner, M. D.; George, S. M. Viscous Flow Reactor with Quartz Crystal Microbalance for Thin Film Growth by Atomic Layer Deposition. *Rev. Sci. Instrum.* **2002**, *73*, 2981–2987.

(25) Bell, F. G.; Ley, L. Photoemission Study of SiO_x ($0< x <2$) Alloys. *Phys. Rev. B: Condens. Matter Mater. Phys.* **1988**, *37*, 8383–8393.

(26) Saito, Y. Adsorption of Anhydrous Hydrogen Fluoride onto Silicon and Native Oxide by XPS. *Surf. Sci. Spectra* **1999**, *6*, 228–236.

(27) Kim, S.; Kim, M. C.; Choi, S. H.; Kim, K. J.; Hwang, H. N.; Hwang, C. C. Size Dependence of Si 2p Core-level Shift at Si Nanocrystal/SiO_2 Interfaces. *Appl. Phys. Lett.* **2007**, *91*, 103113.

(28) DuMont, J. W.; George, S. M. Pyrolysis of Alucone Molecular Layer Deposition Films Studied Using In Situ Transmission Fourier Transform Infrared Spectroscopy. *J. Phys. Chem. C* **2015**, *119*, 14603–14612.

(29) Ferguson, J. D.; Weimer, A. W.; George, S. M. Atomic Layer Deposition of Ultrathin and Conformal Al_2O_3 films on BN Particles. *Thin Solid Films* **2000**, *371*, 95–104.

(30) Queeney, K. T.; Weldon, M. K.; Chang, J. P.; Chabal, Y. J.; Gurevich, A. B.; Sapjeta, J.; Opila, R. L. Infrared Spectroscopic Analysis of the Si/SiO_2 Interface Structure of Thermally Oxidized Silicon. *J. Appl. Phys.* **2000**, *87*, 1322–1330.

(31) Devine, R. A. B. Structural Nature of the Si/SiO_2 Interface Through Infrared Spectroscopy. *Appl. Phys. Lett.* **1996**, *68*, 3108–3110.

(32) Mawhinney, D. B.; Glass, J. A.; Yates, J. T. FTIR Study of the Oxidation of Porous Silicon. *J. Phys. Chem. B* **1997**, *101*, 1202–1206.

(33) Ogata, Y.; Niki, H.; Sakka, T.; Iwasaki, M. Oxidation of Porous Silicon Unver Water Vapor Environment. *J. Electrochem. Soc.* **1995**, *142*, 1595–1601.

(34) Deal, B. E.; Grove, A. S. General Relationship for Thermal Oxidation of Silicon. *J. Appl. Phys.* **1965**, *36*, 3770–3778.

(35) Bhat, V. K.; Pattabiraman, M.; Bhat, K. N.; Subrahmanyam, A. The Growth of Ultrathin Oxides of Silicon by Low Temperature Wet Oxidation Technique. *Mater. Res. Bull.* **1999**, *34*, 1797–1803.

(36) Gupta, P.; Colvin, V. L.; George, S. M. Hydrogen Desorption Kinetics from Monohydride and Dihydride Species on Silicon Surfaces. *Phys. Rev. B: Condens. Matter Mater. Phys.* **1988**, *37*, 8234–8243.

(37) Koehler, B. G.; Mak, C. H.; Arthur, D. A.; Coon, P. A.; George, S. M. Desorption Kinetics of Hydrogen and Deuterium from Si(111)7x7 Studied Using Laser Induced Thermal Desorption. *J. Chem. Phys.* **1988**, *89*, 1709–1718.

(38) Gupta, P.; Mak, C. H.; Coon, P. A.; George, S. M. Oxidation Kinetics of Si(111)7x7 in the Submonolayer Regime. *Phys. Rev. B: Condens. Matter Mater. Phys.* **1989**, *40*, 7739–7749.

(39) Koehler, B. G.; Mak, C. H.; George, S. M. Decomposition of H_2O on Si(111)7x7 Studied Using Laser Induced Thermal Desorption. *Surf. Sci.* **1989**, *221*, 565–589.

(40) Bellamy, L. J. *The Infra-Red Spectra of Complex Molecules*; Chapman and Hall: London, 1975.

(41) Bartram, M. E.; Michalske, T. A.; Rogers, J. W. A Reexamination of the Chemisorption of Trimethylaluminum on Silica. *J. Phys. Chem.* **1991**, *95*, 4453–4463.

(42) Puurunen, R. L.; Root, A.; Haukka, S.; Iiskola, E. I.; Lindblad, M.; Krause, A. O. I. IR and NMR Study of the Chemisorption of Ammonia on Trimethylaluminum-Modified Silica. *J. Phys. Chem. B* **2000**, *104*, 6599–6609.

(43) Yates, D. J. C.; Dembinski, G. W.; Kroll, W. R.; Elliott, J. J. Infrared Studies of Reactions between Silica and Trimethylaluminum. *J. Phys. Chem.* **1969**, *73*, 911–921.

(44) Morrow, B. A.; McFarlan, A. J. Surface Vibrational Modes of Silanol Groups on Silica. *J. Phys. Chem.* **1992**, *96*, 1395–1400.

(45) Grill, A.; Neumayer, D. A. Structure of Low Dielectric Constant to Extreme Low Dielectric Constant SiCOH Films: Fourier Transform Infrared Spectroscopy Characterization. *J. Appl. Phys.* **2003**, *94*, 6697–6707.

(46) Frank, M. M.; Chabal, Y. J.; Green, M. L.; Delabie, A.; Brijs, B.; Wilk, G. D.; Ho, M. Y.; da Rosa, E. B. O.; Baumvol, I. J. R.; Stedile, F. C. Enhanced Initial Growth of Atomic Layer Deposited Metal Oxides on Hydrogen-Terminated Silicon. *Appl. Phys. Lett.* **2003**, *83*, 740–742.

(47) Frank, M. M.; Chabal, Y. J.; Wilk, G. D. Nucleation and Interface Formation Mechanisms in Atomic Layer Deposition of Gate Oxides. *Appl. Phys. Lett.* **2003**, *82*, 4758–4760.

(48) Mackenzie, K. J. Infrared Frequency Calculations for Ideal Mullite ($3Al_2O_3.2SiO_2$). *J. Am. Ceram. Soc.* **1972**, *55*, 68–71.

(49) Kinney, J. B.; Staley, R. H. Reactions of Titanium Tetrachloride and Trimethylaluminum at Silica Surfaces Studies by Using Infrared Photo-acoustic Spectroscopy. *J. Phys. Chem.* **1983**, *87*, 3735–3740.

(50) Barabash, R. M.; Zaitsev, V. N.; Kovalchuk, T. V.; Sfihi, H.; Fraissard, J. Low-Temperature Fluorination of Silica by a Nonaqueous Solution of NH_4F. *J. Phys. Chem. A* **2003**, *107*, 4497–4505.

(51) Gross, U.; Ruediger, S.; Kemnitz, E.; Brzezinka, K.-W.; Mukhopadhyay, S.; Bailey, C.; Wander, A.; Harrison, N. Vibrational Analysis Study of Aluminum Trifluoride Phases. *J. Phys. Chem. A* **2007**, *111*, 5813–5819.

(52) *HSC Chemistry. HSC Chemistry 5.1*; Outokumpu Research Oy: Pori, Finland.

(53) Weidlein, J.; Krieg, V. Vibrational Spectra of Dimethyl and Diethyl Aluminum Fluoride. *J. Organomet. Chem.* **1968**, *11*, 9–16.

(54) Morozova, N. B.; Zherikova, K. V.; Baidina, I. A.; Sysoev, S. V.; Semyannikov, P. P.; Yakovkina, L. V.; Smirnova, T. P.; Gelfond, N. V.; Igumenov, I. K.; Carta, G.; Rossetto, G. Volatile Hafnium(IV) Compounds with Beta-Diketonate and Cyclopentadienyl Derivatives. *J. Phys. Chem. Solids* **2008**, *69*, 673–679.

(55) Mozgawa, W.; Sitarz, M.; Rokita, M. Spectroscopic Studies of Different Aluminosilicate Structures. *J. Mol. Struct.* **1999**, *511*, 251–257.

(56) Lamagna, L.; Wiemer, C.; Perego, M.; Spiga, S.; Rodriguez, J.; Coll, D. S.; Grillo, M. E.; Klejna, S.; Elliott, S. D. Mechanisms for Substrate-Enhanced Growth during the Early Stages of Atomic Layer Deposition of Alumina onto Silicon Nitride Surfaces. *Chem. Mater.* **2012**, *24*, 1080–1090.

(57) Bauer, R. S.; Bachrach, R. Z.; Brillson, L. J. Au and Al Interface Reactions with SiO_2. *Appl. Phys. Lett.* **1980**, *37*, 1006–1008.

(58) Hecht, M. H.; Vasquez, R. P.; Grunthaner, F. J.; Zamani, N.; Maserjian, J. A Novel X-ray Photoelectron Spectroscopy Study of the Al/SiO_2 Interface. *J. Appl. Phys.* **1985**, *57*, 5256–5262.

(59) Tsukada, M.; Ohfuji, S. I. In Situ X-Ray Photoemission Spectroscopic Studies of Al/SiO_2 Interface Formation. *J. Vac. Sci. Technol., A* **1994**, *12*, 1–6.
(60) Wang, L. L.; Munir, Z. A.; Maximov, Y. M. Thermite Reactions - Their Utilization in the Synthesis and Processing of Materials. *J. Mater. Sci.* **1993**, *28*, 3693–3708.
(61) Graham, M. J.; Cohen, M. On the Mechanism of Low Temperature Oxidation (23-450 C) of Polycrystalline Nickel. *J. Electrochem. Soc.* **1972**, *119*, 879–882.
(62) Song, S.; Placido, F. Investigation on Initial Oxidation Kinetics of Al, Ni, and Hf Metal Film Surfaces. *Chin. Opt. Lett.* **2010**, *8* (Suppl.), 87–90.
(63) Derrie, J.; Commandre, M. SiO_2 Ultra Thin Film Growth Kinetics as Investigated by Surface Techniques. *Surf. Sci.* **1982**, *118*, 32–46.
(64) Fehlner, F. P. Formation of Ultrathin Oxide Films on Silicon. *J. Electrochem. Soc.* **1972**, *119*, 1723–1727.
(65) Matsunaga, N.; Rogers, D. W.; Zavitsas, A. A. Pauling's Electronegativity Equation and a New Corollary Accurately Predict Bond Dissociation Enthalpies and Enhance Current Understanding of the Nature of the Chemical Bond. *J. Org. Chem.* **2003**, *68*, 3158–3172.
(66) *NIST X-ray Photoelectron Spectroscopy Database*, version 4.1; National Institute of Standards and Technology: Gaithersburg, MD, 2012; http://srdata.nist.gov/xps/.
(67) Gardella, J. A.; Ferguson, S. A.; Chin, R. L. $\pi^* \leftarrow \pi$ Shakeup Satellites for the Analysis of Structure and Bonding in Aromatic Polymers by X-ray Photoelectron Spectroscopy. *Appl. Spectrosc.* **1986**, *40*, 224–232.
(68) Bose, O.; Kemnitz, E.; Lippitz, A.; Unger, W. E. S. C 1s and Au $4f_{7/2}$ Referenced XPS Binding Energy Data Obtained with Different Aluminium Oxides, -Hydroxides and -Fluorides. *Fresenius' J. Anal. Chem.* **1997**, *358*, 175–179.
(69) Mayer, T. M.; Rogers, J. W.; Michalske, T. A. Mechanism of Nucleation and Atomic Layer Growth of AlN on Si. *Chem. Mater.* **1991**, *3*, 641–646.
(70) Ellingham, H. J. T. Reducibility of Oxides and Sulphides in Metallurgical Processes. *J. Soc. Chem. Ind. London* **1944**, *63*, 125–133.

DOI: 10.1021/acsami.7b01259
ACS Appl. Mater. Interfaces 2017, 9, 10296–10307

2017 Surface Preparation and Cleaning Conference
March 28-29, Austin, Texas, USA

Effect of Additives in Diluted HF Solutions on Removal of Metal Contaminants and Particles on Silicon Wafer

Sung-Hae Jang[a], Hyun-Tae Kim[a], Dong-Hwan Lee[a]
Jae-Hwan Lee[b], Eun-Suck Choi[b] and Jin-Goo Park[a, c, †]

[a] Department of Bio-Nano Technology and
[c] Materials Engineering, Hanyang University, Korea
[b] LG Siltron, Gumi, Korea

†+82(0)31 400 5226
jgpark@hanyang.ac.kr
Hanyang Univ., Ansan, 426-791, KOREA

NEMPL Hanyang University

2017 SPCC, March 28-29, Texas, USA 1

2017 Surface Preparation and Cleaning Conference
March 28-29, Austin, Texas, USA

OUTLINE

1. Research Background
   1) Research motivation
   2) Research objectives
2. Experimental Materials/Procedure
3. Experiment Results and Discussion
   1) Si surface change observation in HF/chelating agent
   2) Metal contamination cleaning by HF-based solution
4. Summary

NEMPL Hanyang University

2017 SPCC, March 28-29, Texas, USA 2

Demands on particle/metal-free wafer in sub 10 nm devices

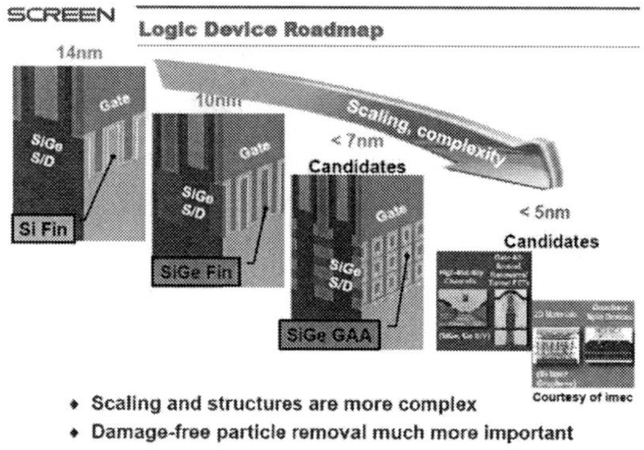

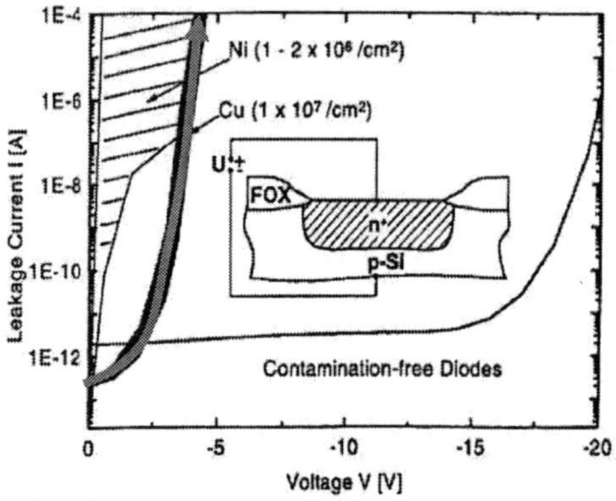

- Electrical device degradation by metal contamination
  - Leakage current ↑ : Threshold voltage(V_T) ↑
- Requiring extremely low metal concentration in FOUP
  - Below 10^8 atoms/cm^2 is required in practice

Gargini, P. The International Technology Roadmap for Semiconductors (2015)
Takiyama, M. (2001). Influence of Metallic Contamination on LSI Fabrication. Nippon Steel Technical Report, 95-99.

Issues in conventional RCA cleaning

SC1 last: Metal ions (<u>10^8 atoms/cm^2</u>)

- **Dissolved metal precipitation** (Al, Cu, Ni…)
- Subsequent metal cleaning process is necessary

SC2 last : Particle redeposition (<50nm, no particles)

- Inefficient for noble metal cleaning (e.g. Cu, Ti, Ni)
  ➔ Remaining residual metal contaminants
- **Particle redeposition** by acidic atmosphere

Simultaneous Removal of both metal and particles,
Chemical Modification of SC1 or SC2

Gottschalk, C., & Schweckendiek, J. (2004). Paper reprinted from MICRO, Canon Communications LLC Publication, March.

Research objective

To control the metallic/pariculate contaminants on the silicon wafer surface by optimizing the concentration of HF, DIO_3 and chelating agent

1. Cu and Al metal contamination method and detection by AFM
2. Evaluation chelating agent and ozone in DHF solutions

This research reports only on metal removal!

Experiment Materials And Procedure

NEMPL Hanyang University

Copper growth on Si wafer

Cu formation :: Electrochemical reaction

1) Presence of Cu^{2+} and Si wafer

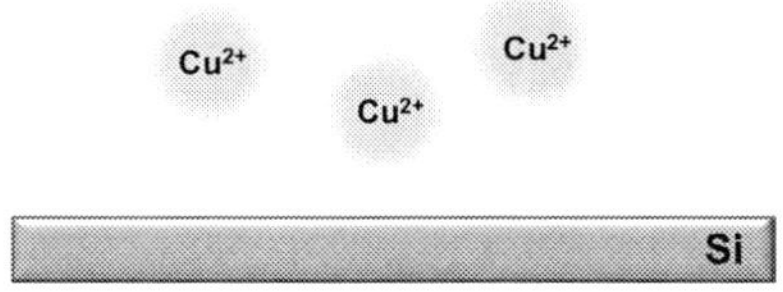

2) Cu nucleation by oxidizing Si through high reduction potential

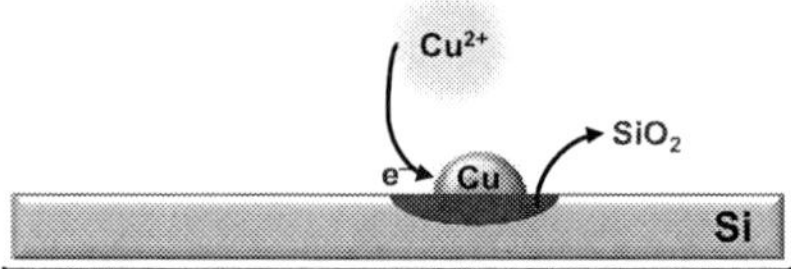

Standard reduction potential (V vs. NHE)	
$Cu^{2+} + 2e^- \rightleftarrows Cu$	0.34
$SiO_2 + 4H^+ + 4e \rightleftarrows Si + 2H_2O$	– 0.84

3) Continuous Cu particle growth

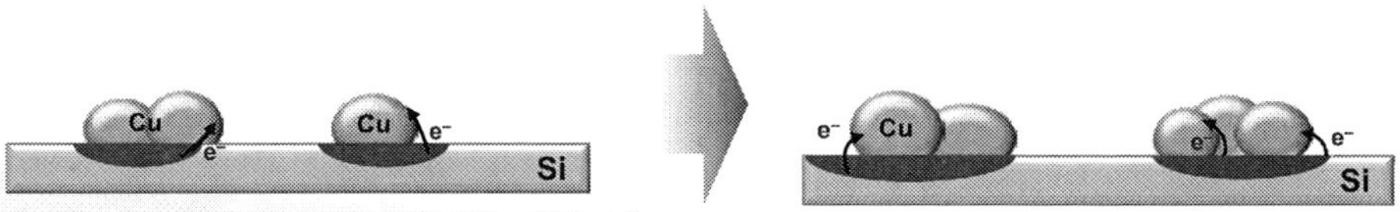

Hitoshi Morinaga, Makoto Suyama, and Tadahiro Ohmi, J. Electrochem., Vol. 141, No. 10 (1994)

Aluminum growth on Si wafer

Aluminum hydroxide formation

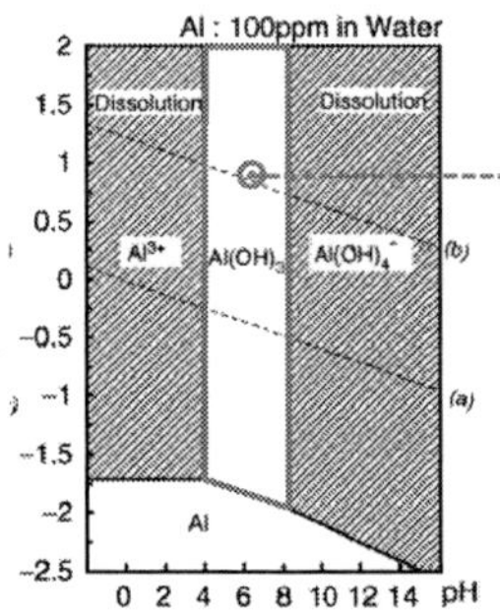

< Aluminum Pourbaix diagram (25 °C) >

- **$Al(OH)_3$** : Stable phase of Al in neutral solution

$$Al\,(s) + OH^- \rightarrow Al(OH)_{ads} + e^-$$
$$Al(OH)_{ads} + OH^- \rightarrow Al(OH)_{2ads} + e^-$$
$$Al(OH)_{2ads} + OH^- \rightarrow Al(OH)_{3ads} + e^-$$

Standard reduction potential (V vs. NHE)	
$SiO_2 + 4H^+ + 4e \rightleftarrows Si + 2H_2O$	– 0.84
$Al_2(OH)_3 + 3e^- \rightleftarrows 2Al + 3OH^-$	– 1.66

Aluminum oxide transformation

- $Al(OH)_3$ transformation into $Al_2O_3{\cdot}3H_2O$ compound in neutral solution
- Precipitation formation within SiO_2 film

$$2Al + 6H_2O \rightarrow 2Al(OH)_{3ads} + 6e^- + 6H^+ \rightarrow Al_2O_3{\cdot}3H_2O + 3H_2$$

Beardmore, J., Lopez, X., Mujika, J. I., & Exley, C. (2016). What is the mechanism of formation of hydroxyaluminosilicates?. Scientific Reports, 6.
Giza, Effect of Aging on the Corrosion of Aluminum Alloy 6061, Faculty of Engineering Cairo University (2010)

Chelating agent

Chelating effect

- **What is chelate?**
  Complex compound formation with metal ion making covalent bonds by polydentate ligand (Electron donors)

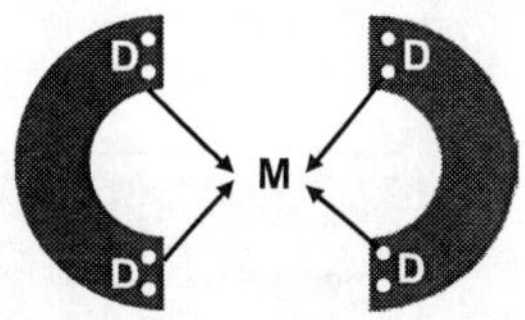

< Formation of chelate by polydentate ligand >

- **Chelating effect**
  1) Stability improvement of complex by chelate rings when metal ion and ligands are combined
  2) The more ligands exist, the more reactions occur

<Chelating agents according to the metal species >

Metal	Chelating reagents
Al	F^-, $C_2O_4^{2-}$, OAc^-, cit, tart, EDTA, OH^-, BAL, NTE
Cu	NH_3, I^-, SCN^-, CN^-, $S_2O_3^{2-}$, TU, EDTA, S^{2-}, DTC, DHG, BAL, $C_2O_4^{2-}$, cit, tart, NTA, TG, NTE, tren, penten

K. Burger, Organic Reagents in Metal Analysis, Vol. 54, p 39-41 (1973)

Metal contamination method

Experiment procedures

Pretreatment **Metal contamination**

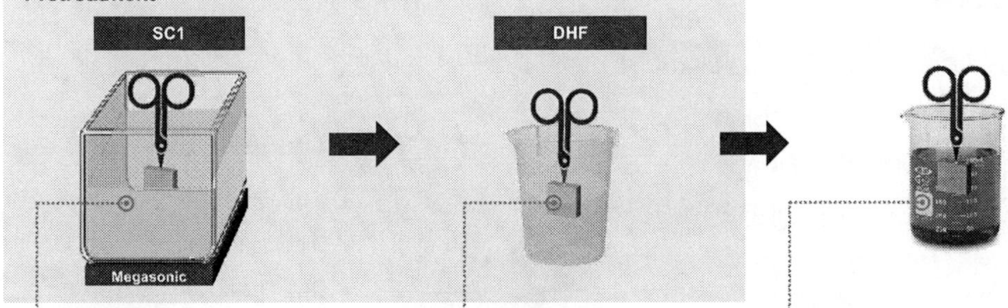

Condition (SC1)	Condition (DHF)	Condition (Metal contamination)
1. **Temperature :** 70°C	1. **Temperature :** 25°C	**1.Cu contamination :**
2. **Time :** 10 min.	2. **Time :** 5 min.	Wafer dipping in $CuCl_2$ 500ppb solution for 1min.
3. **Concentration :**	3. **Concentration :**	**2. Al contamination :**
NH_4OH:H_2O_2:DIW = 1:2:50	HF : DIW = 1:100	Wafer dipping in $AlCl_3$ 100~300ppm solution for 1min.
4. **Megasonic power :** 1MHz, 600W		

AFM images after metal contamination

Cu contamination
(500 ppb, dipping for 1 min.)

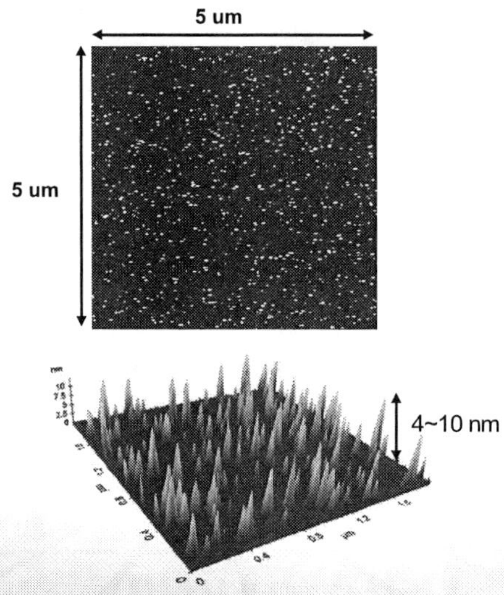

Al contamination
(100~300 ppm, dipping for 1 min.)

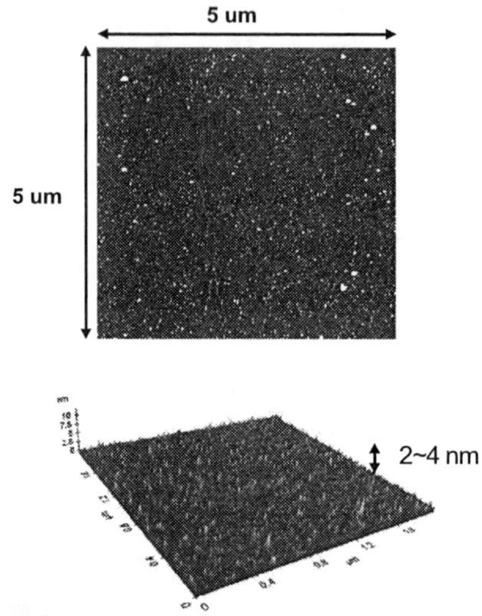

2017 SPCC, March 28-29, Texas, USA 11

Metal removal by chelating agents

Experiment procedures

Metal particle contamination

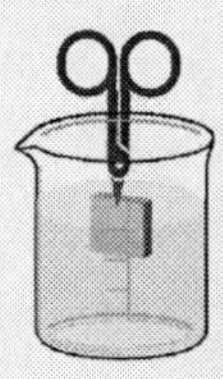

Cu Contamination solution

- Solution : $CuCl_2$ + DIW, 500ppb
- Dipping for **1** min.
- Subsequent analysis using **AFM**

Al Contamination solution

- Solution : $AlCl_3$ + DIW, 100-300ppm
- Dipping for **1** min.
- Subsequent analysis using **AFM**

Contamination cleaning

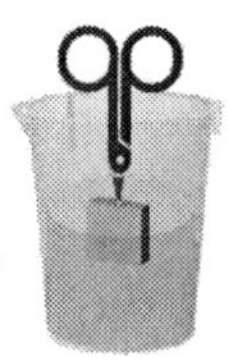

HF + Chelating agent

- 0.05wt% HF + Chelating agent
- Dipping for **5** min.
- DIW rinsing for 1 min.
- Subsequent analysis using **AFM**

2017 SPCC, March 28-29, Texas, USA 12

Metal removal by HF/DIO$_3$

Experiment procedures

Metal particle contamination

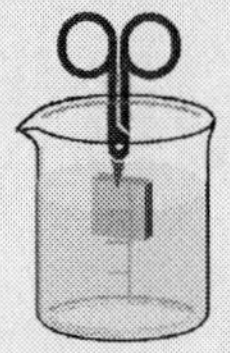

Cu Contamination solution

- Solution : $CuCl_2$ + DIW, 500ppb
- Dipping for **1** min.
- Subsequent analysis using **AFM**

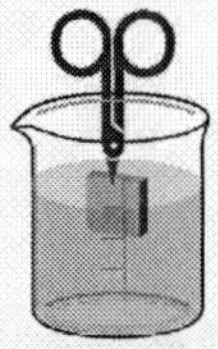

Al Contamination solution

- Solution : $AlCl_3$ + DIW, 100-300ppm
- Dipping for **1** min.
- Subsequent analysis using **AFM**

Contamination cleaning

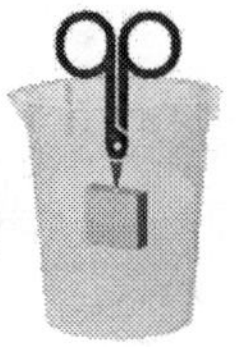

HF + DIO$_3$

- 0.05wt% HF + 10ppm DIO_3
- Dipping for **5** min.
- Etch rate = 2.3 Å/min.
- Subsequent analysis using **AFM**

2017 Surface Preparation and Cleaning Conference
March 28-29, Austin, Texas, USA

Experiment Results and Discussion

NEMPL Hanyang University

Effect of Chelating agent on Contact angle

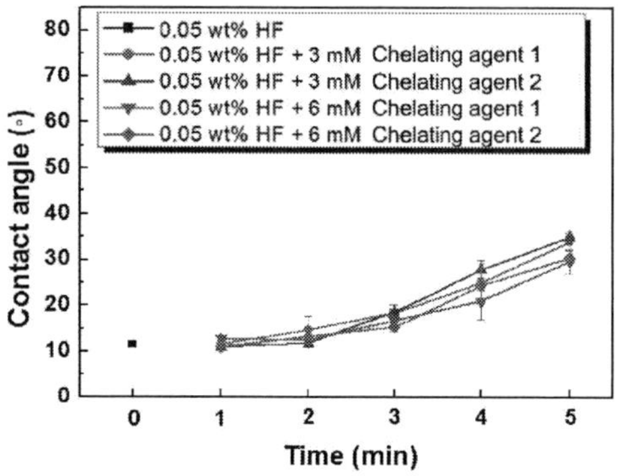

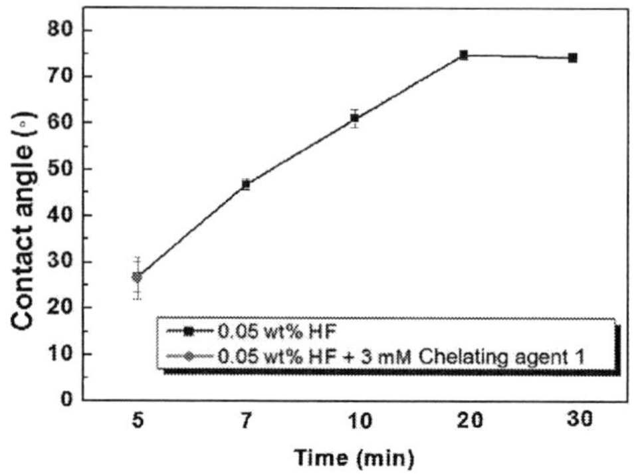

- No contact angle change by chelating agents

  ➔ **No surface modification effects** by chelating agent species and concentration

- Hydrophobic change by applying HF based solution

  ➔ Residual SiO_2 exists until 5 min : Minimum cleaning condition to maintain **hydrophilicity**

Surface roughness change

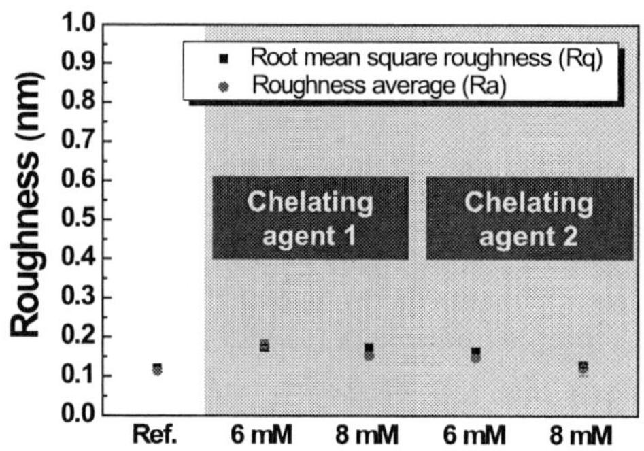

In 1 um X 1 um (in 0.05 wt% HF, 5 min)

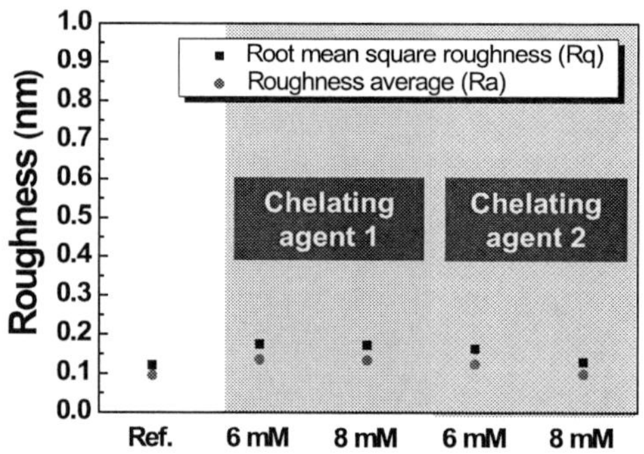

In 5 um X 5 um (in 0.05 wt% HF, 5 min)

- Roughness measurement after treating the surface with HF/chelating agent using AFM
- **No roughness change** despite the chelating agent species and concentration change

Metal removal by Chelating agents

Metal removal efficiency evaluation

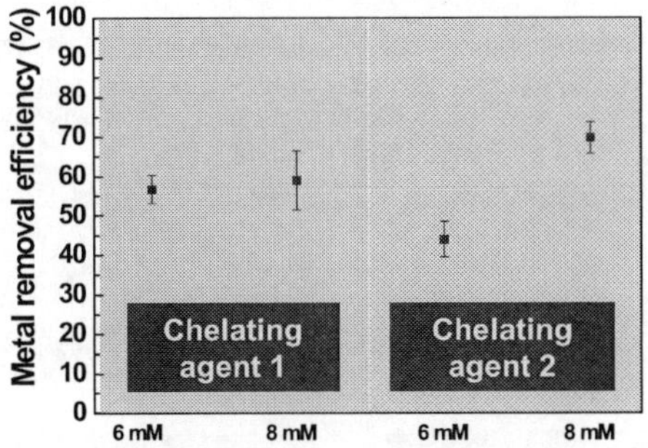

Cu removal in 0.05 wt% HF, 5 min

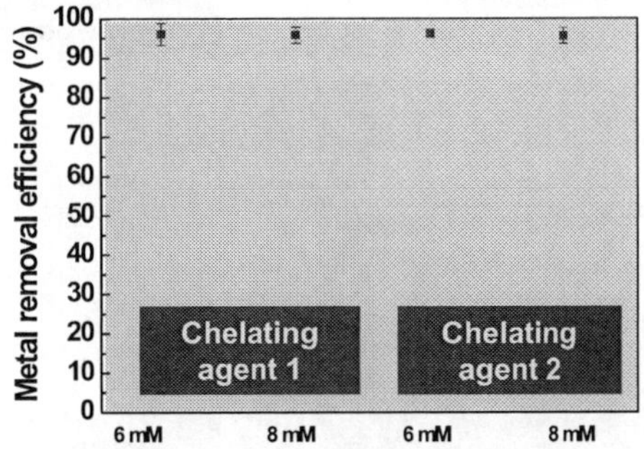

Al removal in 0.05 wt% HF, 5 min

- Cu MRE at 6 mM : **Chelating agent 1 (57%)** > Chelating agent 2 (44%)
  → More functional groups in chelating agent 1 ⇔ Higher metal removal efficiency
- Cu MRE at 8 mM : Chelating agent 1 (59%) < **Chelating agent 2 (70%)**
  → Less dissociation in low pH ⇔ Lower metal removal efficiency
- Al removal : Showing over **90%** of MRE supported by both chelating agent and HF

AFM images after metal contamination removal

Cu contamination cleaning
(in 0.05 wt% HF)

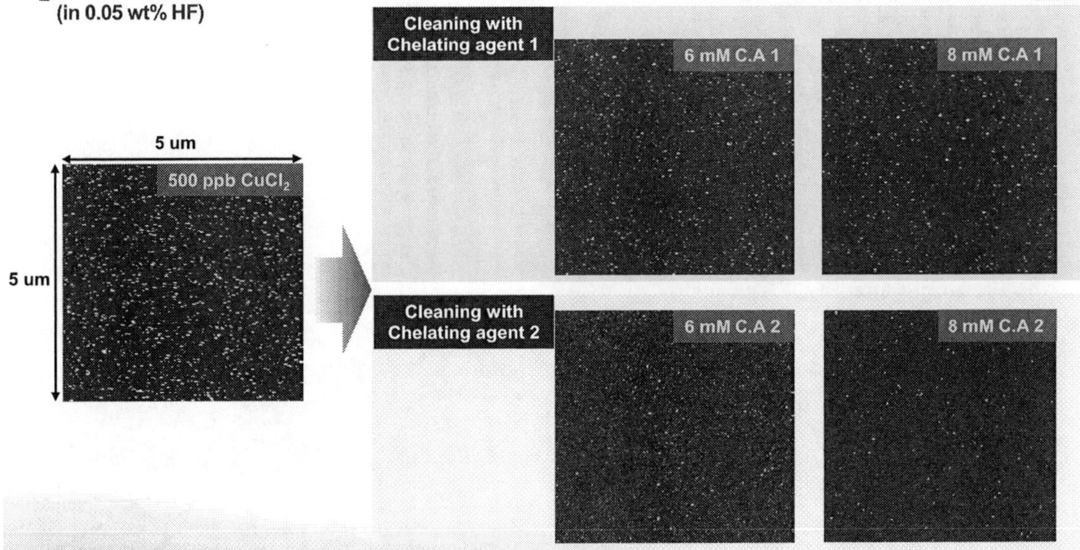

AFM images after metal contamination removal

Al contamination cleaning
(in 0.05 wt% HF)

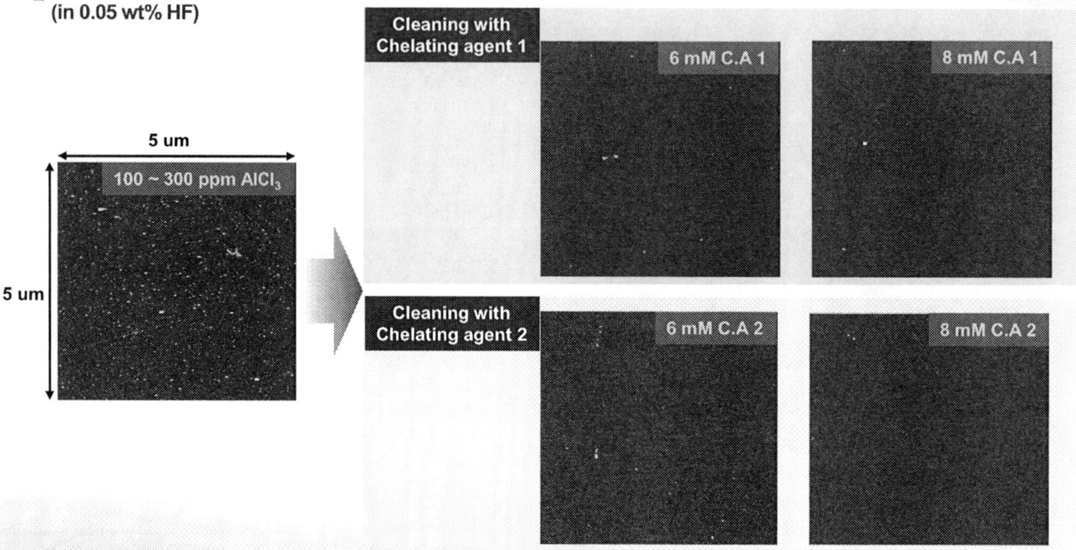

Metal removal by HF/DIO_3

Metal removal efficiency evaluation

- Showing over 90% of MRE for Cu and Al both
- Much higher MRE compared to chelating agent

➔ Expected to show higher MRE than any conditions of HF/DIO_3

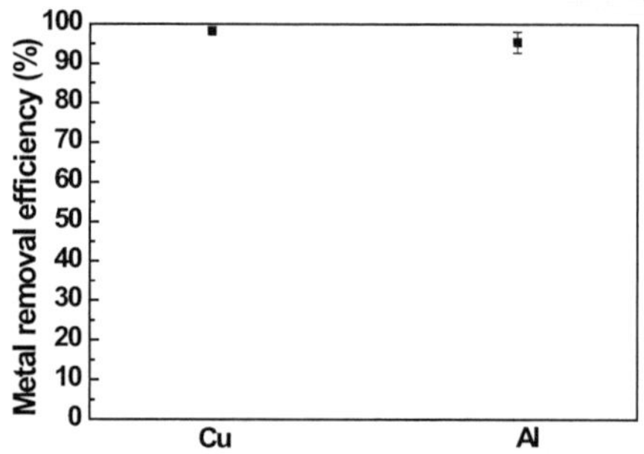

AFM images after HF/DIO_3 cleaning

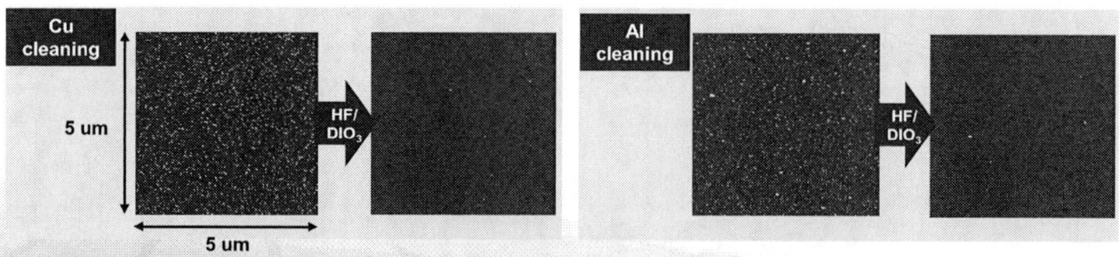

HF/Ozone solution

Metal removal by ozone

- Metal oxidation by O_3

1) Metallic impurity can be removed by oxidation by O_3
2) Ionization ↑ in **acidic** region :
   Complex formation of $[M(H_2O)_x]^+$ ↑

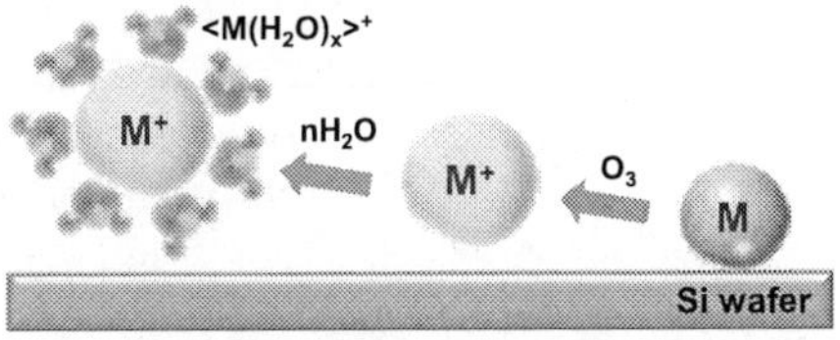

< Metal oxidation & removal mechanism by O_3 >

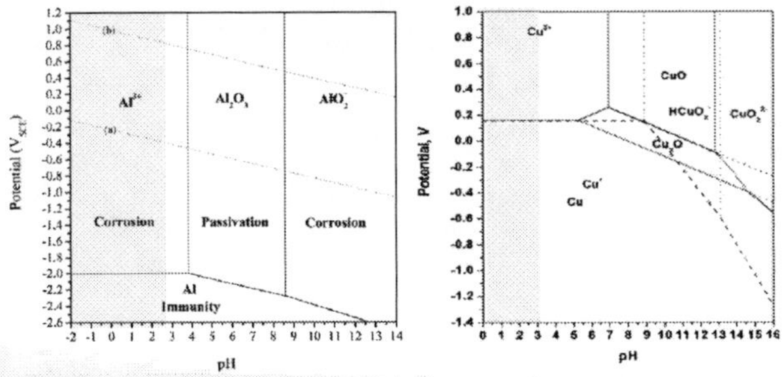

< Pourbaix diagram for Al and Cu >

HF	DIO_3	pH
0.05 wt%	0 ppm	3.18
	10 ppm	3.03
	20 ppm	2.96
	30 ppm	3.03
0.1 wt%	0 ppm	3.07
	30 ppm	3.08
Red-ox potential	$Cu^{2+} + 2e^- \rightleftarrows Cu$ / 0.34V	
	$Al_2(OH)_3 + 3e^- \rightleftarrows 2Al + 3OH^-$ / -1.66V	

Hyoseob Y, Kunkul R. A study of Si-wafer cleaning by electrolyzed water. Korea journal of materials research. Vol. 11, No.4 (2001)
Oh, Y. J., Park, G. S., & Chung, C. H. (2006). Journal of the Electrochemical Society, 153(7), G617-G621.
Sukiman, N. L., Hughes, A. E., Thompson, G. E., Mol, J. M. C., Birbilis, N., Garcia, S. J., & Zhou, X. (2012), techniques and developments. INTECH Open Access Publisher.

HF/Ozone solution

SiO_2 growth in DIO_3

- SiO_2 formation by higher red-ox potential of O_3

$$Si + 2O_3 \rightarrow SiO_2 + 2O_2$$
$$Si + O_2 \rightarrow SiO_2$$

Standard reduction potential (V vs. NHE)	
$O_3 + 2H^+ + 2e^- \rightleftarrows O_2 + H_2O$	2.076
$SiO_2 + 4H^+ + 4e \rightleftarrows Si + 2H_2O$	– 0.84

SiO_2 removal by HF

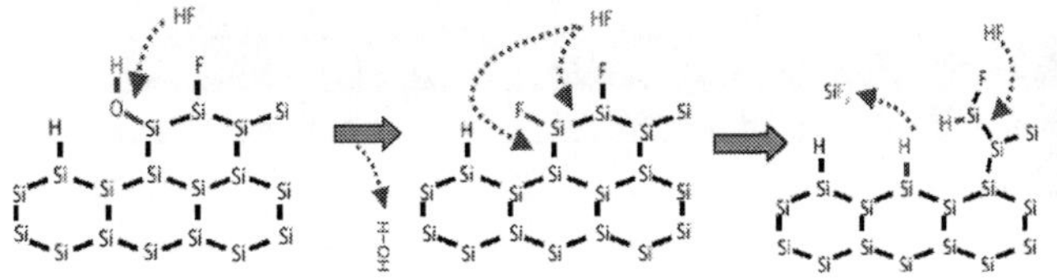

$$SiO_2 + 6HF \rightarrow H_2SiF_6 + 2H_2O$$
$$H_2SiF_6 \rightarrow 2HF + SiF_4$$

< SiO_2 etching mechanism by hydrofluoric acid >

- Stable Si-F bond formation breaking Si-O bond
- Volatile SiF_x formation by high electronegativity of fluorine ion(F^-) → **SiO_2 / Contaminant removal**

Yokota, Y. et al. (2011). U.S. Patent No. 7,972,441. Washington, DC: U.S. Patent and Trademark Office
Knotter, D. M. (2000).Journal of the American Chemical Society, 122(18), 4345-4351.
Michalak, D. J. et al. (2010).. Nature materials, 9(3), 266-271.

HF/Ozone solution

Si etch rate change by HF/DIO_3 concentration

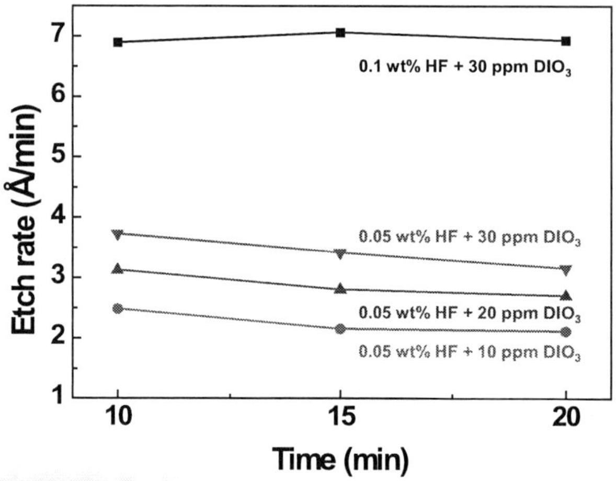

Cleaning conditions		
DHF	DIO_3	Etch rate (Å/min)
0.05 wt%	10 ppm	2.249
	20 ppm	2.877
	30 ppm	3.434
0.1 wt%	30 ppm	6.96

Summary

Metal removal using chelating agents

- Cu removal
  1) Chelating agent 1 : More ligands but less dissociation in acidic region
  → Proper for high pH or less chelating agent concentration for metal removal
  2) Chelating agent 2 : Less ligands but better dissociation in acidic region
  → Proper for low pH or high chelating agent concentration for metal removal
- Al removal
  1) High MRE supported by HF in addition to the chelating agents

	6mM		8mM	
	Chelating agent 1	Chelating agent 2	Chelating agent 1	Chelating agent 2
Cu MRE	57%	44%	59%	70%
Al MRE	Over 90%			

Metal removal using HF/DIO_3

- > 90% of MRE for Cu and Al by repeating the oxide film formation and removal

"Higher metal contamination removal efficiency in HF/O_3 Chemistry"

HANYANG UNIVERSITY
한양
1939
THANK YOU
NEMPL
NEMPL
Hanyang University
2017 SPCC, March 28-29, Texas, USA

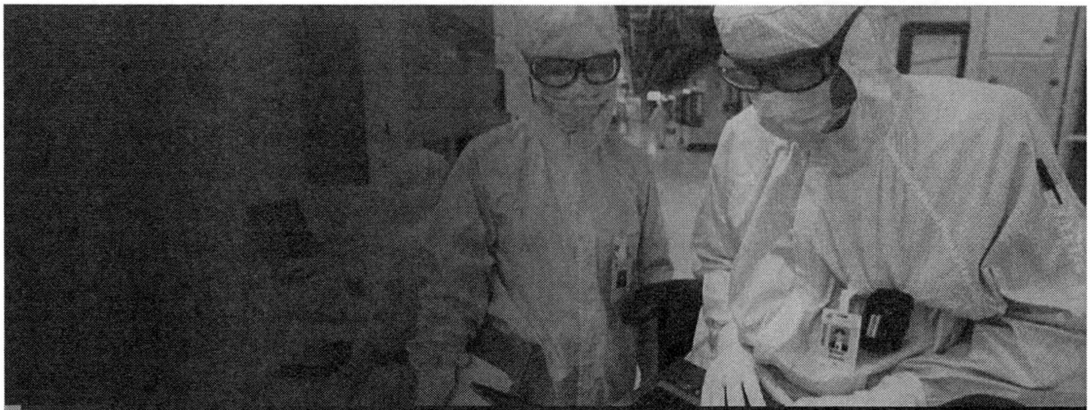

Post CMP in-situ cleaning for 14/7nm transistor scaling: a crucial process for yield enhancement at advanced node device fabrication

Tae Hoon Lee
Advanced Module Engineering
GLOBALFOUNDRIES®

CMOS Transistor Scaling & New Device Platform

Improving performance is about shrinking and expanding

BEOL materials innovations in the 'good old days'
Device and channel levels drive future improvements.
Novel materials innovations = device performance improvements.

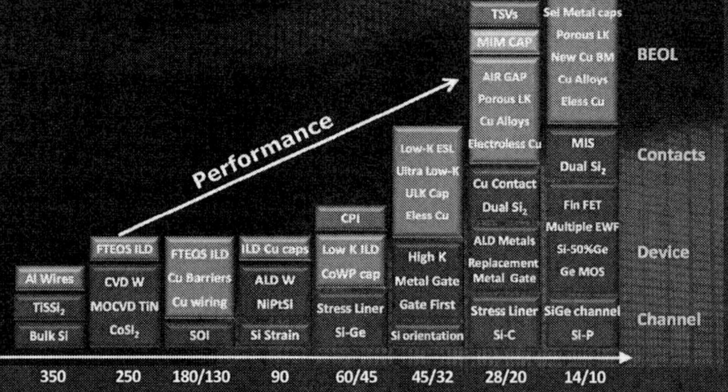

Materials Innovation of Technology Node. Source: Bartlett, GLOBALFOUNDRIES, SEMI Strategic Materials Conference

Scaling continues by structural changes and new materials

2

CMP Process Challenges

Updated from 2015 SEMATECH

	20nm	14nm	7nm
Selectivity/ Materials	FEOL: SiN, Ox MOL: SiN, Ox, W, TiN, Al BEOL: Cu, Ta, TaN, TiN	FEOL: Si, SiN, Ox, a-Si MOL: W, SiN, Ox, poly Si, TiAl, TiC BEOL: Cu, Ta, TaN, TiN	FEOL: Si, SiN, Ox, ***low k*, *Lower Density*** MOL: W, SiN, Ox, poly, ***low k spacer, SiOC, Co*** BEOL: Cu, Ta, TaN. TiN, ***Ru/Co***, BDIII
Dishing/ Erosion	**Higher PD <200A**	**Higher PD <150A**	**Higher PD <100A**
Uniformity	**3sigma <120A**	**3sigma <100A**	**3sigma <80A**

- New materials expected at future nodes to meet both process and device requirements
- More than 50% newer materials at advanced technology node

GLOBALFOUNDRIES Confidential 3

Gate height control – Defects

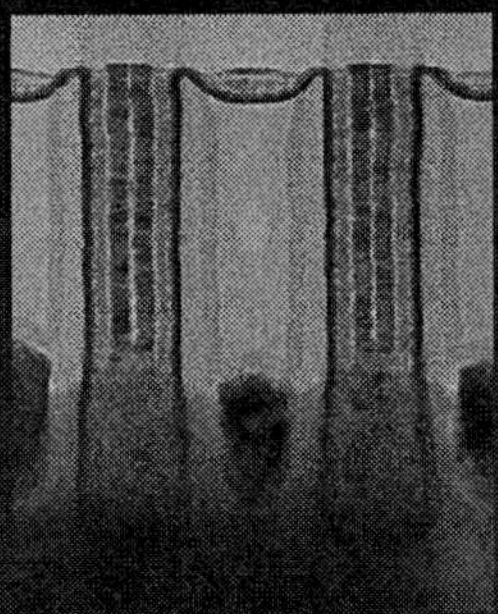

Nominal device

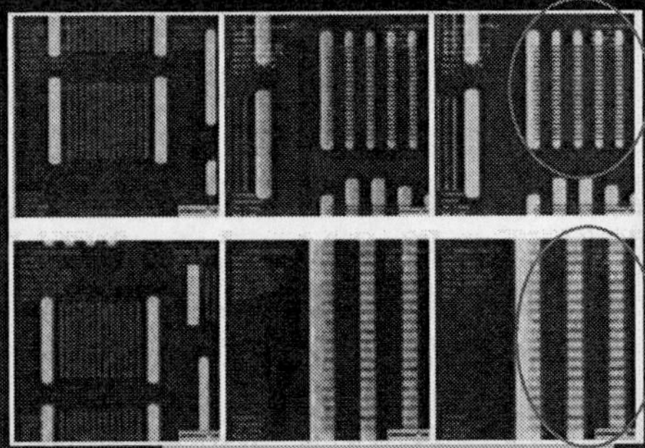

Wide gate >150nm

Taller gate → residues → potential shorts/Open
Shorter gate → Fin exposure → Opens

- Defects can be impacted by gate height control
- Residues and overpolish
  - Accumulation of topography from prior layers
  - Integration scheme
  - Selectivity between different materials
- Very narrow process window → Control is the key!!

4

Impacts of CMP Cleaning on Yield Detract

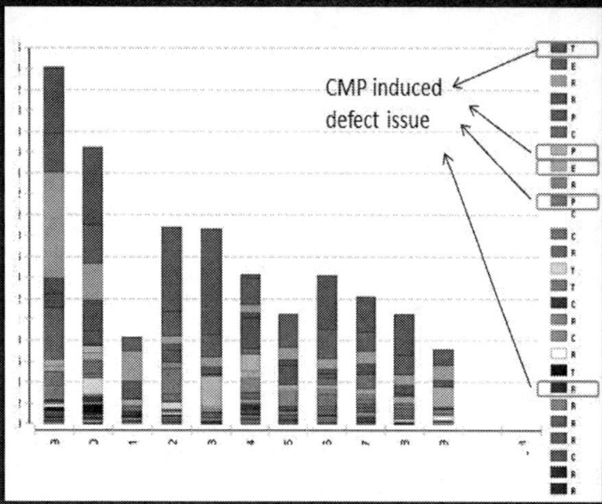

Defects	Failure Mode
Metallic Particle	Leakage Failure
Non-metallic Particle	Open Contact (Block Etch)
Organic Residue	Leakage Path
Microscratch	Pattern Damage
Metal Corrosion	Resistance

Post CMP cleaning becomes one of the key yield impact detractors

Challenges in CMP In-situ Cleaning

Sources	Challenges & Opportunities
Equipment (Cleaner Module)	• High performance cleaning implement (ex. spray jet) • Closed chamber for various chemical application • Tool cleanliness
Megasonic	• Control acoustic bubble motion • High physical force (ex. gigasonic) implementation
Cleaning Chemical	• Limited chemicals available (highly diluted acid, NH_4OH, SC1) • Minimize surface damage (ex. roughness, charge, hydrophobicity) • Non-hazardous and cost effective
Brush	• Nodule design and control of contact area • Brush material for improved cleaning efficiency • Nodule height uniformity control at manufacturing • Pre-broken conditioned brush • Cleanliness of brush at manufacturing stage
Cleaning Recipe	• Optimize cleaning efficiency with minimize cross contamination • Minimize throughput impact
Dryer	• End point detection at drying point

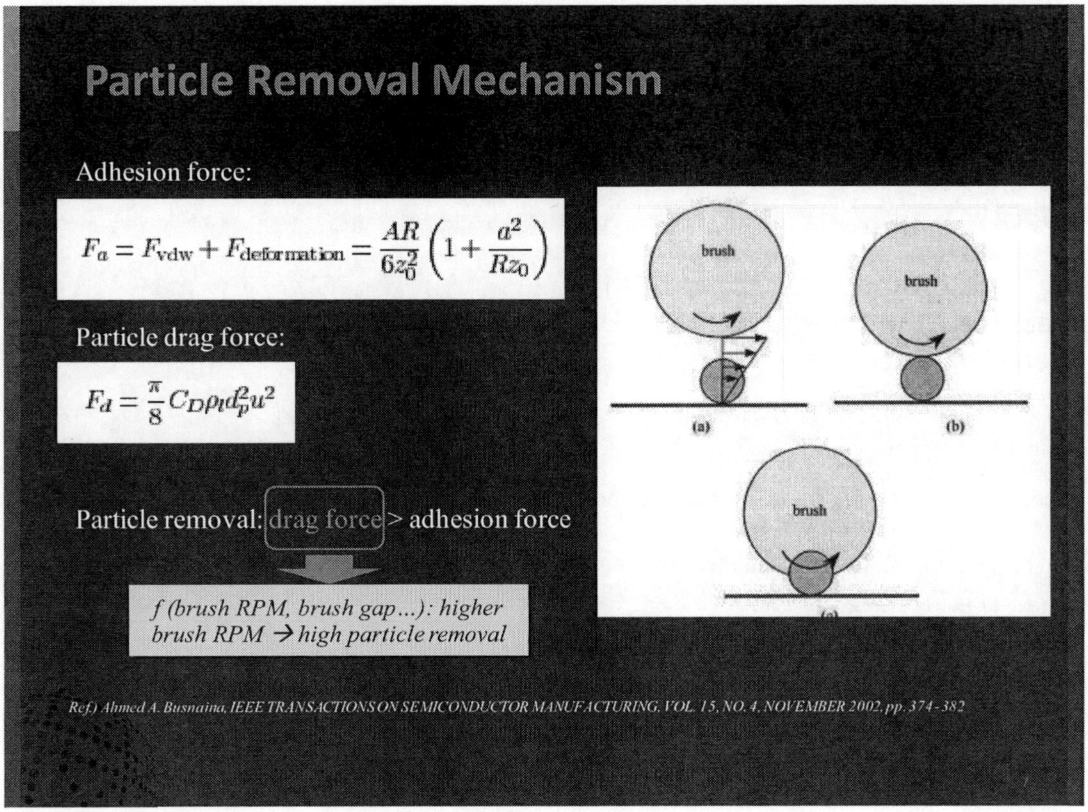
Particle Removal Mechanism
Adhesion force:
$F_a = F_{\text{vdw}} + F_{\text{deformation}} = \frac{AR}{6z_0^2}\left(1+\frac{a^2}{Rz_0}\right)$
Particle drag force:
$F_d = \frac{\pi}{8} C_D \rho_l d_p^2 u^2$
Particle removal: drag force > adhesion force
f (brush RPM, brush gap…): higher brush RPM → high particle removal
brush
brush
brush
(a)
(b)
Ref.) Ahmed A. Busnaina, IEEE TRANSACTIONS ON SEMICONDUCTOR MANUFACTURING, VOL. 15, NO. 4, NOVEMBER 2002, pp. 374-382

Thank You!!
Taehoon.Lee@globalfoundries.com

Entegris
50 YEARS OF PURE ADVANTAGE

SPCC POST-CMP CONFERENCE

High Performance, Ceria Post-CMP Cleaning Formulations for STI/ILD Dielectric Substrates

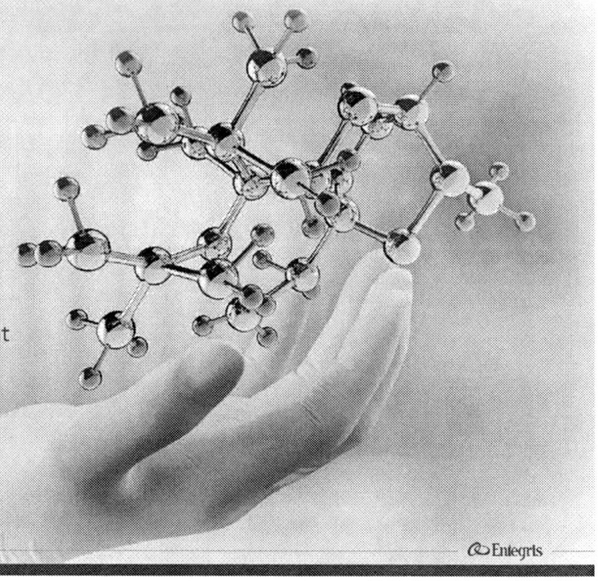

Daniela White*, PhD – Sr. Principal Scientist
Atanu Das, PhD – Scientist
Thomas Parson, PhD – Scientist
Michael White, PhD – Director, Post-CMP Development

ENTEGRIS Inc., R&D Surface Preparation and Integration
7 Commerce Drive, Danbury CT 06810
Ph: 203-739-1470, daniela.white@entegris.com

Entegris

OUTLINE

- Development of a low pH and high pH family of efficient ceria cleaners for STI and ILD dielectric surfaces (PETEOS, SiN, SiC, thermal oxide, etc.), PlanarClean® AG Ce-XXXX
- Mechanistic considerations specific to PlanarClean AG Ce-XXXX formulation design
  - Understanding CeO_2 surface chemistry and CeO_2-SiO_2 interactions: Raman, FTIR
  - Ce^{4+}/Ce^{3+} oxidation state characterization: UV-VIS, Raman, potentiometric titrations
  - $Ce^{4+/3+}$ impact on the ceria cleaning mechanism
- Ceria particle defect count results by SEM, Dark field Microscopy (DFM) and ICP data on dielectric surfaces cleaned with PlanarClean AG Ce-XXXX formulation
- Conclusions and path forward

2 Entegris

WHY FORMULATED CERIA CLEANERS VS. COMMODITY?

1. EHS/Safety concerns with traditional cleans (hot SPM, dHF, SC-1, TMAH + dHF)
2. One-step clean process requirement for throughput improvement
3. Need for improved particle removal
4. Need for improved metal removal
5. No damage to dielectric substrates

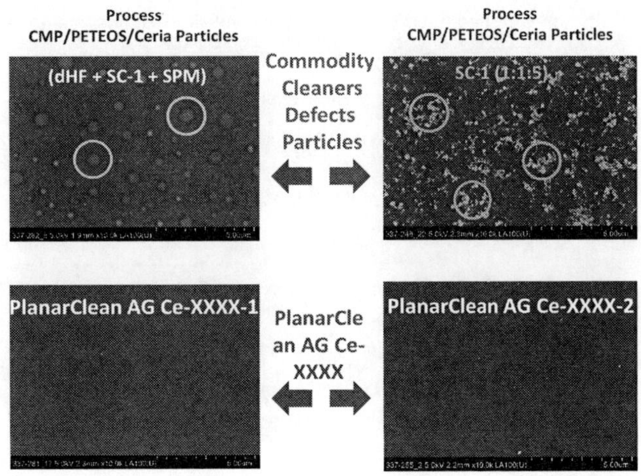

CERIA POST-CMP CLEANING FORMULATION – MECHANIC DESIGN CONCEPT

Formulation Design Options

1. High pH hydrolysis of -Ce-O-Si- bonds by HO^- nucleophilic attack to Ce^{4+} plus additives needed to stabilize –Ce-OH species and prevent re-deposition
2. High pH partial etch/dissolution of the surface –Si-O-Ce- groups plus re-deposition prevention
3. Bond-breaking additives, followed by CeO_2 complexation, particles stabilization and dispersion

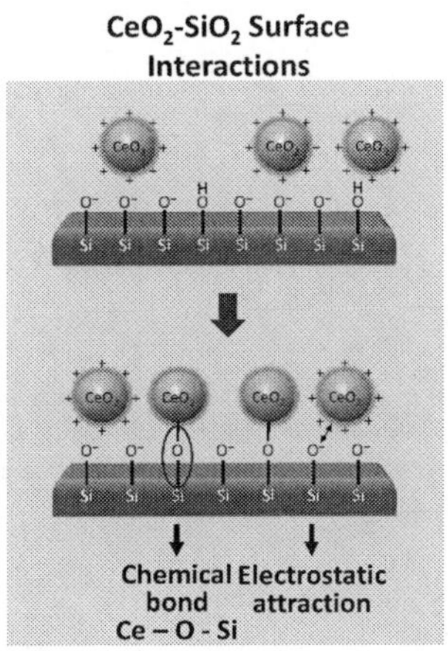

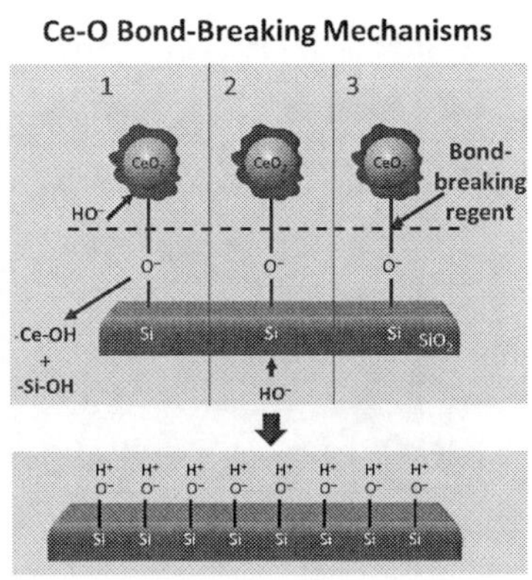

INTERACTIONS OF ORGANIC MOLECULES WITH CERIA SURFACE AND SURFACE CHARGE VS. PH

- Surface modified CeO_2 particles in the CMP slurries: positive or negative surface charge
- Particle size: 15–200 nm
- Need to understand CeO_2 surface chemistry and types of interactions with the dielectric surfaces
- Six different types of CeO_2 particles tested – can we design a universal cleaner?

Surface-Modified Ceria in CMP Slurries

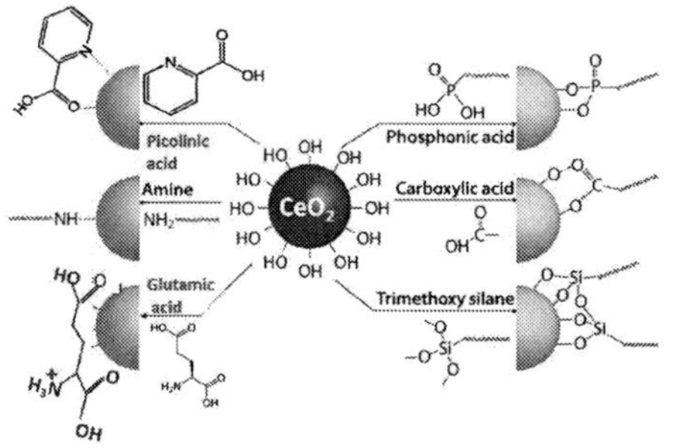

Various CeO_2 Particles (A-F) Tested

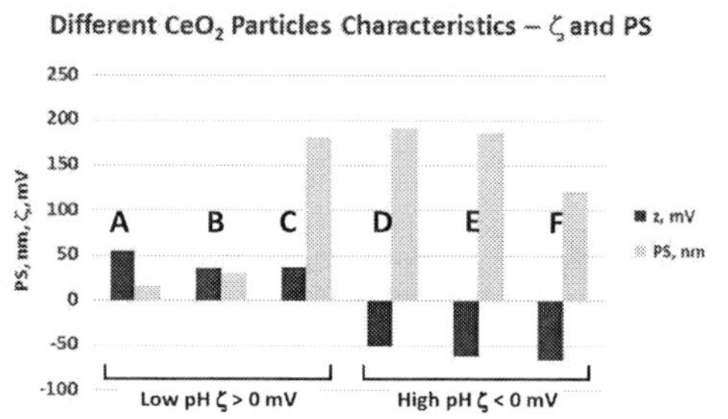

 Entegris

TYPES OF CeO_2 SURFACE GROUPS

A. Hydroxyl groups by FTIR[1] and titration

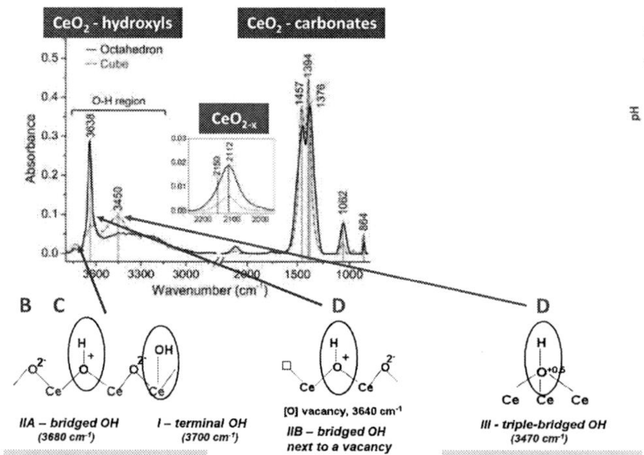

Basic sites, pK_a = 7–10

Acidic sites, pK_a = 3–4

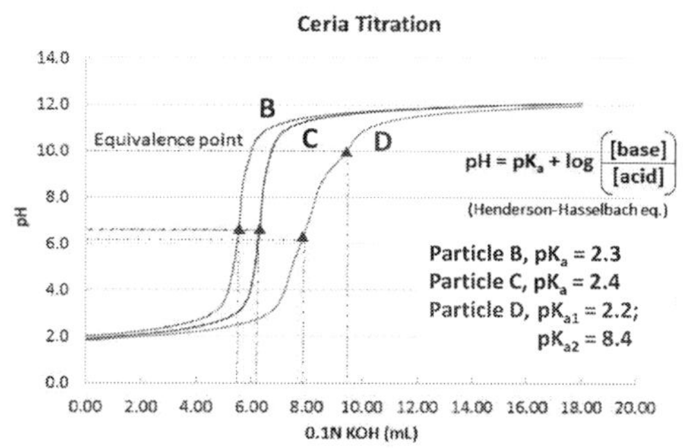

-Ce- HO_2 + OH ⟷ -Ce – OH + H_2O (I)	PK_{a1}: 8–10 most basic
-Ce- HO_2 + OH ⟷ -Ce – OH + H_2O (IIA, IIB)	PK_{a2}: 4–7
-Ce- HO_2 + OH ⟷ -Ce – OH + H_2O (III)	PK_{a3}: <3- most acidic

 1. Christoph T. Nottbohm, Christian Hess, Catalysis Communications 22 (2012) 39–42 Entegris

TYPES OF CeO_2 SURFACE GROUPS

B. Carbonates on reduced and stoichiometric ceria nanoparticles[2] (RAMAN)

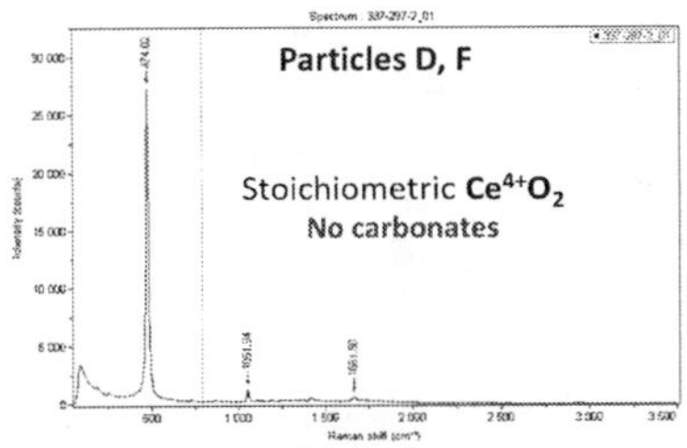

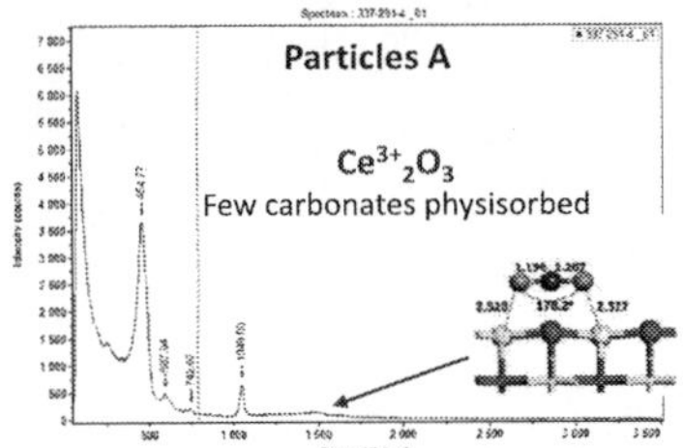

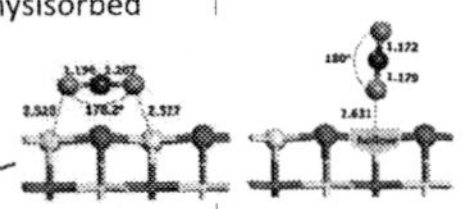

Expect different CMP (RR) and post-CMP cleaning behavior!

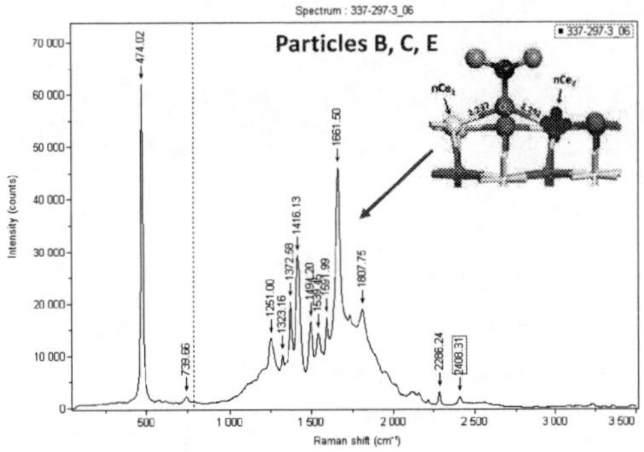

Reduced $CeO_x - CO_2$ chemisorption → unidentate/bidentate carbonates

7 2. E. E. Benson, C. P. Kubiak, A. J. Sathrum, and J. M. Smieja, Chemical Society Reviews 38, 89 (2009).

Entegris

DIFFERENT CLEANING FORMULATIONS FOR DIFFERENT CERIA SURFACE CHEMISTRIES?

Particles E, F

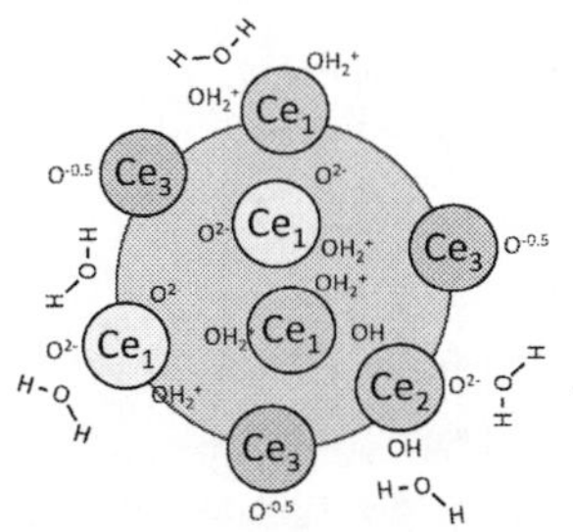

- More acidic surface, partially hydroxylated
- Small amount of surface water H-bonded
- Surface exposed –OH for -Si-O- condensation
- Stronger Ce-O-Si bonds, difficult to break/clean

Particles B, C, D

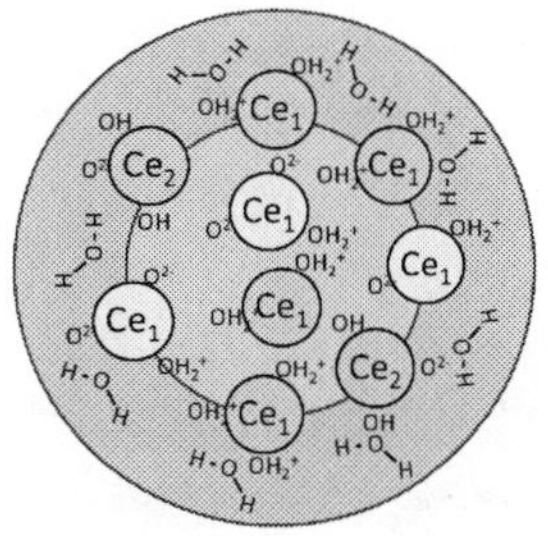

- More basic surface, more hydroxylated
- Outer-sphere shell of H-bonded water
- Reduced surface reactivity
- Weaker Ce-O-Si bonds, easier to break

Based on the FTIR-ATR UV-VIS and tritation experiments data

8

Entegris

CERIA REACTS WITH H_2O_2 BY BOTH REDUCTION AND OXIDATION MECHANISMS

$CeOH^{3+} + H^+ + e \rightarrow Ce^{3+} + H_2O$ ... Ered = +1.715 V
$H_2O_2 \rightarrow 2\ H^+ + O_2 + 2\ e$ ... Eox = -0.695 V
Ered + Eox > 0, reaction can proceed

Reduction of Ce^{4+} to Ce^{3+}
Oxidation of H_2O_2 to O_2

$Ce^{3+} + H_2O \rightarrow CeOH^{3+} + H^+ + e$ ... Eox = -1.715 V
$H_2O_2 + 2\ H^+ + 2\ e \rightarrow 2\ H_2O$ ... Ered = +1.776 V
Eox + Ered > 0, reaction can proceed

Oxidation of Ce^{3+} to Ce^{4+}
Reduction of H_2O_2 to H_2O

Commodity cleaners as controls

- **SC-1** - H_2O:$\mathbf{H_2O_2}$:NH_3 (1:1:5)
- **SPM** – H_2SO_4:$\mathbf{H_2O_2}$ (1:4), T > 100°C

Why $\mathbf{H_2O_2}$?

RAMAN SPECTRA FOR BIG CERIA PARTICLES C (>100 nm) IN REACTION WITH H_2O_2/SC-1

No changes on the surface ration Ce^{4+}/Ce^{3+} upon addition of SC-1/H_2O_2

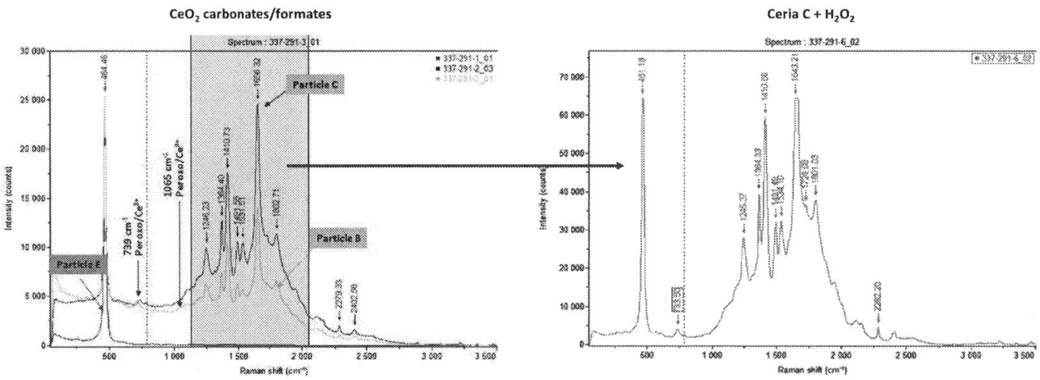

$Ce^{4+} + H_2O_2 \not\longrightarrow Ce^{3+}$

RAMAN SPECTRA FOR SMALL CERIA PARTICLES A (15 nm) IN REACTION WITH H_2O_2/SC-1

Before H_2O_2 addition

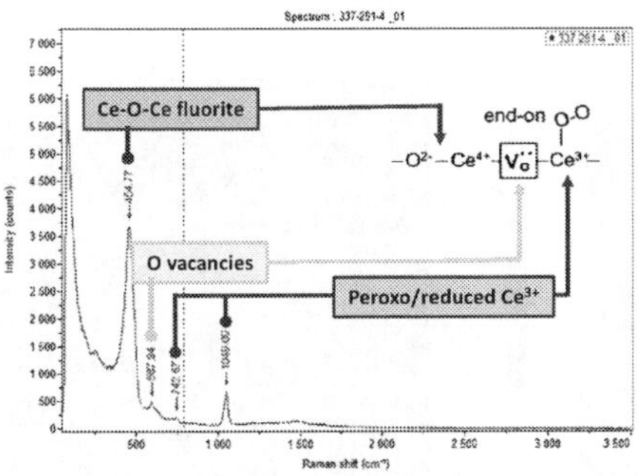

After H_2O_2 addition

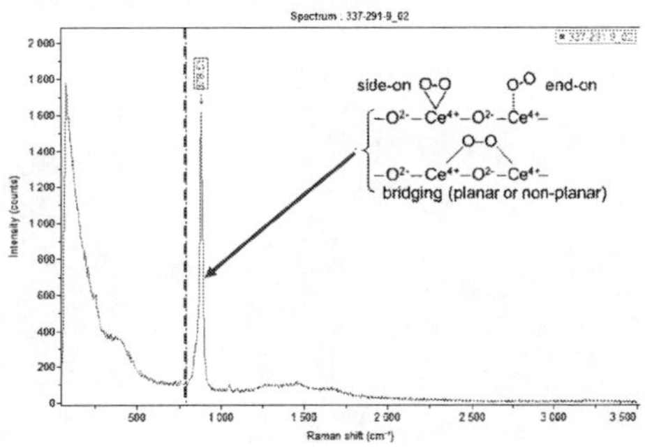

- Reduced fluorite $Ce^{4+}O_2$ surface species
- Vo.. Vacancies
- Peroxo/reduced Ce^{3+} species

+ H_2O_2

$Ce^{3+} \rightarrow Ce^{4+}$

- Fully oxidized Ce^{3+} to Ce^{4+} (only peroxo-Ce^{4+} species)
- No Vo.. vacancies
- No remaining Ce4+-O2--Ce^{4+} fluorite structure
- No remaining reduced peroxo-Ce^{3+} species

THE NEGATIVE EFFECT OF H_2O_2 IN COMMODITY CLEANERS ON DIELECTRIC SUBSTRATES

PETEOS/Ceria E/ PlanarClean AG Ce-XXXX-3

PETEOS/Ceria E/ PlanarClean AG Ce-

PETEOS/Ceria E/ (dHF + SC-1 + SPM)

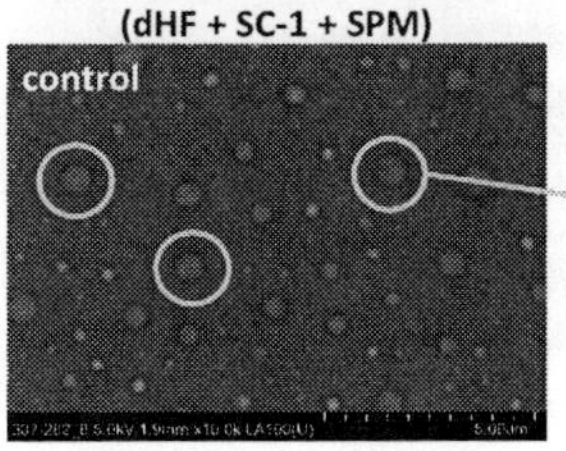

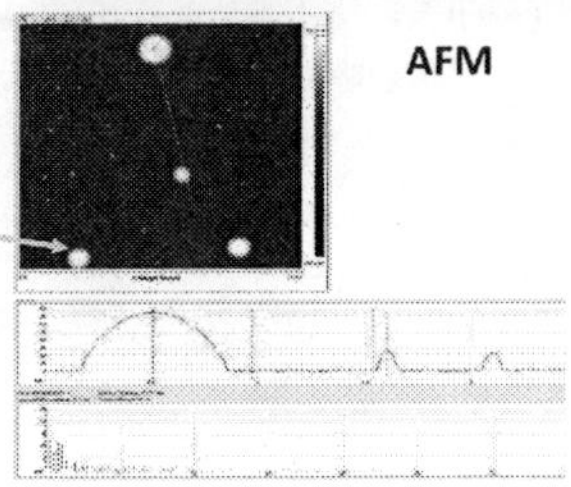

AFM

SEM-EDX

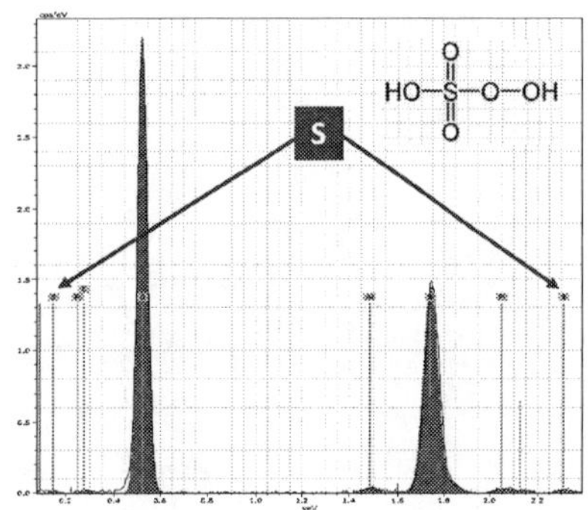

- S spherical deposits from peroxymonosulfuric acid
- PETEOS surface damaged after H_2O hot rinse
- PlanarClean AG Ce-XXXX-1 and -3 leave very clean, undamaged PETEOS surfaces

PETEOS SURFACE DAMAGE POST-SPM AND SC-1 CLEANED & POST-RINSED WITH HOT WATER

Damaged/etched surface

PETEOS/Ceria E/(dHF + SC-1 + SPM)

Agglomerated ceria particles

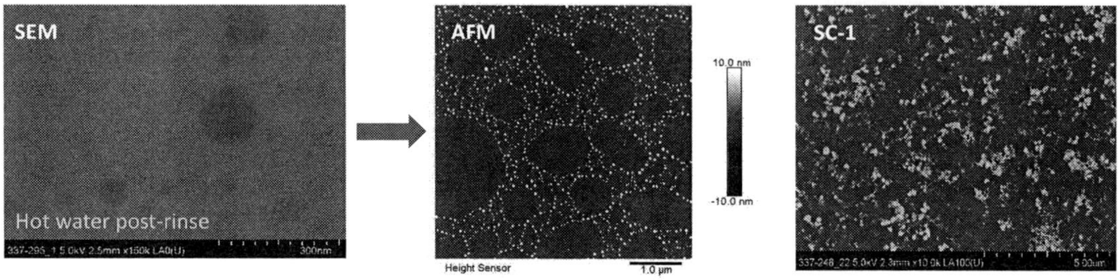

Commodity cleaners such as hot SPM and SC-1 potentially damage and leave agglomerated ceria particles and residue on dielectric surfaces

PLANARCLEAN AG CE-XXXX FORMULATION ADDITIVES LIST – FUNCTION AND MECHANISM

Component	Function	Mechanism
A	Non-TMAH pH adjustor	◦ Provides the hydroxyl anions and adjust pH needed for surpassing CeO_2 pH_{IEP} ◦ Ensures negative surface charge on both dielectric surface and ceria & organic contamination, by being adsorbed on inorganic and organic residues.
B	Complexing reagents Package	◦ Adsorption at the ceria surface ◦ Stabilization of ceria particles via electrosteric repulsion, preventing agglomeration and re-precipitation
C	Bond-breaking reagent	- Ce - O - Si – C - **Ce – OH + HO –** Si -
D	Cleaning Additives Package	◦ Interacts with particles and dielectrics surfaces to prevent particles aggregation and organics re-deposition

EXPERIMENTAL PROCEDURE

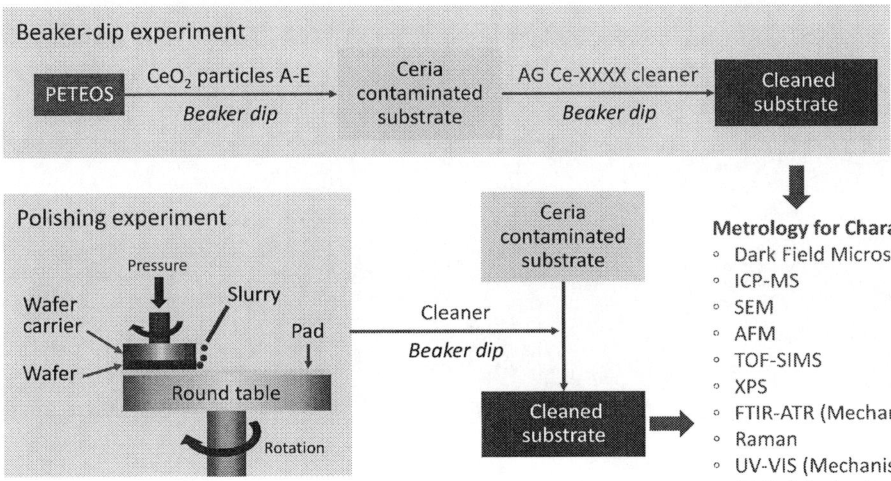

Metrology for Characterization

- Dark Field Microscopy (DFM)
- ICP-MS
- SEM
- AFM
- TOF-SIMS
- XPS
- FTIR-ATR (Mechanism)
- Raman
- UV-VIS (Mechanism)
- NMR (Mechanism)

FORMULATION DEVELOPMENT (PERFORMANCE = CERIA CONTAMINATION AFTER CLEANING)

Dark Field Microscopy Data

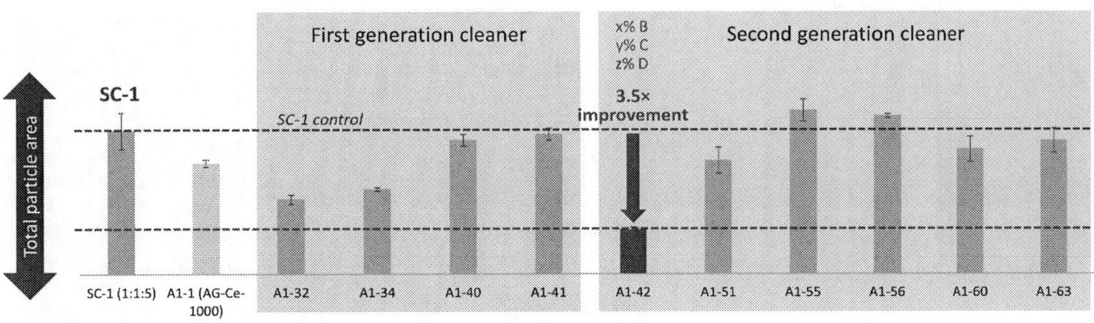

2nd generation cleaner: 3.5× more efficient than SC-1 and 2× than the 1st generation

3RD GEN PLANARCLEAN AG CE-XXXX-2 – 150× BETTER CLEANING PERFORMANCE VS. SC-1 (1:1:5)

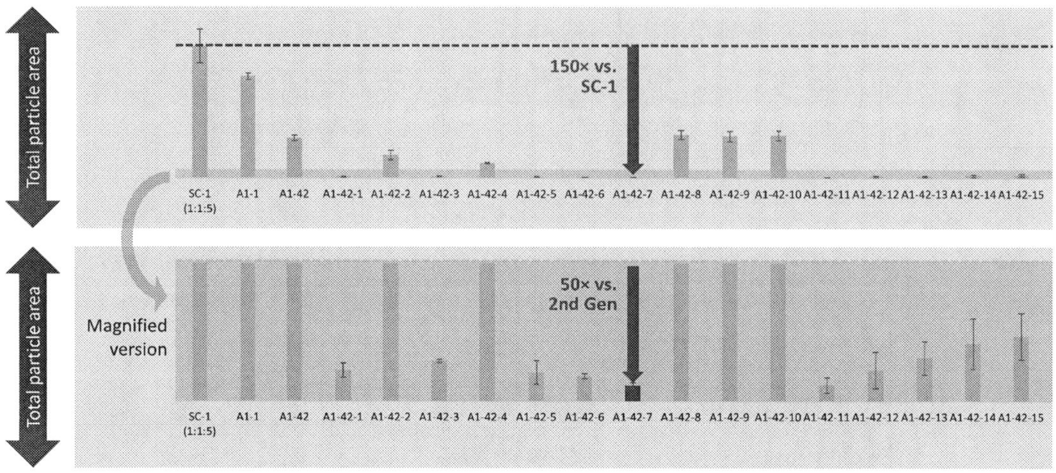

Developed cleaner showed extremely good performance – 150× better than SC-1

SEM AND ICP-MS CHARACTERIZATION

Process: Beaker dip/PETEOS/Particle C

Dark field microscopy

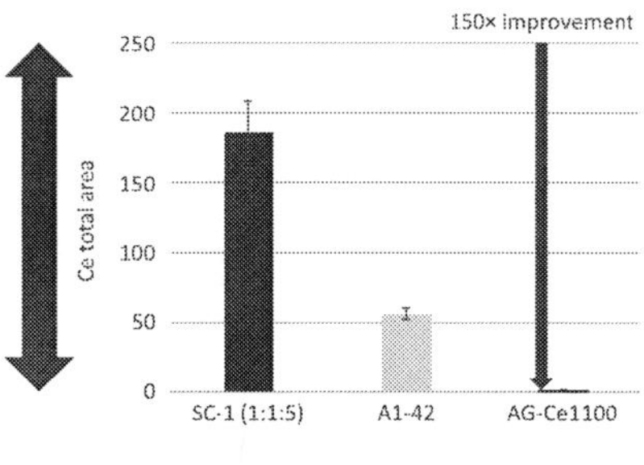

ICP-MS data

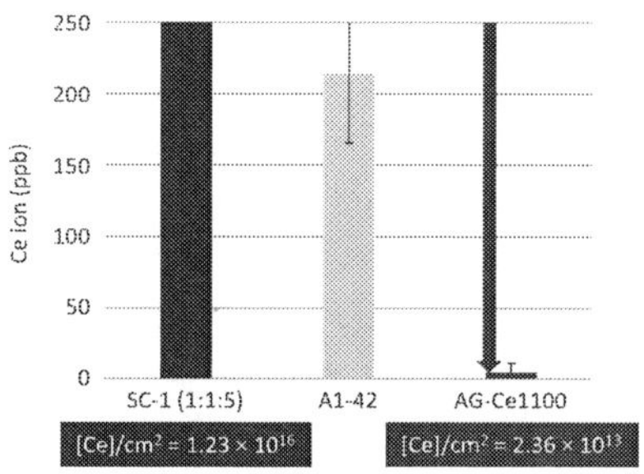

CMP/Ceria Slurry/
PlanarClean AG Ce-XXXX-2

- ICP-MS supports the dark field microscopy data and it shows >150× improvement over SC-1
- SEM data strongly supports DFM and ICP data

CONCLUSIONS

- Several low-pH and high-pH high-performance ceria cleaning formulation PlanarClean AG Ce-XXXX were developed at Entegris based on in-depth mechanistic understanding on silica-ceria surface interactions
- All ceria cleaning formulations contain complexing reagents, silica-ceria bond breaking reagents and dispersing reagents for particles agglomeration and re-deposition prevention
- Low-pH PlanarClean AG Ce-XXXX-2 and PlanarClean AG Ce-XXXX-3 formulations perform well on dielectric surfaces polished with both low- and high-pH ceria dispersions
- High-pH PlanarClean AG Ce-XXXX-1 ceria cleaning formulations perform best on dielectric surfaces polished with high-pH ceria dispersions
- We demonstrated that commodity cleaners such as hot SPM and SC-1 are the root cause for defective and damaged dielectrics surfaces, also highly contaminated with ceria aggregated particles
- PlanarClean AG Ce-XXXX formulations show improved ceria particles removal vs. commodities by as much as 150× (ICP-MS)

19 Entegris

ACKNOWLEDGEMENTS

- Robin Van Den Nieuwenhuizen – brainstorming and financial support
- Emanuel Cooper – brainstorming and consulting
- Changfeng Chen – Raman Spectroscopy
- Michele Stawasz – AFM and SEM-EDX
- Mike Deangelo – SEM Characterization
- Wonlae Kim – ICP-MS measurements
- Mike Owens – lab formulations, contact angle measurements
- Cuong Tran – marketing advice
- Fadi Coder – sales and marketing advice

20 Entegris

Entegris

Entegris
50 YEARS OF PURE ADVANTAGE

Mechanistic and Electrochemical Aspects of Copper Post CMP Cleaners for 5-7 nm Nodes

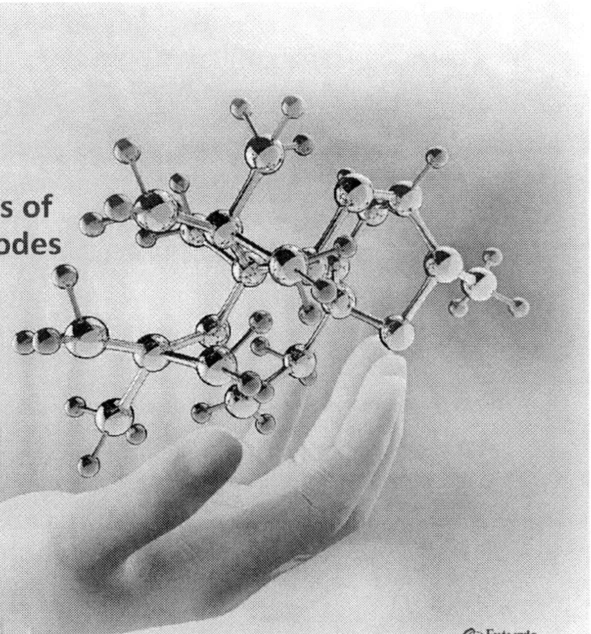

Michael White, Daniela White, Volley Wang, Jun Liu, Elizabeth Thomas, Don Frye, Ruben Lieten, Thomas Parson and Atanu Das

Entegris Surface Preparation and Integration
7 Commerce Dr.
Danbury, CT 06877
michael.white@entegris.com

WHAT IS CMP (CHEMICAL MECHANICAL PLANARIZATION)?

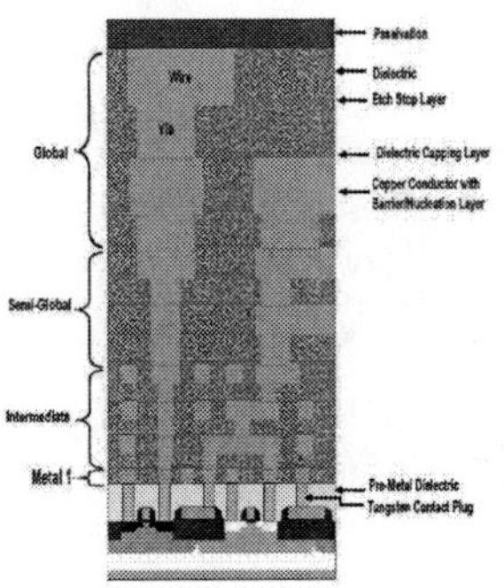

Source: International Roadmap for Semiconductors ITRS press conference, Dec 2004, 25.

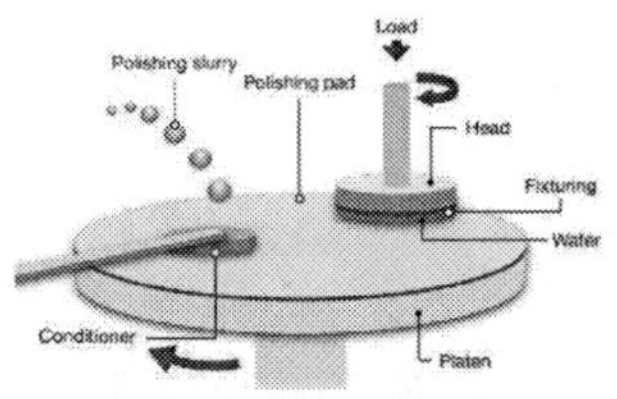

Source: http://www.azom.com/article.aspx?ArticleID=12527

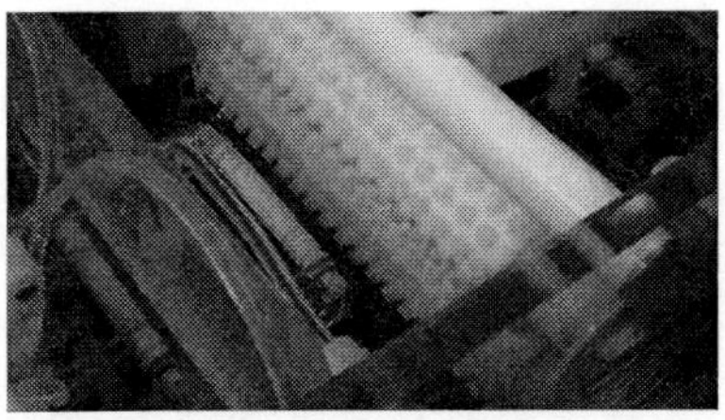

Layers/Materials that need to be Cleaned

1. Copper/barrier
   Ta, TaN, TiN, Co, Ru, Mn
2. Tungsten
3. Cobalt (bulk)
4. Aluminum
5. Dielectrics (including CeO_2 polishing)
   - TEOS, Si_3N_4, Low-k dielectric, SiC (SiOC, SiON, ...), Polysilicon, Single crystal silicon (wafer, various doping)

Challenges in Post CMP Cleaning

1. Variety of CMP slurry particle types
   - Silica (native, surface treated + or -)
   - Al_2O_3, CeO_2, TiO_2, ZrO_2, SiC, diamonds
2. Organic residue
   1. Corrosion inhibitors (BTA, ...)
   2. Dishing/erosion/selectivity additives (polymers, surfactants, small molecules)
   3. Rate additives
3. Pad debris
4. Plating additives

PERFORMANCE GOALS FOR POST-CMP CLEANERS

1. Best in class defectivity
    1. Very low particle defects (silica, ceria, alumina, ...)
    2. Greater challenges arising as particle sizes decrease
    3. Low organic residue (Cu-BTA or other thick film formers, W or other metal inhibitors, pad debris, plating additives, ...)
2. Very low or no interfacial or surface metal/barrier corrosion or recess
    1. Advanced nodes <10 nm
    2. Low galvanic corrosion
3. Uniform, smooth etching with low roughness
    1. Affects thresholds for defectivity measurements
    2. No attack on low-k dielectric/dielectric loss
4. Low metallic impurities on wafer (<10^{10} atoms/cm^2) on dielectrics
5. Good buffering/no brush interactions to avoid ring scratches

THE RATIONAL DESIGN OF A POST CMP CLEANER PLANARCLEAN® AG COPPER CLEANING

PlanarClean® AG – Advanced Generation Copper Cleaning Mechanism

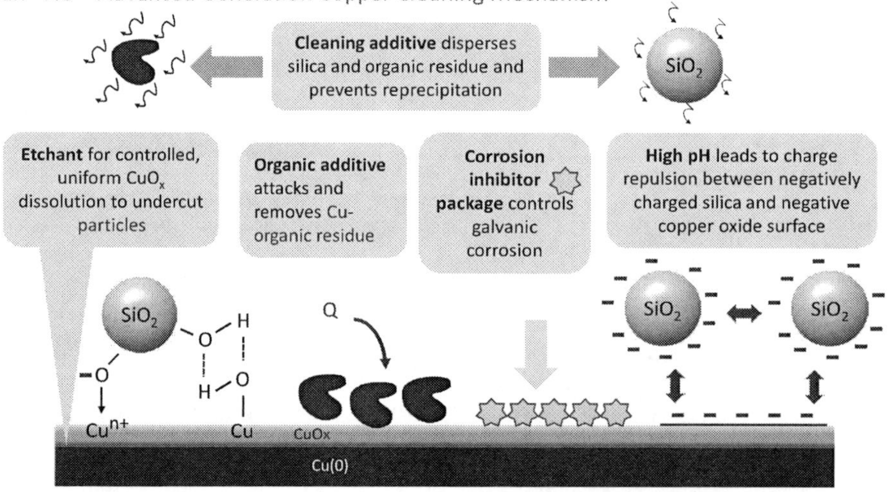

PLANARCLEAN® AG PREVENTS SILICA AGGREGATION

- Particle adhesion mechanisms
  - **Physisorption** (van der Waals attraction – increases with PS)
  - **Electrostatic** attraction or repulsion (zeta potential)
  - **Chemisorption** (chemical reaction particle-surface)
  - **Capillary condensation**

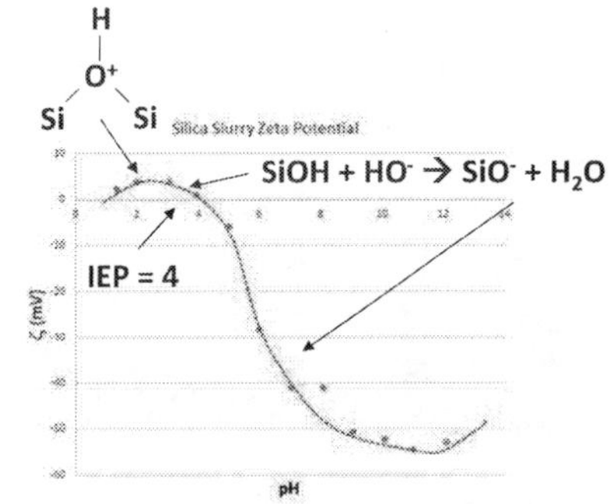

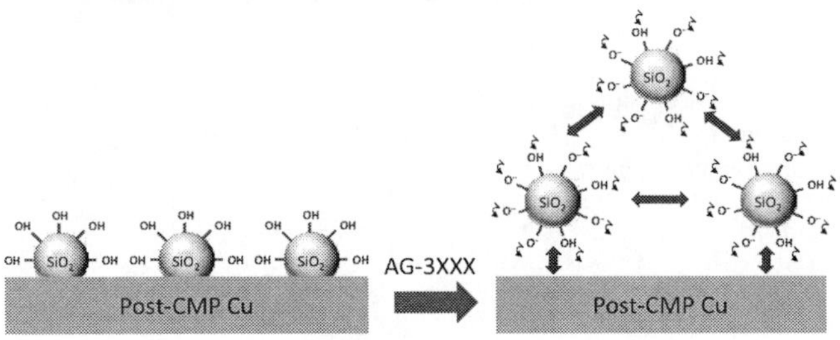

Zeta Potential

$\zeta = 4\pi\gamma(\nu/E)/\varepsilon$

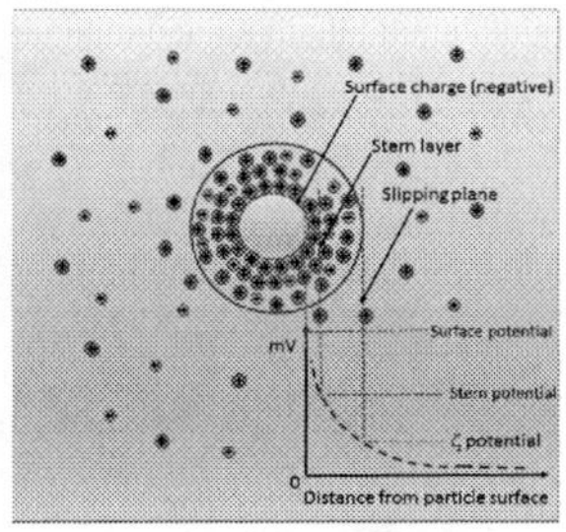

DEFECTIVITY CORRELATED TO CHARGE REPULSION BETWEEN SILICA PARTICLES

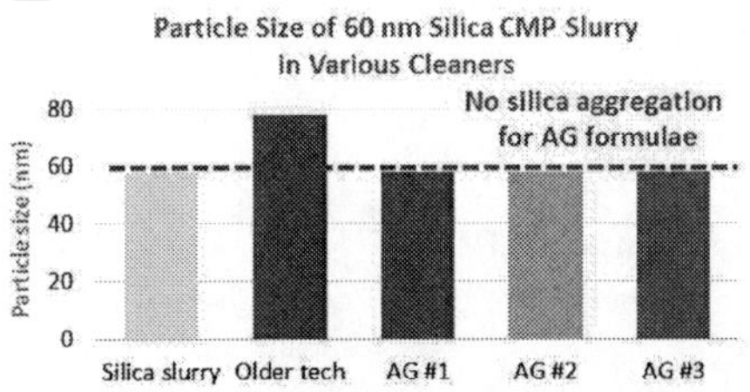

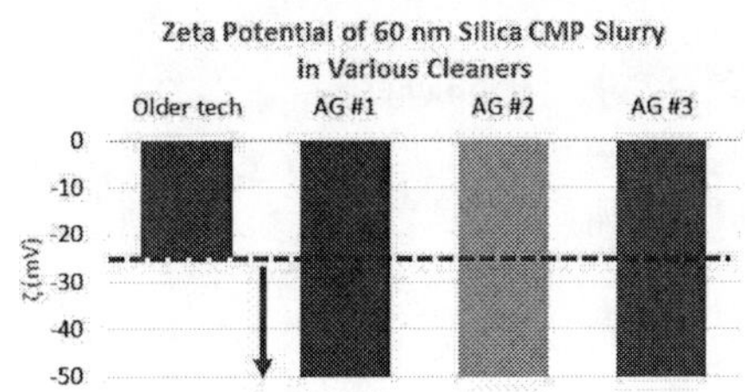

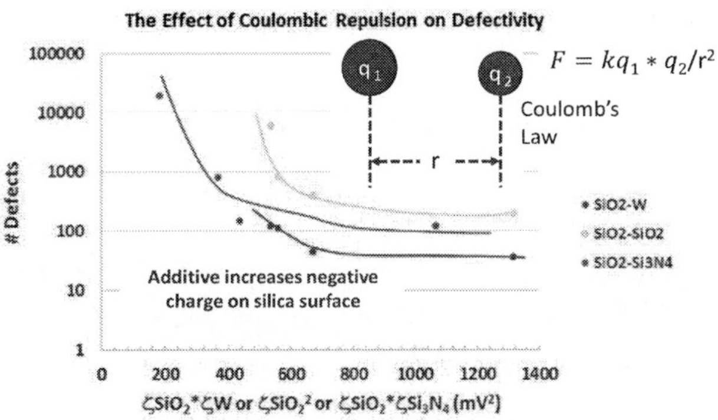

Zeta potential of all AG formulations is highly negative

1. White, M. L. et al, Mater. Res. Soc. Symp. Proc. 991, 0991-C07-02 (2007)
2. Hedge, S. and Babu, H. V. 2Eelectrochem. Soc. St. Lett. V7, pp. 316-318 (2008)
3. White, M. L. et al. Mat. Sc. For. 1249 E04-07 (2010).

BREAK-UP AND DISPERSION OF Cu(I)-BTA COMPLEXES

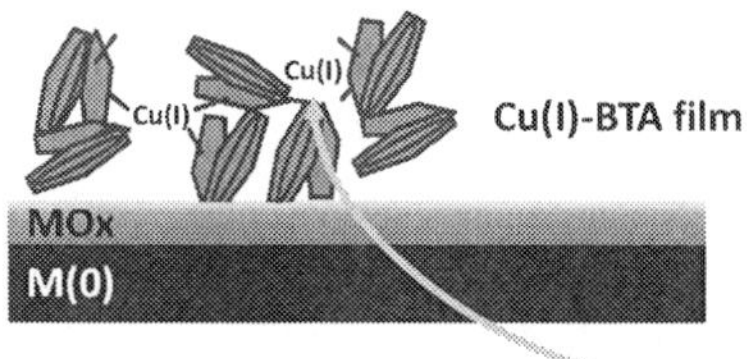

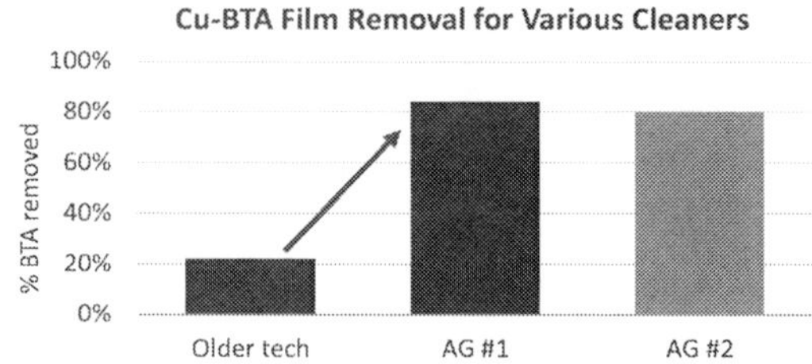

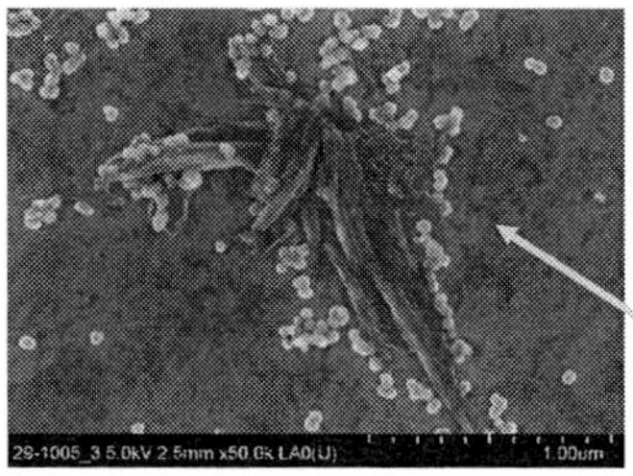

Additive tailored to attack Cu(I)BTA and similar films

- Fast kinetics
- Thermodynamically favored
- (Higher Cu binding constant than BTA)

Cu(I)-BTA can redeposit if Cu is not properly complexed or dispersed

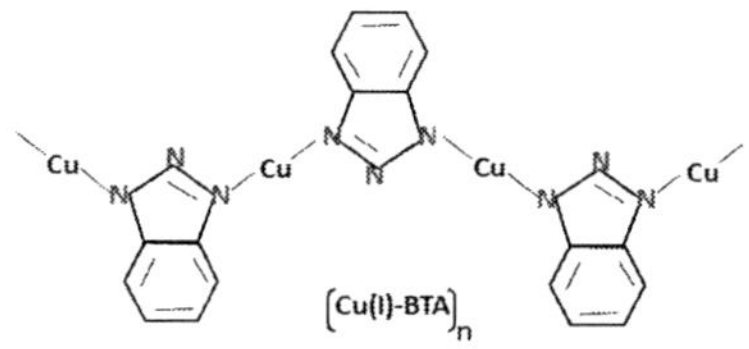

UV-VIS USED TO PREDICT OPTIMUM COMPLEXANT AND LIGAND CONCENTRATION

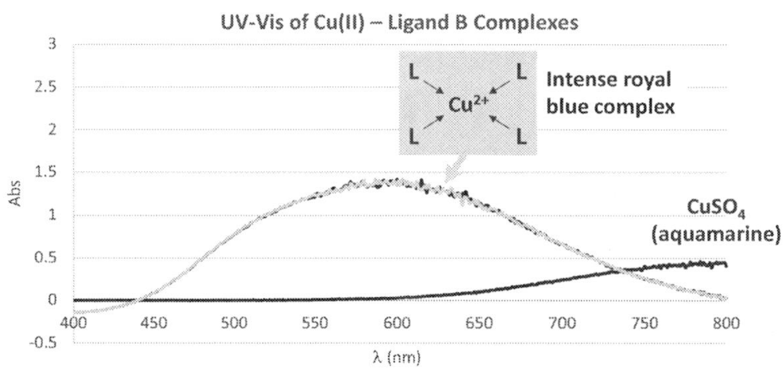

Ligands complex copper as soluble Cu(II) and prevent redeposition as CuOx defects or reprecipitating BTA

PLANARCLEAN AG: LOWER DEFECTS THAN COMPETITORS

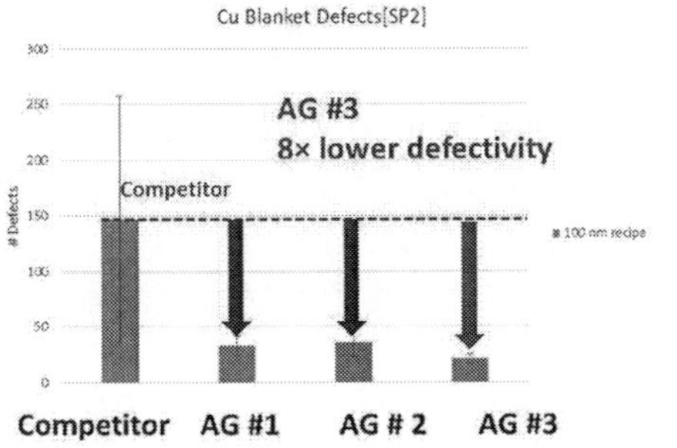

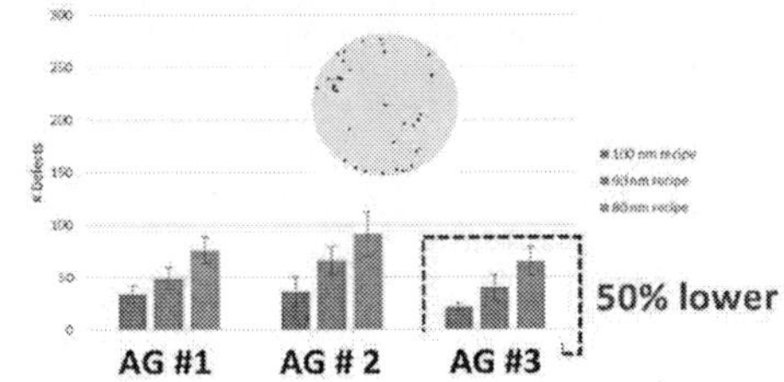

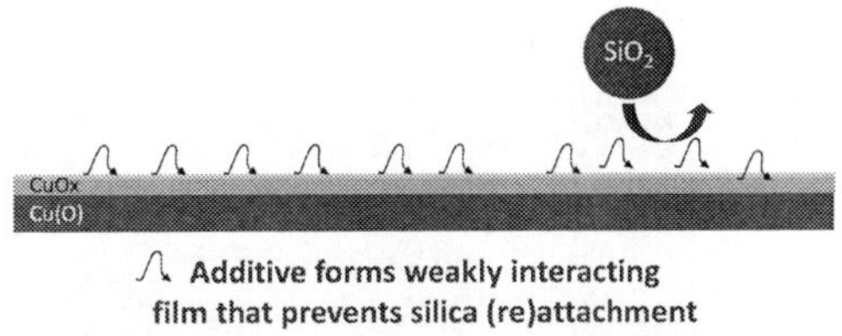

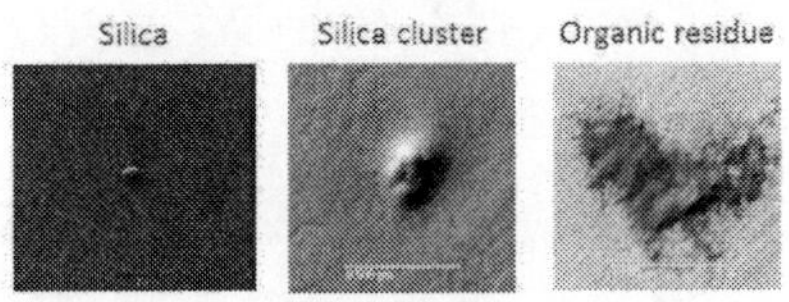

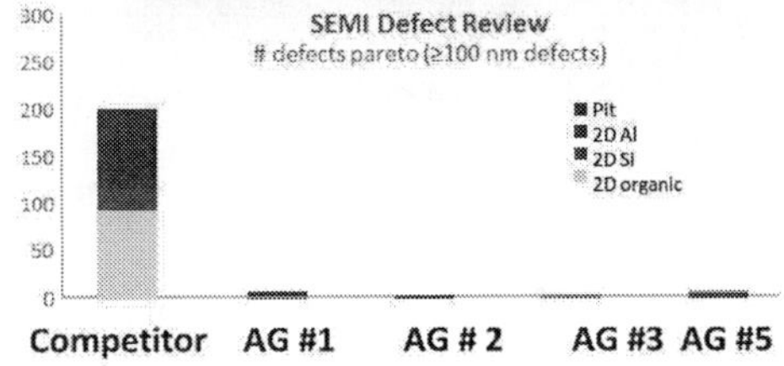

Entegris

ELECTROCHEMISTRY REVEALS PLANARCLEAN® AG EXHIBITS IMPROVED CORROSION PERFORMANCE

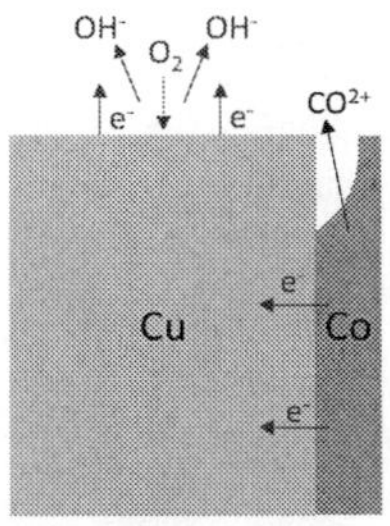

$$Co \rightarrow Co^{2+} + 2e^-$$

$$0.5O_2 + H_2O + 2e^- \rightarrow 2OH^-$$

Co OCP < Cu OCP-:
Co not protected

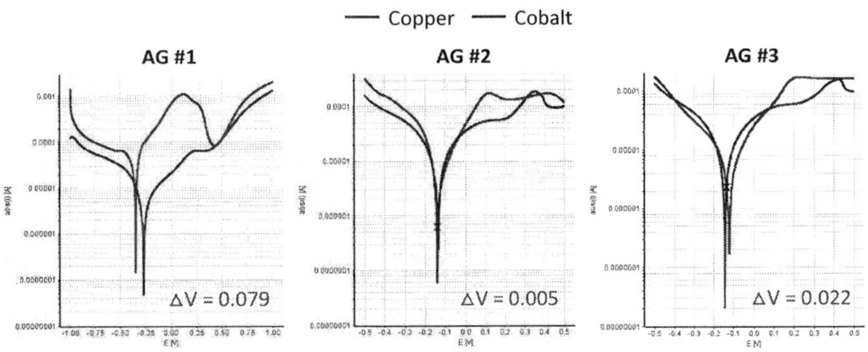

AG #1	AG #2	AG #3
Co protected (Cu 1.210 Å/min)	Nearly zero galvanic corrosion (Cu 0.078 Å/min)	Very low galvanic corrosion (Co 0.270 Å/min)

Controlled Electrochemical properties
- ✓ **Ligands to control potential gap**
- ✓ **Passivation to modify resistivity**

Entegris

IMPEDANCE SPECTROSCOPY SHOWS THAT AG COPPER INHIBITOR IMPROVES Cu PASSIVATION

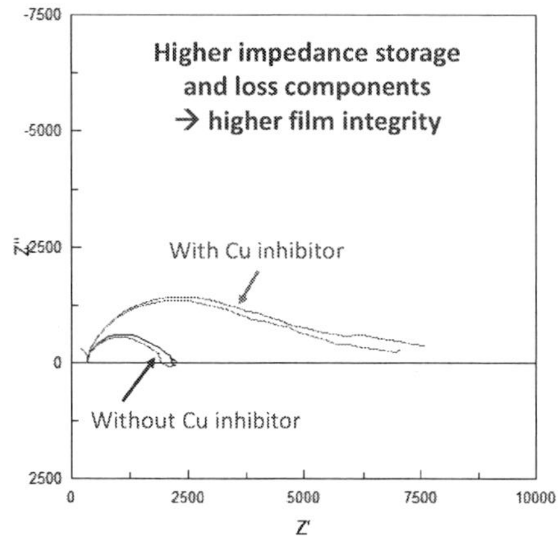

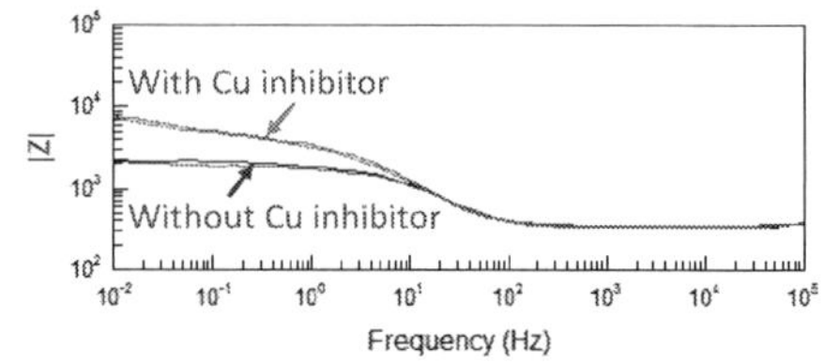

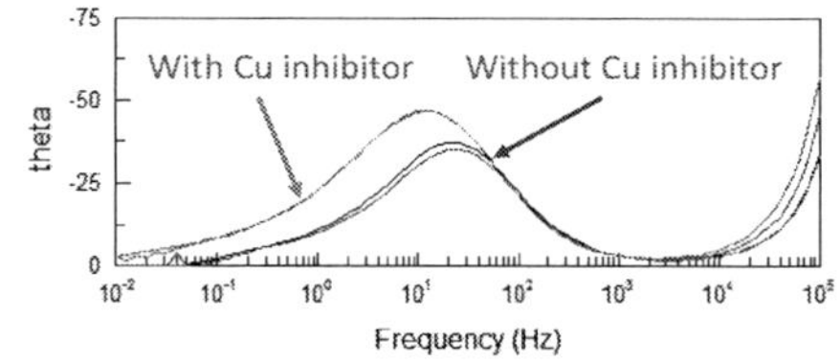

Entegris

PLANARCLEAN® AG FORMULATIONS PROVIDE BETTER PASSIVATION ON BOTH Cu AND Co

Impedance Spectroscopy

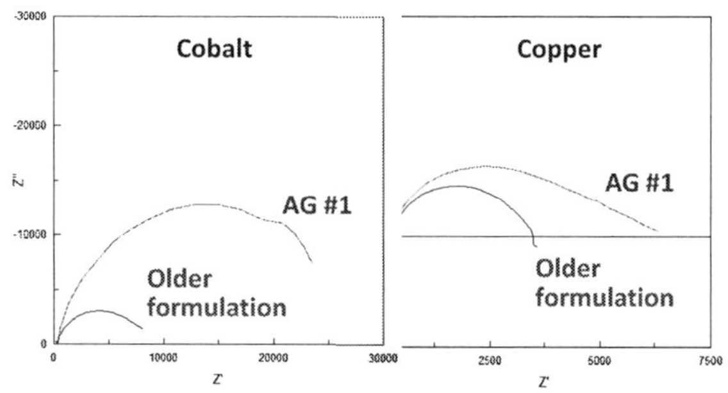

Additional Novel Cu Inhibitor Improves Cu Passivation

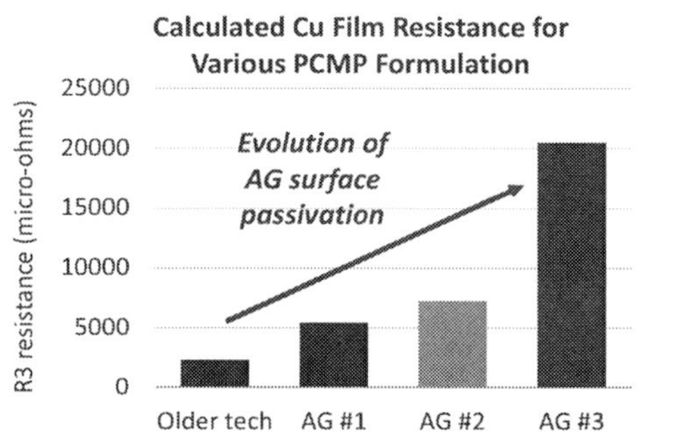

When ω -> 0

$$Z' = R_{\Omega} + \frac{R_{ct} + \sigma\omega^{-1/2}}{(\sigma\omega^{1/2}C_{dl}+1)^2 + \omega^2 C_{dl}^2 (R_{ct} + \sigma\omega^{-1/2})^2}$$

$$Z'' = -\frac{\omega C_{dl}(R_{ct} + \sigma\omega^{-1/2})^2 + \sigma^2 C_{dl} + \sigma\omega^{-1/2}}{(\sigma\omega^{1/2}C_{dl}+1)^2 + \omega^2 C_{dl}^2 (R_{ct} + \sigma\omega^{-1/2})^2} a$$

Ref:
1. Wang, et al. SPIE Beijing 2016 Conf. Proc.
2. Bard, A. J. Faulkner, L. R. Electrochemical Methods: Fundamentals and Applications; Wiley and Sons 2001

Entegris

ELECTRON MICROSCOPY SHOWS SIGNIFICANTLY IMPROVED Cu/Co CORROSION PERFORMANCE FOR PLANARCLEAN® AG

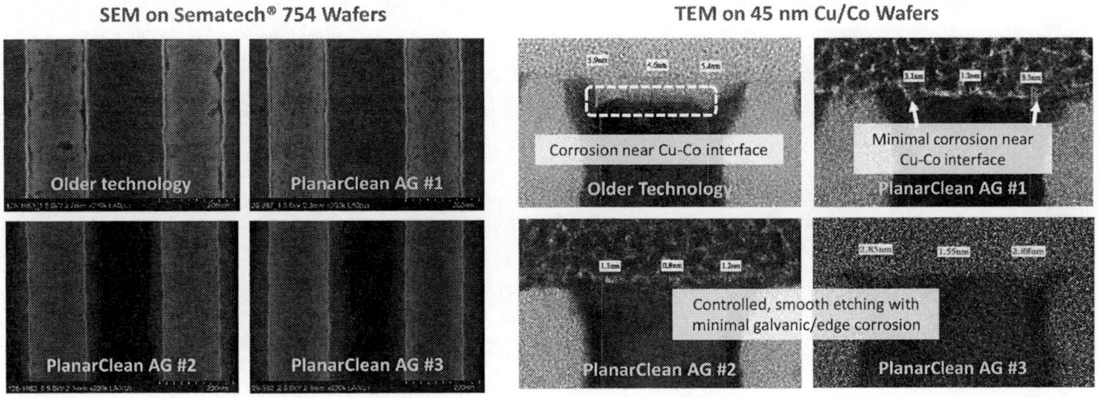

Sematech® is a trademark of Sematech, Inc.

IMPROVED CORROSION ADDITIVES REDUCE SURFACE ROUGHNESS BY 2-4×

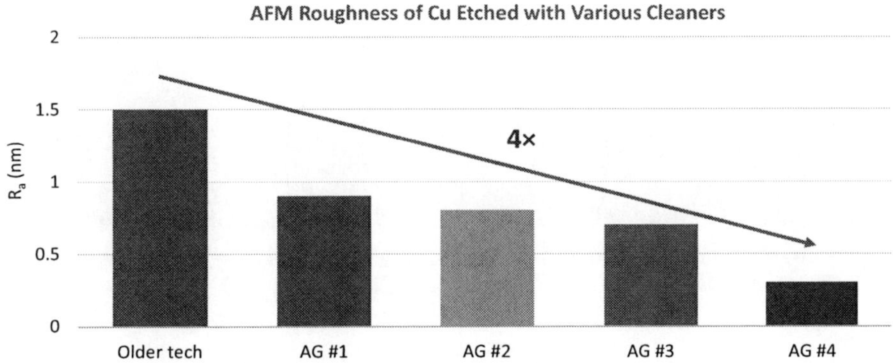

CONCLUSIONS

- Charge repulsion shown to be a key driver towards cleaning performance
- Rate of attack on Cu(I)-BTA polymer, dispersion and complexation important for removal and preventing re-deposition of organic residues and particles
- OCP gap must be minimized by optimal ligand selection to minimize galvanic corrosion
- Impedance spectroscopy and Tafel plots have been used to Optimize corrosion inhibiting package
- Very low Cu roughnesses can be obtained with the correct inhibitor (4 Å)

Chemical Mechanical Planarization
STACK TRECK

SPCC 2017 | Viorel Balan
Viorel.balan@cea.fr

> 50 years of Moore's Law
Stacking Is The New Scaling

Lithography Enables Scaling / CMP Enables Stacking

Building Stacked Devices by Hybrid Direct Bonding

Challenges

Direct Bonding → Roughness & Clean SiO_2 & Cu Non-Patterned Wafers

Hybrid Direct Bonding → Patterned Wafers Topography: Local/Global

Conclusion

---

leti ceatech

Stacking Heterogenous Devices

Application to 3D Stacked Back Side Imagers

ST life.augmented

3D stacked BSI Imager

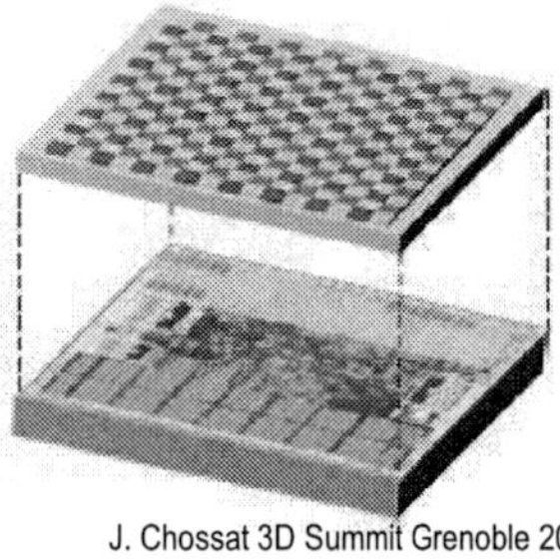

J. Chossat 3D Summit Grenoble 2016

Top die

- Pixels

Bottom die

- Analog & Readout
- Sensor Logic
- And more !!!

Wafer to Wafer Stacking by Hybrid Direct Bonding Technology

→ Top Die Contains only Pixels

→ Passive Substrate replaced by Advanced Digital CMOS wafer

Manufacturing Flexibility →

Different Technology Nodes (**Design Rules**) for different components

Stacking Heterogenous Devices

3D Stacked BSI by Hybrid Direct Bonding

J. Chossat 3D Summit Grenoble 2016

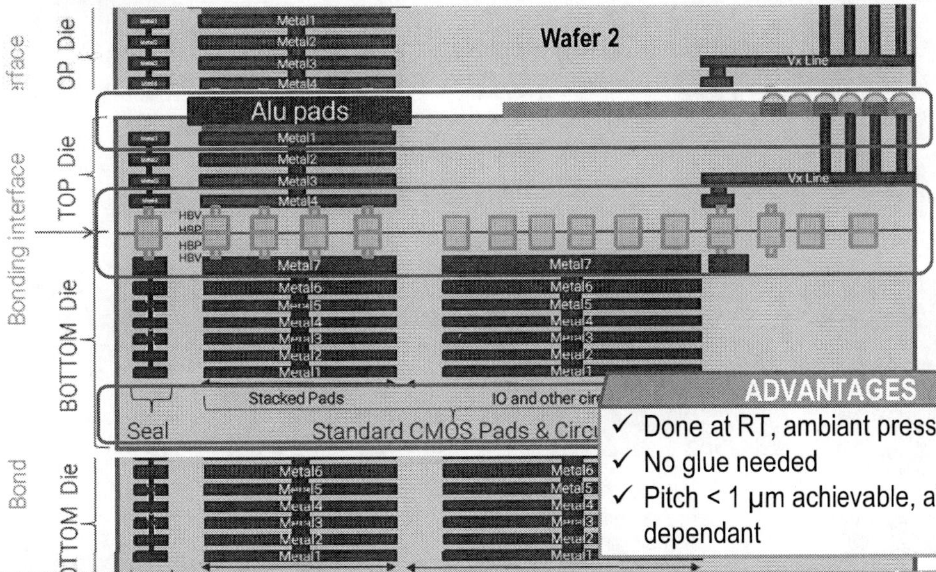

ADVANTAGES

- ✓ Done at RT, ambiant pressure
- ✓ No glue needed
- ✓ Pitch < 1 µm achievable, alignment dependant

The key technology step for 3D stack is the interconnection of the upper and lower part of the circuit.

From Direct Bonding

To Hybrid Direct Bonding

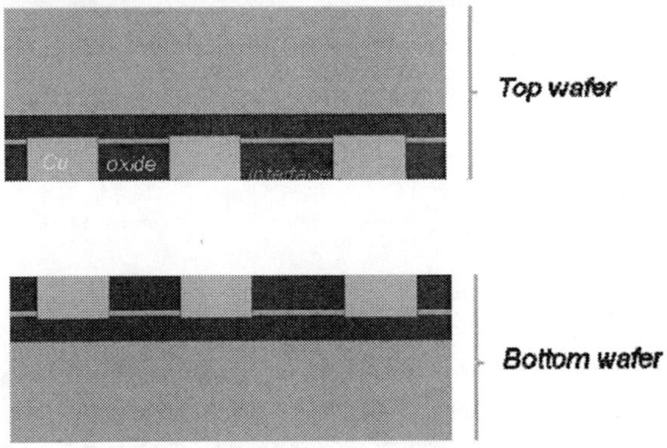

Dielectric
Mechanical Strength
SiO_2

+

Metal
Electrical connection
Cu: microelectronic

- Cu → most promising 3D integration candidate
  - Main metal used for CMOS interconnects
  - Cost of ownership
- Copper/oxide surface direct bonding advantages:
  - Very high interconnect density possible.
  - Low-temperature process
  - Compatibility FEOL/BEOL requirements for a sequential approach

Chemical Mechanical Planarization

Challenges for Recreating The Bulk from 2 **Mixt** Surfaces

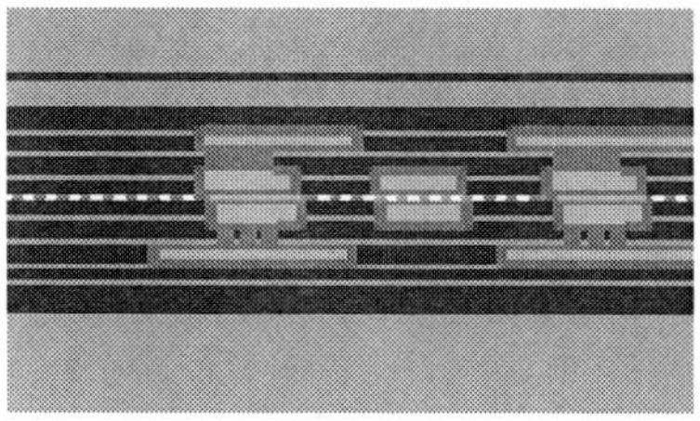

Bond 2 Surfaces with
- Cu
- Barrier
- Dielectric
- → Manage Topography

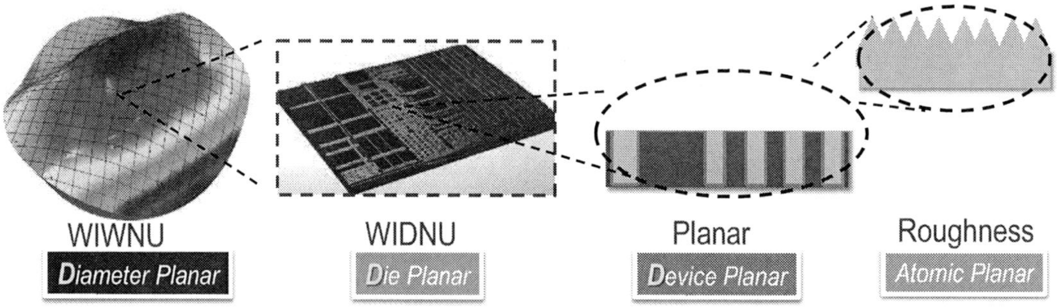

Intimate Surface Contact Needed → Planarization @all Spatial λ

Chemical Mechanical Planarization

Challenges for Recreating The Bulk from 2 Surfaces

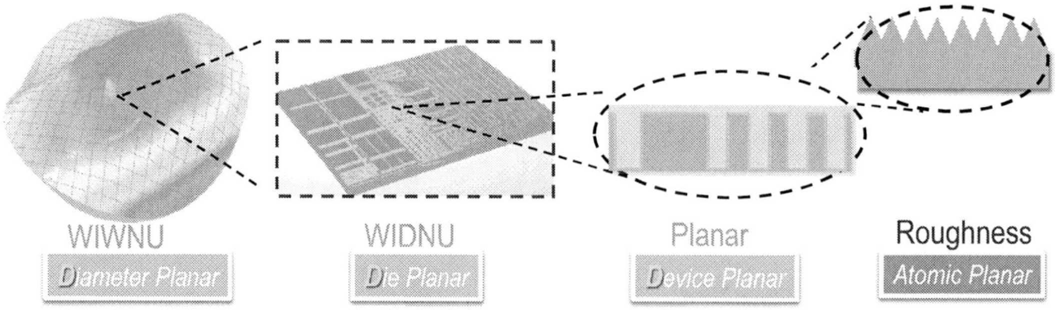

Direct Bonding SiO_2

Roughness Impact

F. Rieutord, ECS 2006

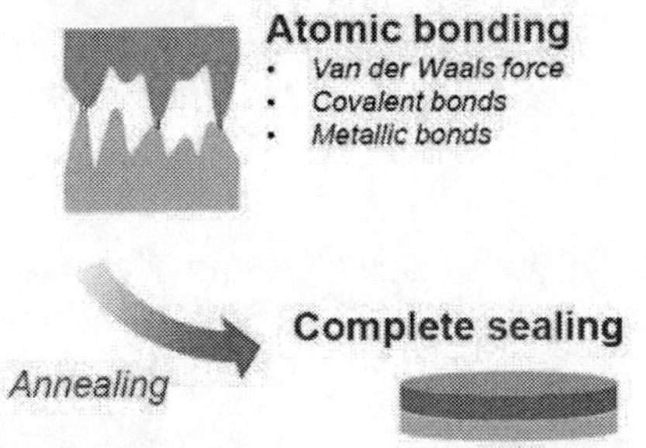

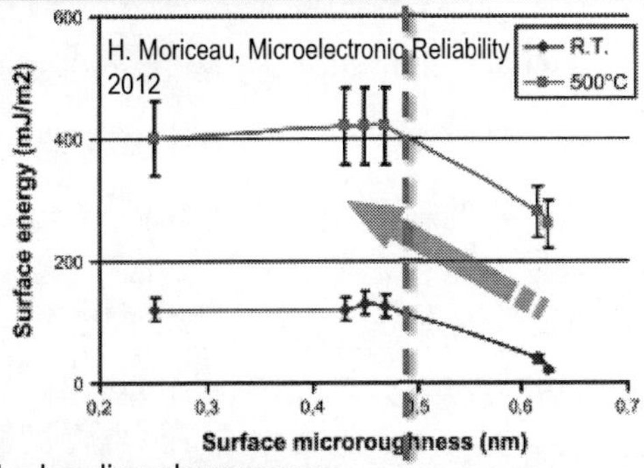

Surface high frequency micro-roughness → key role in the bonding phenomenon,
Low frequency roughness can be accommodated by deformation of each substrate (elastic energy price)

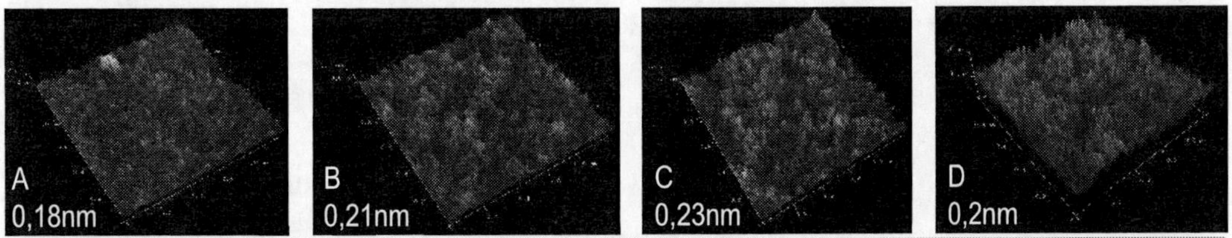

SiO_2/SiO_2 Blanket Bonding RMS<0.5nm

Direct Bonding

Cu-Cu

Roughness

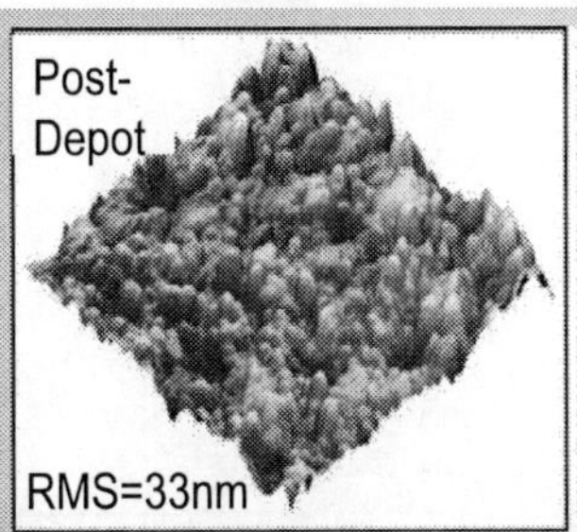

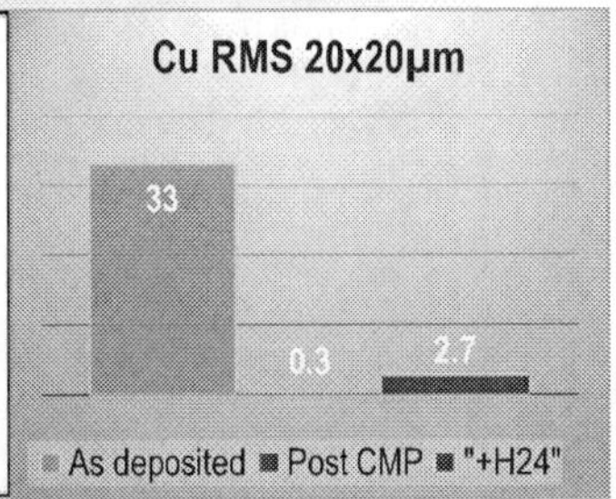

Hydrophilicity

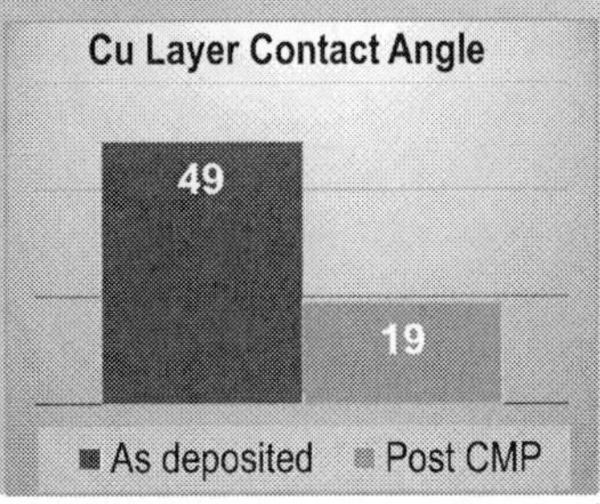

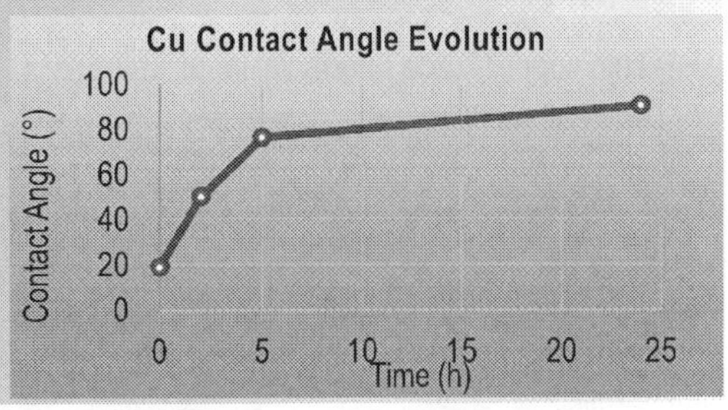

P. Gueguen ECS 2008

RMS & Contact Angle Increase → Q Time Impact

Direct Bonding

Cleaning Influence

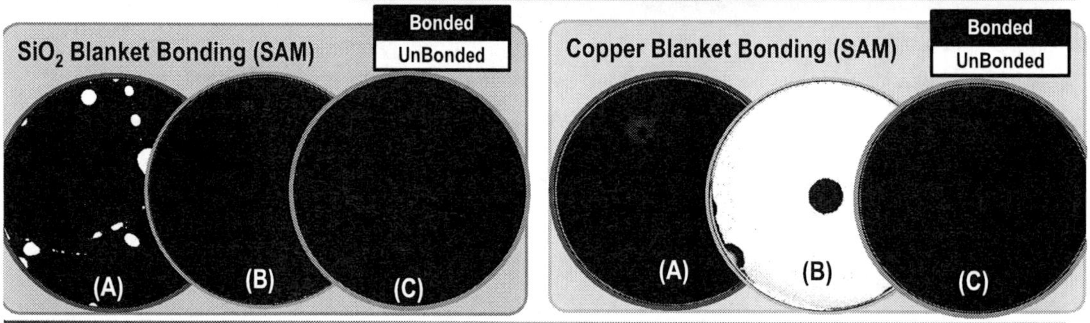

Cleaning Solution Adapted both to Cu & SiO_2

Direct Bonding

Cu-Cu Interface Evolution=f(T°C)

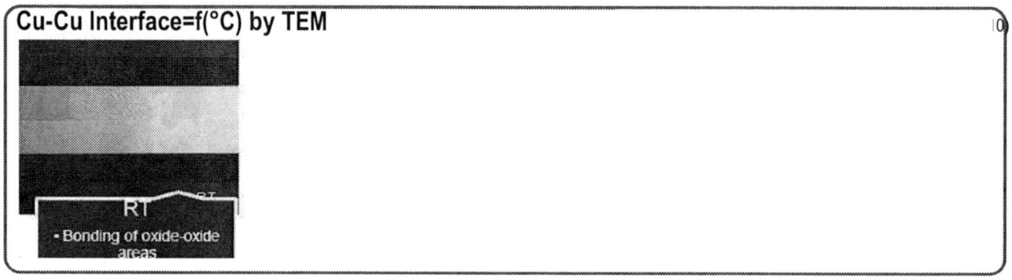

- Thin copper oxide interfacial layer at the bonding interface
- 200°C, this copper oxide becomes thermodynamically unstable → grain growth→ sealing: XRR
- Bonding interface turns into a grain boundary with a high bonding energy

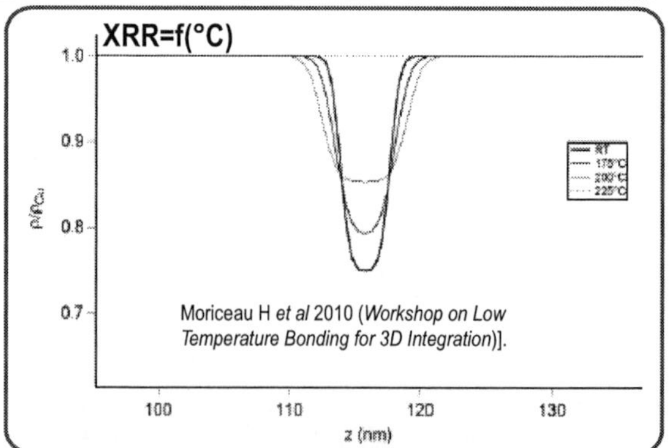

Chemical Mechanical Planarization

Challenges for Recreating The Bulk from 2 Surfaces

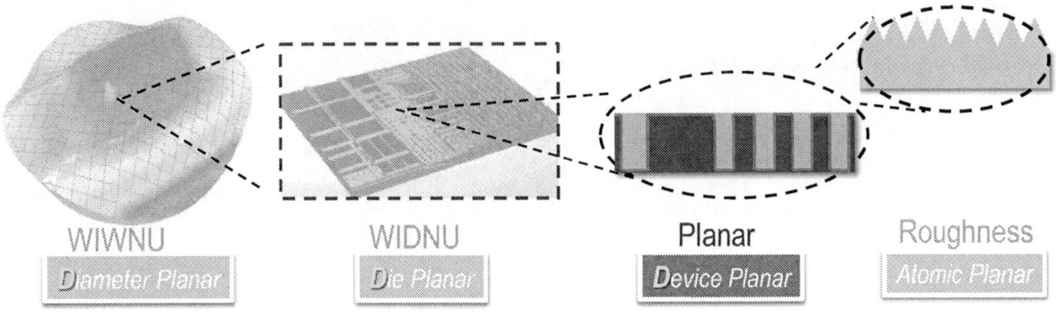

Hybrid Direct Bonding

Local Topography Impact Modelisation

- Dishing critical: topography doubled → negative impact electrical contact

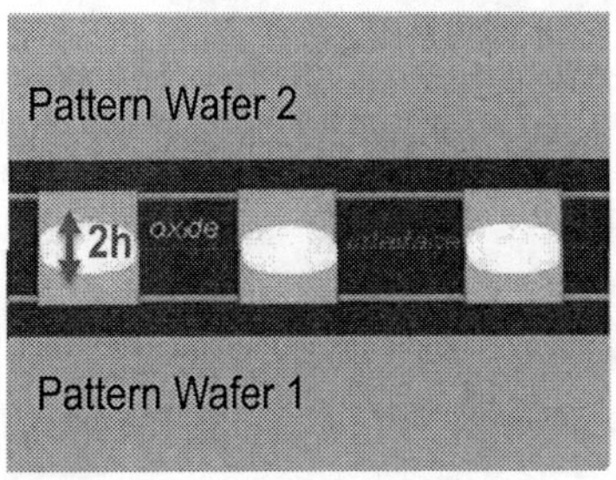

- Cohesive tractions effective below threshold value

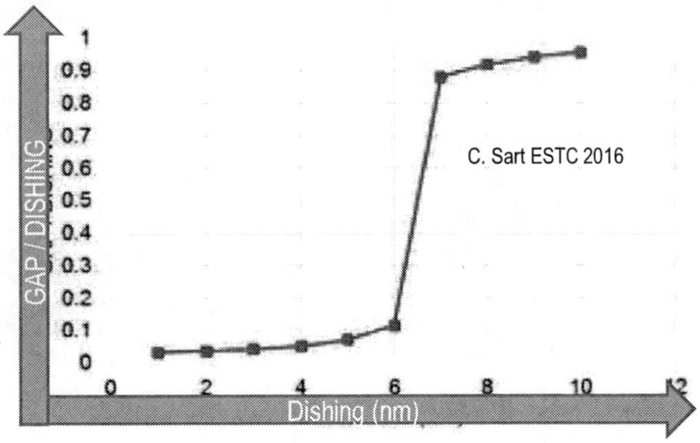

Hybrid Direct Bonding

Local Topography Impact Modelisation

- Dishing critical: topography doubled → negative impact electrical contact

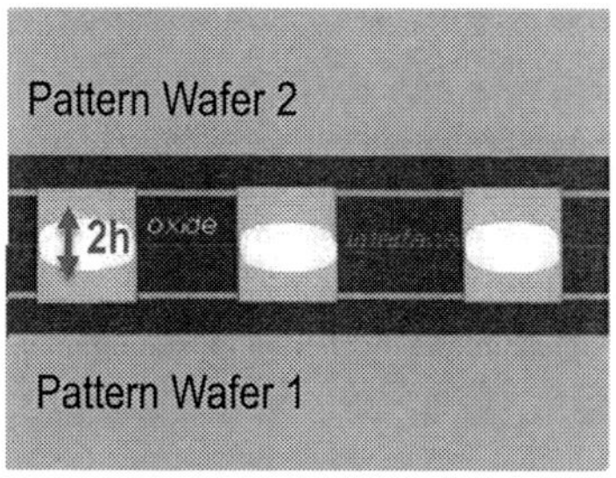

- Small dishing of Cu pads overcome during post bond annealing even @ low T°C: 200°C

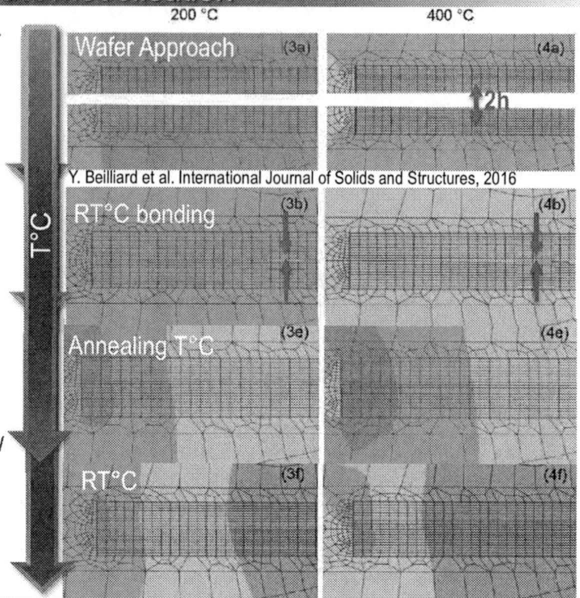

Cu/Cu Interface Remains Closed after Cooling Down to Room Temperature Agreement w/ Experimental Results

Hybrid Direct Bonding

Predictive Model

Model

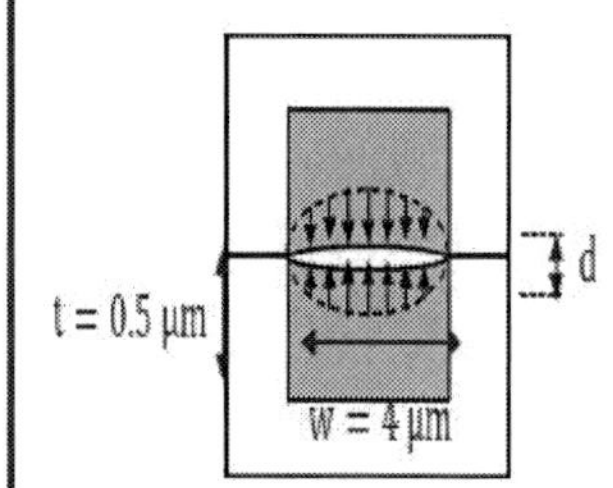

In case of initial dishing in Cu pads

► Definition of the gap after thermal budget

Thermal budget post bonding

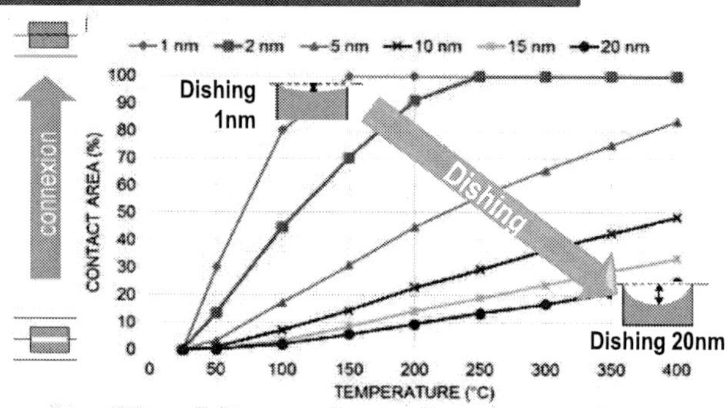

Forecast on % pad closure depending on anneal temperature

- Clues for process adaptation with dishing,
- Further models with introduction of grain size and crystal orientation.

S. Lhostis European 3D Summit 2017 C. Sart ESTC 2016

Predictive Model for the Process of Hybrid Bonding Layer

Hybrid Direct Bonding

Patterned Wafers DOE

- >250 process conditions used → pads, slurries, %abrasif %/oxydant, V,P, flow, time…

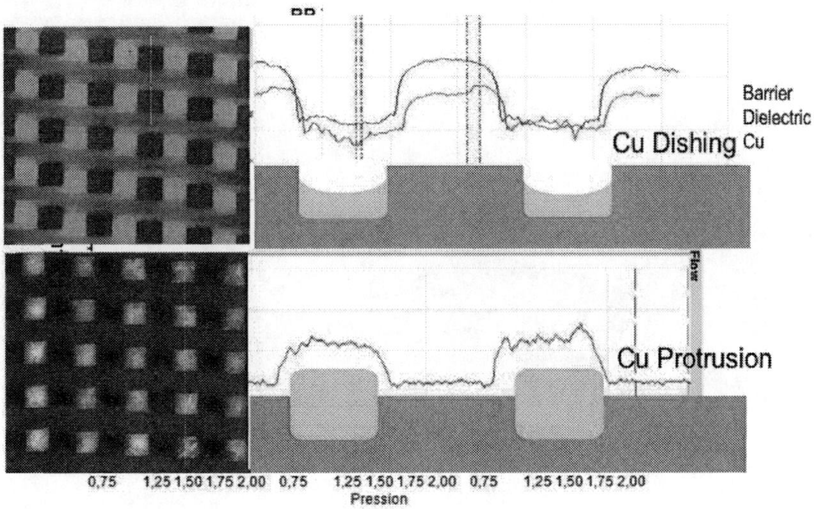

Hybrid Direct Bonding

Process Optimization

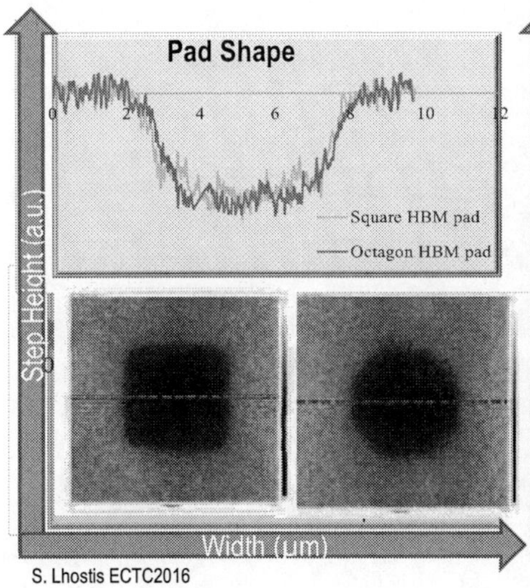

S. Lhostis ECTC2016

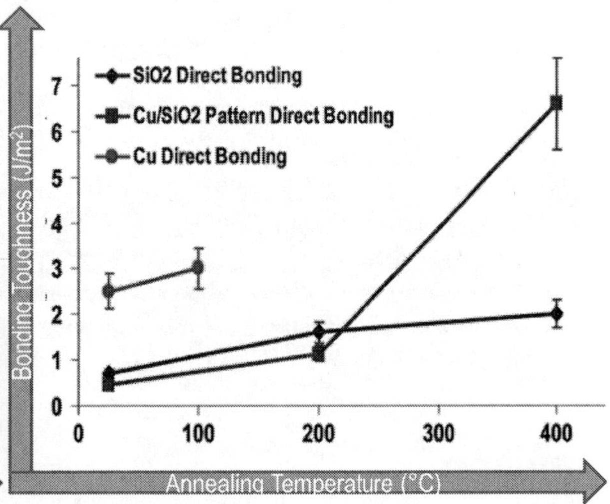

L. Di Cioccio J. Electrochem. Soc,158, 2011

Very Low Dishing, Uniform vs Cu Pad Shapes & Sizes
High Bonding Toughness Obtained

Chemical Mechanical Planarization

Challenges for Recreating The Bulk from 2 Surfaces

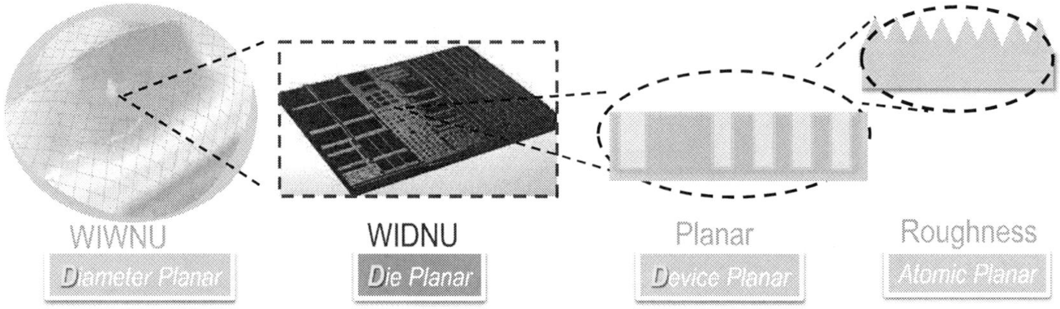

Stacking Heterogenous Devices

Dimension & Density Impact

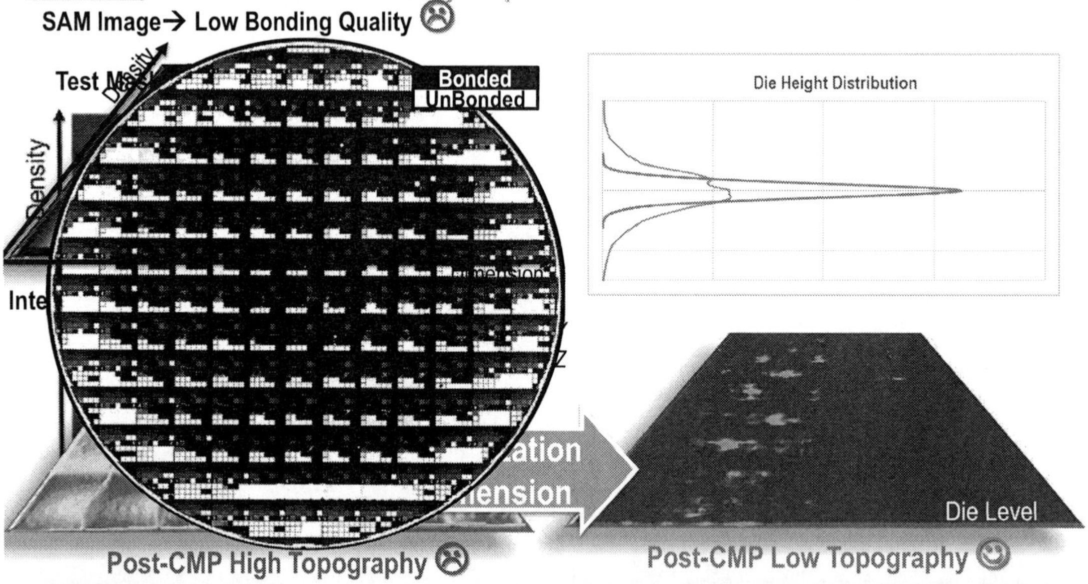

Atomically Flattened Surfaces thru Design Optimization
→Excellent Hybrid Direct Bonding Quality

Chemical Mechanical Planarization

Challenges for Recreating The Bulk from 2 Surfaces

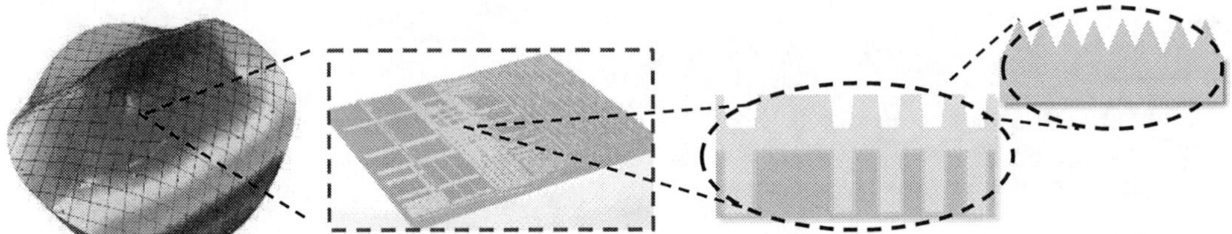

WIWNU	WIDNU	Planar	Roughness
Diameter Planar	*Die Planar*	*Device Planar*	*Atomic Planar*

Hybrid Direct Bonding

Patterned Wafers DOE

- Pattern to pattern → 3 wafers/investigated point: 2 bonded, 1 characterized →Pattern (PW) to non-patterned (NPW) bonding →validation PW to PW

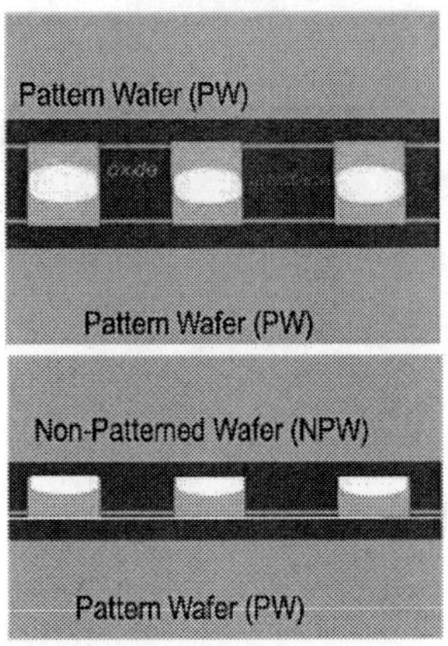

Process Conditions Adjustement

VP/Flow	Time	SiO2/Cu Selec	Post-CMP Topo	PW/NPW	PW/PW
h/a/z	a	a	c		
	b		a		
	c		b		
k/b/y	d	b	d		
	e		a		
k/c/y	f	c	e		
	e		a		
l/b/y	a	d	c		
	b		b		
l/c/y	b	e	c		
k/b/z	c	f	a		
	d		c		

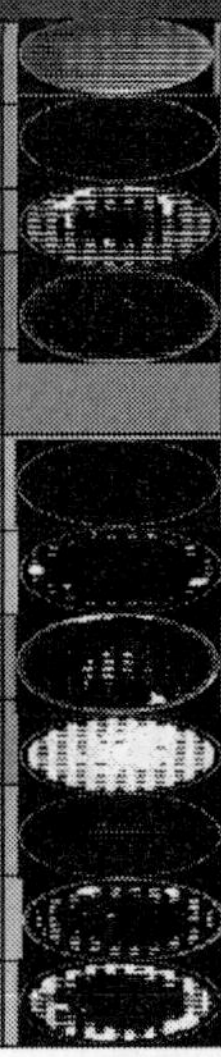

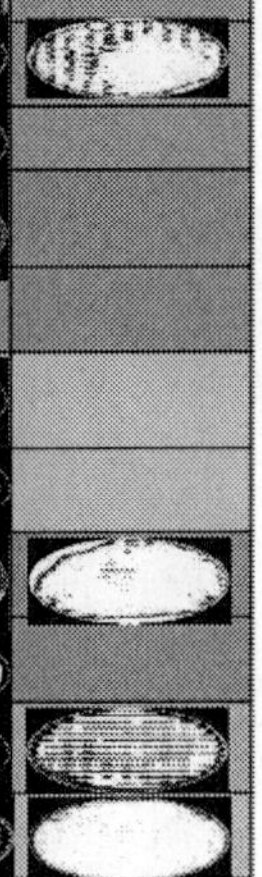

Costly Process in Patterned Wafers

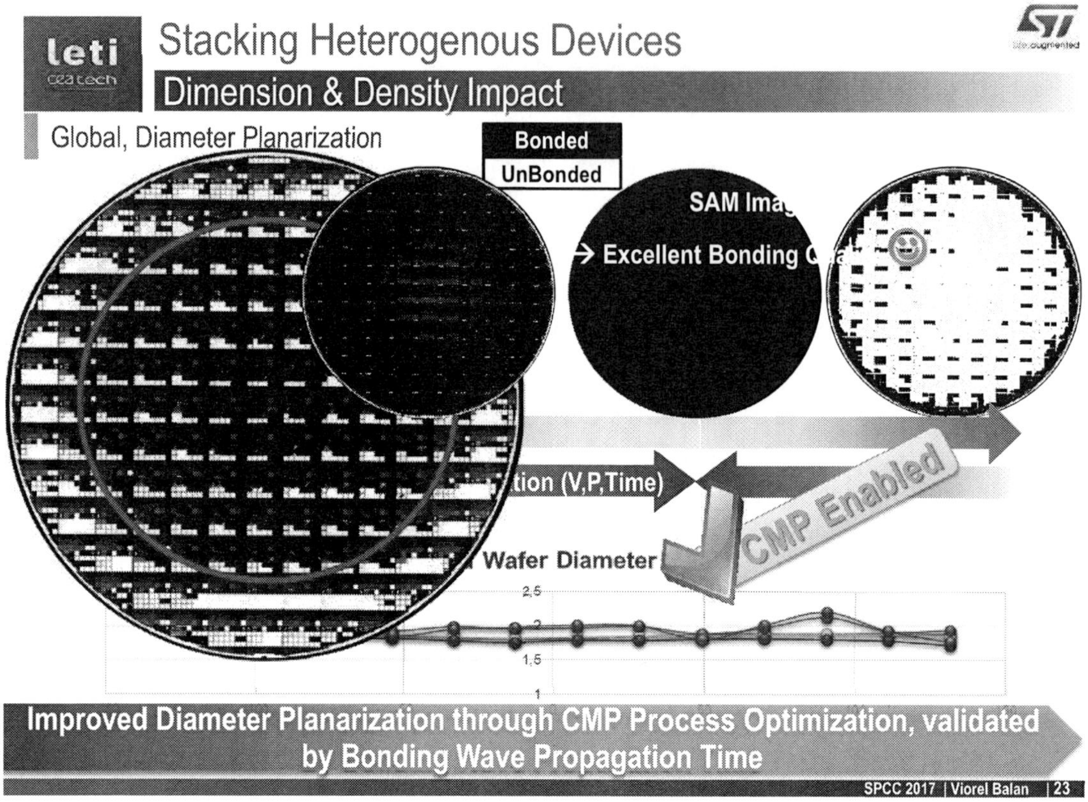
leti
cea tech
Stacking Heterogenous Devices
Dimension & Density Impact
Global, Diameter Planarization
Bonded
UnBonded
SAM Ima
→ Excellent Bonding
tion (V,P,Time)
Wafer Diameter
CMP Enabled
2,5
2
1,5
1
Improved Diameter Planarization through CMP Process Optimization, validated by Bonding Wave Propagation Time
SPCC 2017 | Viorel Balan | 23

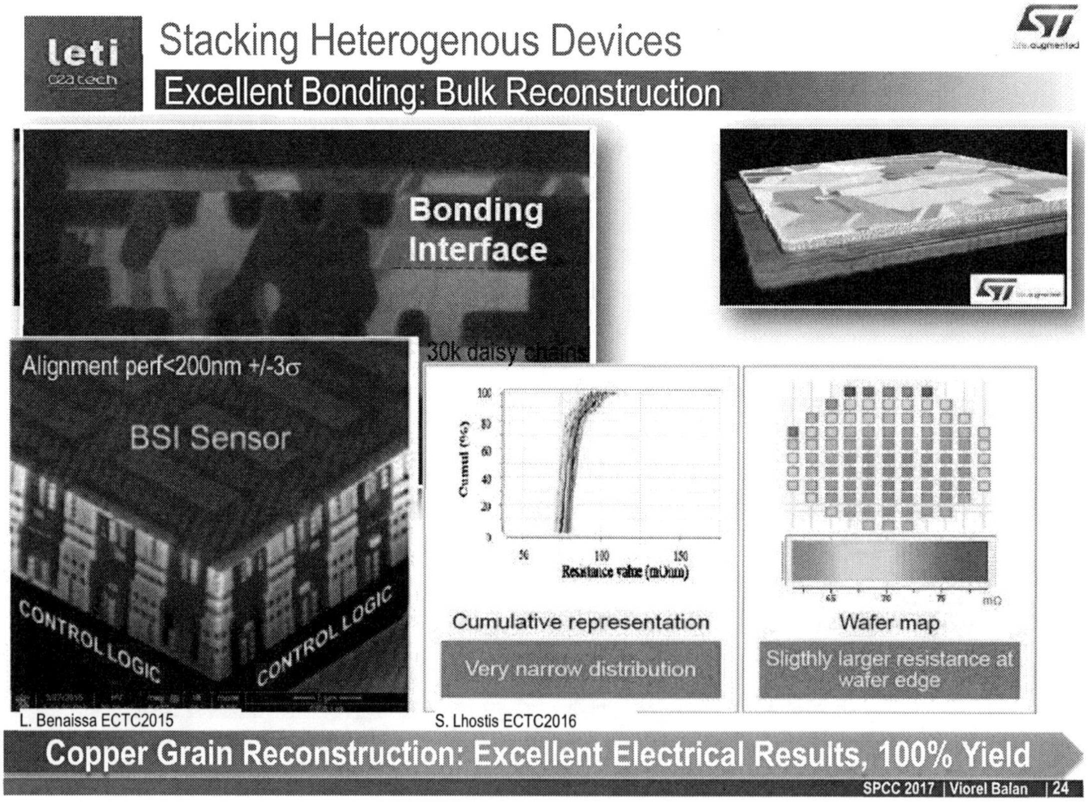
leti
cea tech
Stacking Heterogenous Devices
Excellent Bonding: Bulk Reconstruction
Bonding
Interface
30k daisy chains
Alignment perf<200nm +/-3σ
BSI Sensor
CONTROL LOGIC
CONTROL LOGIC
Cumul (%)
Resistance value (mOhm)
Cumulative representation
Very narrow distribution
Wafer map
Slightly larger resistance at wafer edge
L. Benaissa ECTC2015
S. Lhostis ECTC2016
Copper Grain Reconstruction: Excellent Electrical Results, 100% Yield
SPCC 2017 | Viorel Balan | 24

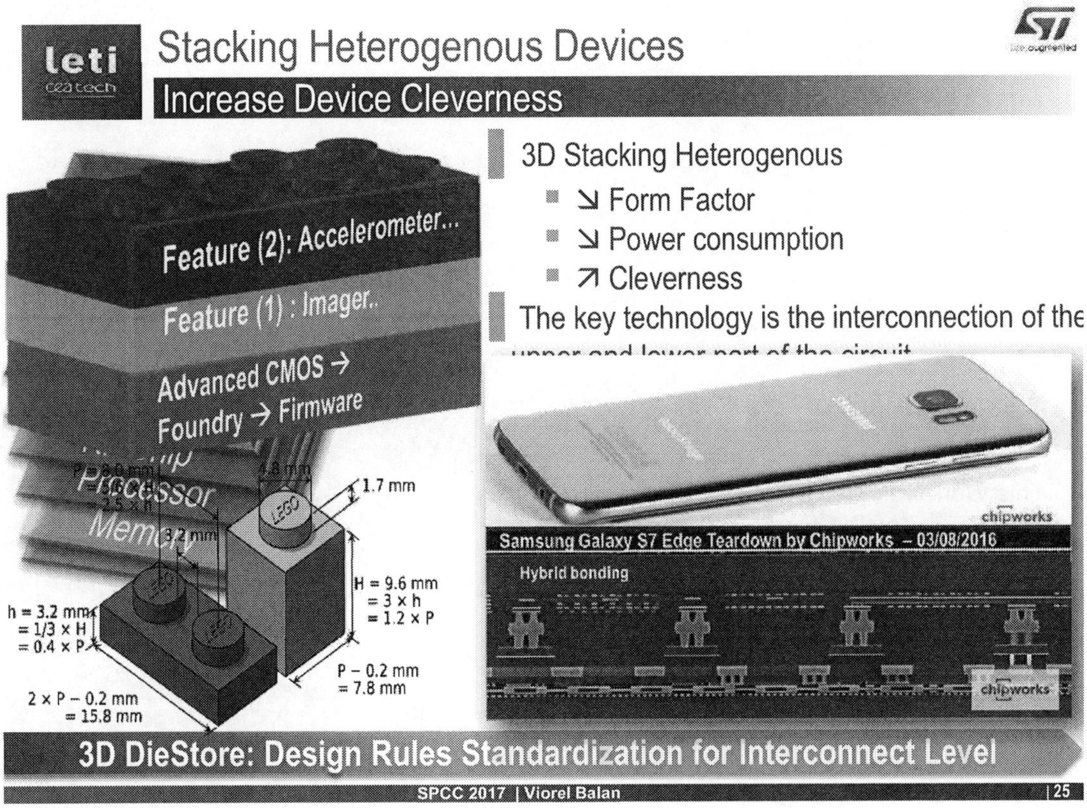

leti
cea tech
Stacking Heterogenous Devices
Increase Device Cleverness
3D Stacking Heterogenous
↘ Form Factor
↘ Power consumption
↗ Cleverness
The key technology is the interconnection of the
Feature (2): Accelerometer...
Feature (1) : Imager..
Advanced CMOS →
Foundry → Firmware
Processor
Memory
4.8 mm
1.7 mm
3.2 mm
H = 9.6 mm
= 3 x h
= 1.2 x P
h = 3.2 mm
= 1/3 x H
= 0.4 x P
P − 0.2 mm
= 7.8 mm
2 x P − 0.2 mm
= 15.8 mm
chipworks
Samsung Galaxy S7 Edge Teardown by Chipworks – 03/08/2016
Hybrid bonding
chipworks
3D DieStore: Design Rules Standardization for Interconnect Level
SPCC 2017 | Viorel Balan
| 25

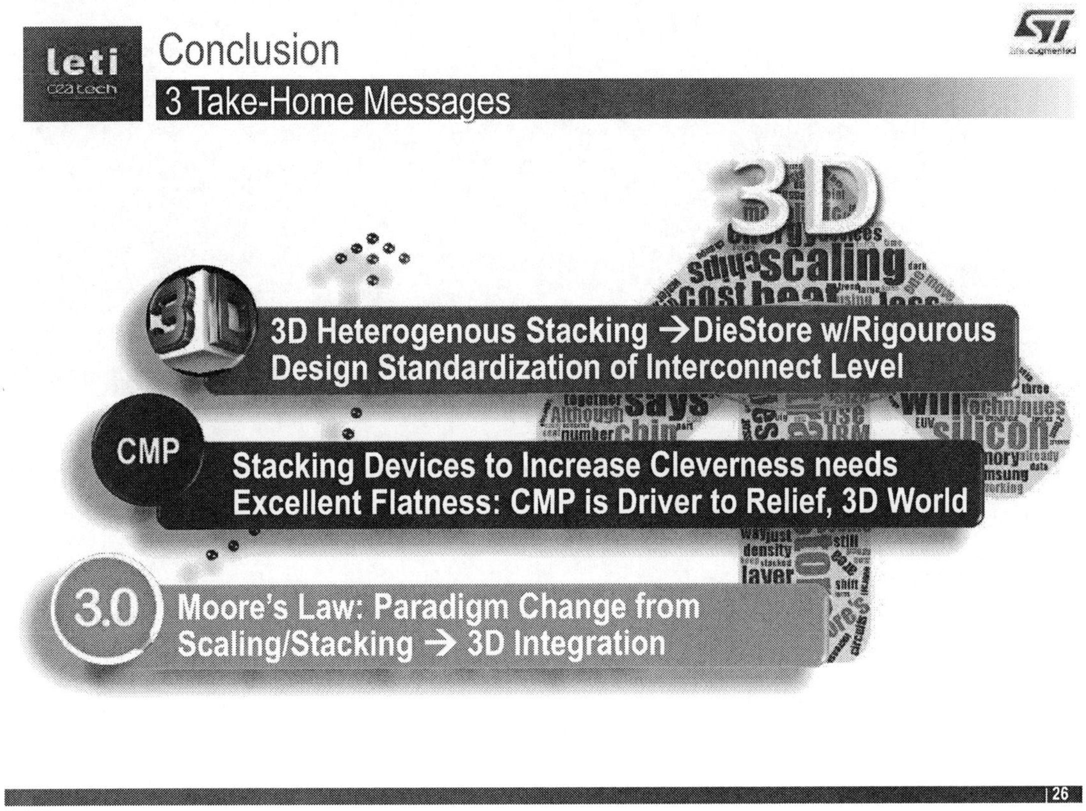

leti
cea tech
Conclusion
3 Take-Home Messages
3D
3D Heterogenous Stacking →DieStore w/Rigourous
Design Standardization of Interconnect Level
CMP
Stacking Devices to Increase Cleverness needs
Excellent Flatness: CMP is Driver to Relief, 3D World
3.0
Moore's Law: Paradigm Change from
Scaling/Stacking → 3D Integration
| 26

This Work was performed at STMicroelectronics conjointly with CEA-LETI, EVG and MCL.

Thanks to all the 3D integration team at STMicroelectronics and CEA-LETI.

This work was conceived within, and supported by "Pilot Optical Line for Imaging and Sensing" (POLIS), an ENIAC Joint Undertaking project.

POLIS

eniac

Thank You

Leti, technology research institute
Commissariat à l'énergie atomique et aux énergies alternatives
Minatec Campus | 17 rue des Martyrs | 38054 Grenoble Cedex | France
www.leti.fr

INSTITUT CARNOT CEA LETI

Chemical/Mechanical Balance Management through Pad Microstructure in CMP

R. Yim[1,2,5], C. Perrot[2], V. Balan[1], P-Y. Friot[3], B. Qian[3], N. Chiou[3], G. Jacob[3], E. Gourvest[2], F. Salvatore[4], S. Valette[5]
[1] CEA, LETI, MINATEC, Univ. Grenoble Alpes, 17, rue des Martyrs, 38054 Grenoble Cedex 9, France, ratanak.yim@cea.fr
[2] STMicroelectronics, 850 rue Jean Monnet, 38926 Crolles, France, ratanak.yim@st.com
[3] Dow Electronic Materials, 451 Bellevue Rd., Newark, DE 19713, USA
[4] ENISE, 58, rue Jean Parot, 42023 Saint-Etienne cedex2, France
[5] Univ. Lyon, Ecole Centrale de Lyon, Laboratoire de Tribologie et Dynamique des Systèmes, UMR 5513, 36 avenue Guy de Collongue, 69134 Ecully, France

Introduction

CMP = SYNERGY between a Chemical and a Mechanical action

The slurry chemically modify the surface while the pad with the abrasive particles mechanically remove the modified surface. The **pad** plays a key role in the CMP process, it helps both the chemical and mechanical actions.

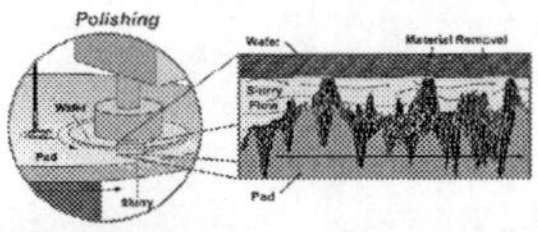

Pores + Grooves → Slurry Transport

Pores + Asperities → Wafer contact

CMP

Pad's pores and grooves allow the slurry transport while the contact with the wafer is determined by its pores and asperities. In this study we focused on the pores which act on the two sides of the balance.

Experimental and Methodology

Blanket Si Wafers
polished on two kinds of
Pads with different pore sizes

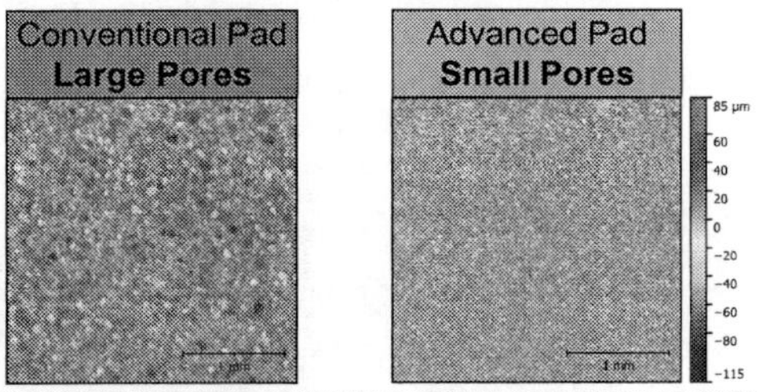

Pads Characterization :
- Roughness Parameters (Texture Open),
- Height Distributions
- Asperities Properties

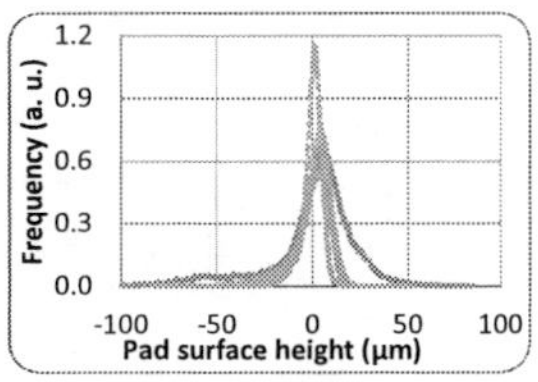

Wafers Characterization :
- Surface Roughness and Height Distributions (AFM),
- Defect Levels
- Si removal rates

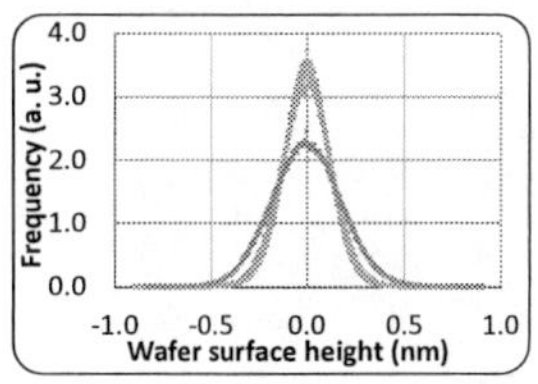

Results and Discussion

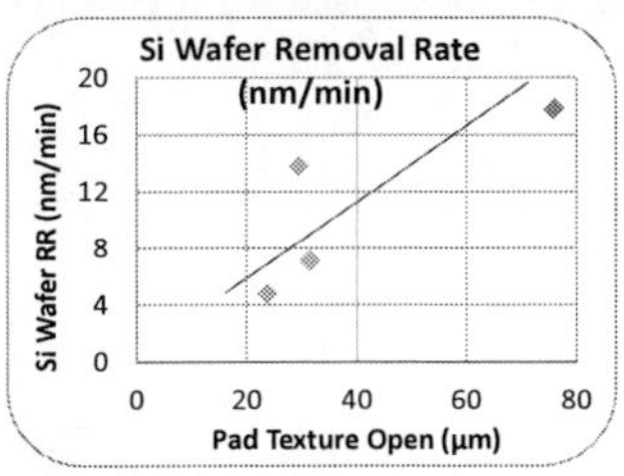

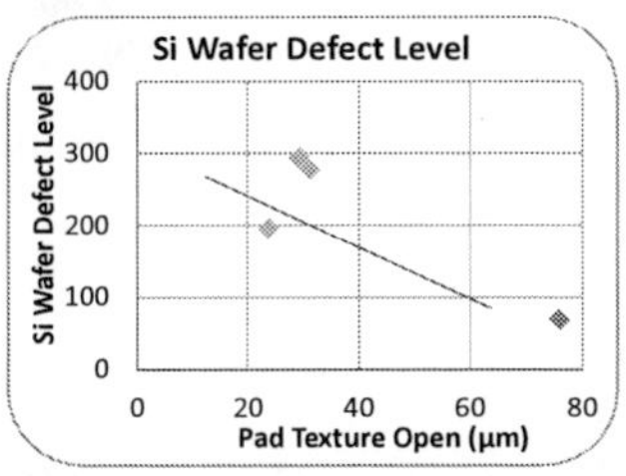

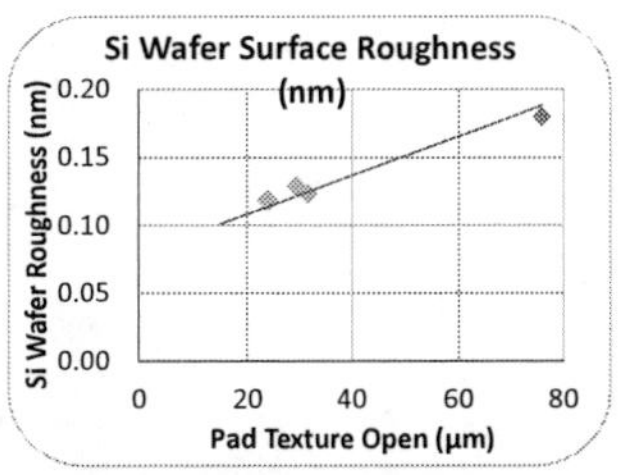

$$\text{Texture Open} = \frac{\text{Empty Volume}}{\text{Projected Area}}$$

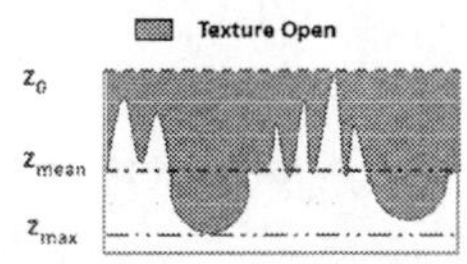

As **Si CMP** is a more chemical process [1], removal rate is very dependent on slurry transport ability and consequently on texture open. **Higher pore size** increases the texture open and leads to higher removal rate. **Larger pores** also allow a better evacuation of the byproducts, resulting in better defect level.
On the contrary, decreasing the texture open favor the mechanical action of the process, that's why the planarization is better.

Large pore size → Higher RR & Lower Defectivity Level

Small pore size → Better Surface Roughness

Chemical/Mechanical Balance can be managed through **Pad Microstructure.** Advanced Pad & Wafer Characterization are required to better understand **Roughness Transfer from Pad to Wafer** and to achieve **better CMP Performances.**

[1] Y. Liu *et al.*, Microelectronic Engineering, Volume 66, Issues 1–4, April 2003, Pages 438-444

Process Model Identification for Optimizing Particle-to-Part Selectivity during Ultrasonic Parts Cleaning

Osama Khalil, John Deem, Iordan Iordanov, Ardeshir Sidhwa Ph.D., Matthew House, David Zuck

QUANTUM GLOBAL® TECHNOLOGIES, LLC

Phoenix, AZ ✧ Quakertown, PA

Introduction

- Meeting defect limitations is a persistent challenge in chamber component parts cleaning as technology nodes continue to scale down. Particles, which are a target of the final stages of cleaning, create killer defects at the component level, and can be expected to severely impact product yields at sub-10 nm technology nodes.
- Ideal final stage cleaning removes particles that are loosely or partially adhered to a part substrate or coating, without generating particles from pitting the substrate material itself.
- Realistically, ultrasonic cleaning can quickly remove adhered particles but can adversely impact the substrate material.

The present work deals with characterizing ultrasonic cleaning by:

1. Mapping spatial acoustic energy distributions under different frequencies, supplied power, measurement probes, and tank designs
2. **Automating experiments for process model identification of particle removal dynamics**
3. Computer simulation-based design of ultrasonic tanks

Process Modeling

The goal of "grey" process model system identification is to identify the relationship between inputs and outputs of a system by applying both first principles and experimental data. In understanding particle removal dynamics during ultrasonication, the modeling variables will be defined as:

y =output variable=particle count by liquid particle counter (LPC)

u =input variable=ultrasonic power (on/off)

Data indicates 2nd-order and single zero behavior [1].

Ordinary differential equation in time domain:

$$T_D{}^2\frac{d^2y(t)}{dt^2}+2\xi T_D\frac{dy(t)}{dt}+y(t)=K\left(\alpha\frac{du(t)}{dt}+u(t)\right)$$

Following the Laplace Transform to the gain-time constant form for a second-order linear system in the complex s domain gives the transfer function g(s) for the system:

$$\frac{u(s)}{y(s)}=g(s)=\frac{K(\alpha s+1)}{T_D^2s^2+2\xi T_D s+1}$$

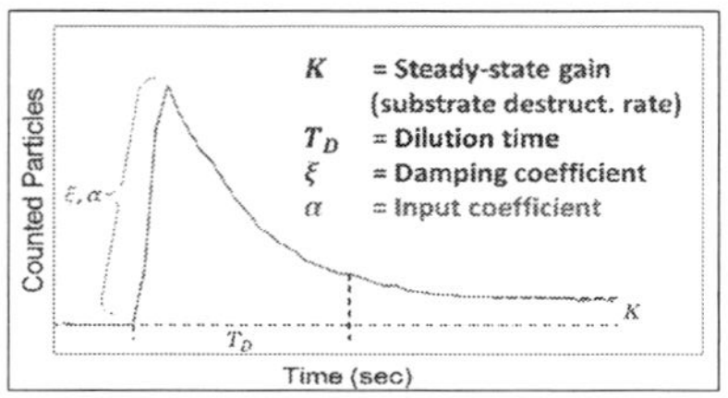

Figure 1: Sample LPC data demonstrating how transfer function parameters indicate measurable quantities (i.e. substrate destruction rate, tank dilution time, maximum particle removal rate)

Results

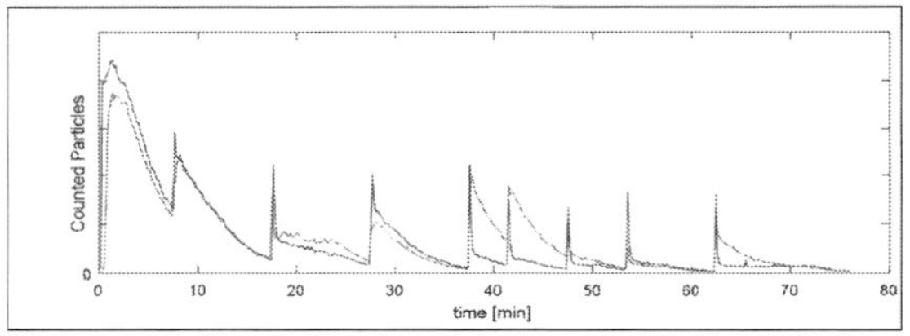

Figure 2: LPC data for two replicate ultrasonic pulse experiments using the same pseudo-random binary sequence of ultrasonic power signals on a textured aluminum sample.

Partial results are shown in Figure 2, which plots particles counted as the output to an input sequence of ultrasonic power cycles. Figure 2 also demonstrates the remarkable repeatability of this experiment between two entirely independent replicates.

Transfer functions can only be used to describe a process that is linear time-invariant (LTI)[2]:

1. The relationship between input and output is a linear map
2. If we apply an input to the system now or T seconds from now, the outputs would be identical

In our case, peaks diminish with each successive ultrasonic power activation:

- The LPC-ultrasonic system is not LTI, so no single model can accurately represent serial cleans
- Repeatability testing demonstrates that reliable models could be generated for every nth ultrasonic power activation

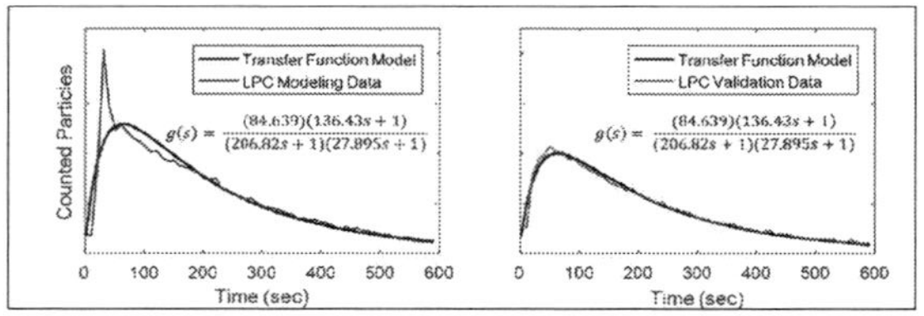

Figure 3: Modeling data (left) and validation data (right) extracted from data in Figure 2 for ultrasonic power activation n = 4. The same transfer function generated from modeling data is applied to the validation and demonstrates a strong fit.

Strong results from repeatability studies and model validation led to the design of an automated experiment with which large set of data are captured from the LPC-ultrasonic system. Captured data is for extensive modeling and model validation to create a large library of models to have an operating model to better describe particle removal dynamics under universal conditions.

Conclusions

Understanding particle removal dynamics will be critical in continuing to meet the ever-tightening defect limitations of shrinking semiconductor node dimensions. Data from the present work have shown that particle removal is repeatable and can be modeled using system identification methodology. Automating these experiments help build a large library of models to have an operating model to better describe particle removal dynamics under universal conditions. These models will facilitate design optimization of ultrasonic rinse tanks as a part of QuantumClean's continued mission to keep pace with Moore's Law.

Related Work

A full characterization of ultrasonication for final parts cleaning goes beyond particle removal dynamics. The next fundamental component of characterization is understanding what factors affect the spatial uniformity of ultrasonic energy density. Figure 4 shows two examples of ultrasonic energy mapping data from a QuantumClean ultrasonic rinse tank. Future work will be to use models from this study and energy density distribution data to perform computer simulation-based pilot designing of ultrasonic tanks to optimize bulk particle removal (Figure 5), and energy density uniformity.

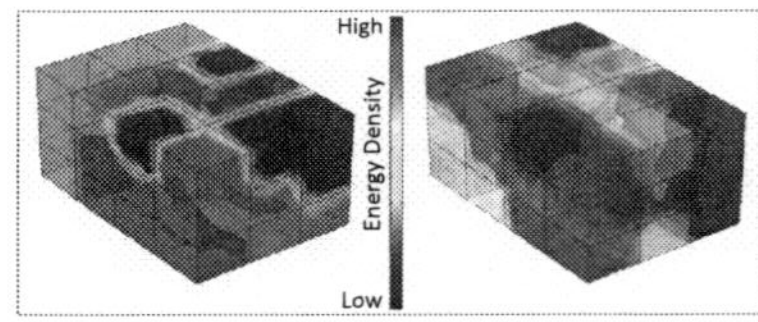

Figure 4: Three-dimensional representation of ultrasonic (acoustic) energy density throughout the volume of an ultrasonic rinse tank

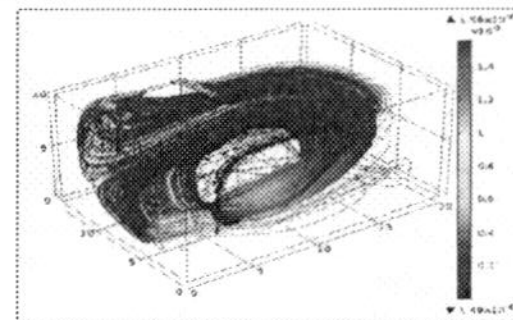

Figure 5: A time-variant finite element simulation of particle trajectories within the flowing fluid space of an ultrasonic rinse tank

References

[1] Bequette, B.W. (2002). Process Control: Modeling, Design and Simulation.

[2] Hespanha, J.P. (2009). Linear Systems Theory, 21-25.

Contact

Osama Khalil
osama.khalil@quantumclean.com
Quantum Global Technologies, LLC
Phoenix-East Facility
3925 E. Watkins St., Ste 100
Phoenix, AZ 85034

QUANTUM GLOBAL® TECHNOLOGIES, LLC

QUANTUM CLEAN®

SPCC 2017
Surface Preparation and Cleaning Conference
March 28-29 Austin, TX

Advantages of Using RGA as Part of Atomically Clean Surface (ACS™) for Sub-10 nm Parts Cleaning Process

Matthew House, Osama Khalil, John Deem, Ardeshir Sidhwa Ph.D., David Zuck

QUANTUM GLOBAL® TECHNOLOGIES, LLC

Phoenix, AZ ✧ Quakertown, PA

Introduction

The ability to detect and mitigate moisture, hydrocarbons and other types of contamination from cleaned parts is becoming more critical as semiconductor technology nodes progress to and beyond the 10 nm threshold. QuantumClean® incorporates residual gas analyzers (RGAs) for Atomically Cleaned Surface™ (ACS™) process development for critically cleaned chamber components. ACS™ processes provide solutions for critical parts cleaning and address critical defect concerns such as the detection and elimination of moisture, metallic, ionic, organic, particulate contamination. Component outgassing rate affects post-PM recovery time, which is important to end users. Decreasing recovery time increases product throughput and decreases production cost. Materials used during cleaning and packaging, including gloves and wipes, are also factors in cleaned part final cleanliness. The present work addresses some of the challenges, solutions, and learning afforded by RGA use at various stages of the parts cleaning development process.

Background / Process Analysis

RGAs are mass spectrometers that measure the chemical composition of gasses and are often used as leak detectors in various high vacuum systems operated at pressures less than 10^{-5} mbar. There are three main components of an RGA: an ion source, a quadrupole mass analyzer and a detector. The detector measures the mass-to-charge ratio (M/z) as ion current. Sweeping through a 1-200 amu (atomic mass unit) range of M/z ratios, one can determine molecular concentrations of individual gas mixture components in a vacuum environment. Equation 1 defines component outgassing rate $Q_x(t)$ of species x (H_2O or C_xH_y volatile or C_xH_y non-volatile). The output data is plotted as a bar graph and is referred to as a mass spectrum (Figure 1).

$$Q_x(t) = I_x(t) \times \text{Ratio}(t) \times S_{\text{eff}} \qquad \text{Eqn. 1}$$

Where,

$$I_{H_2O}(t) = I_{18}(t) \qquad I_{C_xH_y,\text{v}}(t) = \sum_{\text{amu}=45}^{100} I_{\text{amu}}(t) \quad I_{C_xH_y,\text{nv}}(t) = \sum_{\text{amu}=101}^{200} I_{\text{amu}}(t)$$

$$Ratio(t) \equiv P_{\text{vac}}\left(\sum_{\text{amu}=1}^{200} I_{\text{amu}}(t)\right)^{-1}$$

and effective pumping speed $S_{\text{eff}} = Q_{\text{leak}}/\Delta P$,
where Q_{leak} is the calibrated leak rate
and $\Delta P = P_{leak} - P_{no\ leak}$.

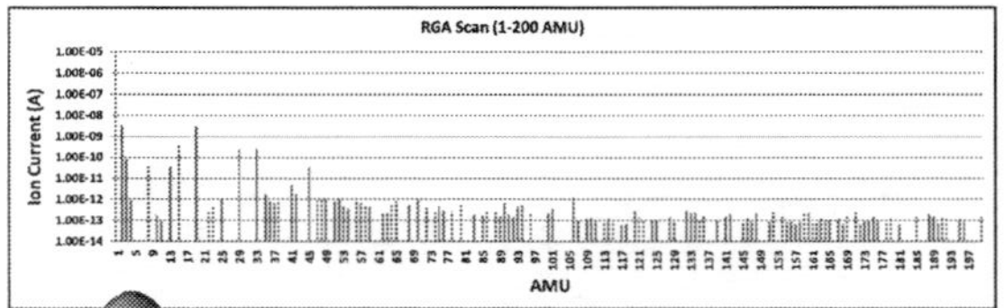

Figure 1: RGA Mass Spectrum

Kit Part Analysis and Comparison

Figure 2 shows an outgassing summary of six different ACS™ cleaned kit parts from a single PVD shield set and one new buy supplier part for comparison.

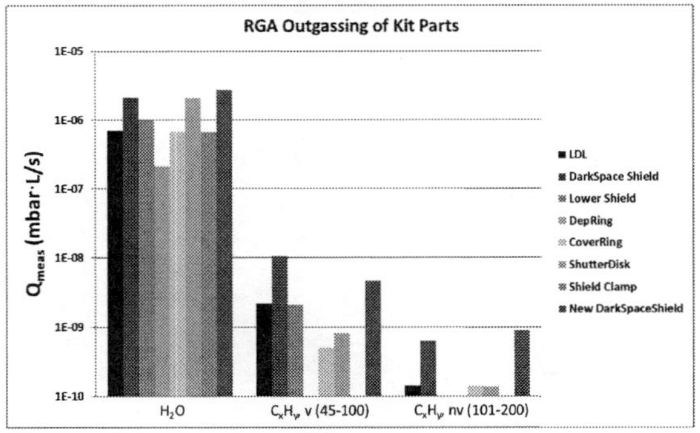

Figure 2: Cleaned Part Outgassing Comparison

Bagging Material Comparison

Figure 3 illustrates RGA testing results of polyethylene packaging from six different suppliers and compares results for residual moisture, volatile hydrocarbons and non-volatile hydrocarbons.

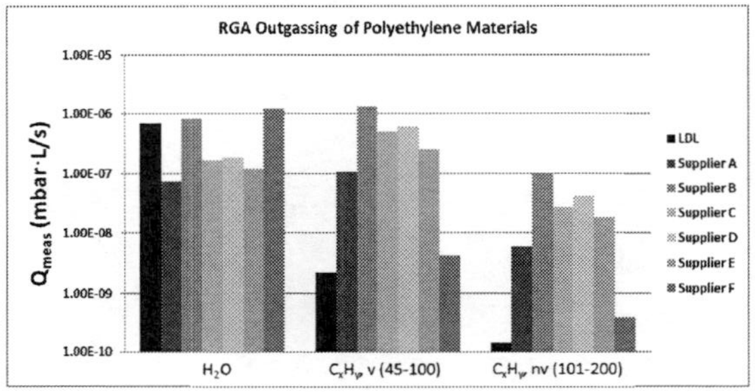

Figure 3: Polyethylene Bag Material Outgassing Comparison

Glove Comparison

Figure 4 shows RGA results of cleanroom gloves from four different suppliers and is used to recommend supplier and product preferences for ACS™ processing.

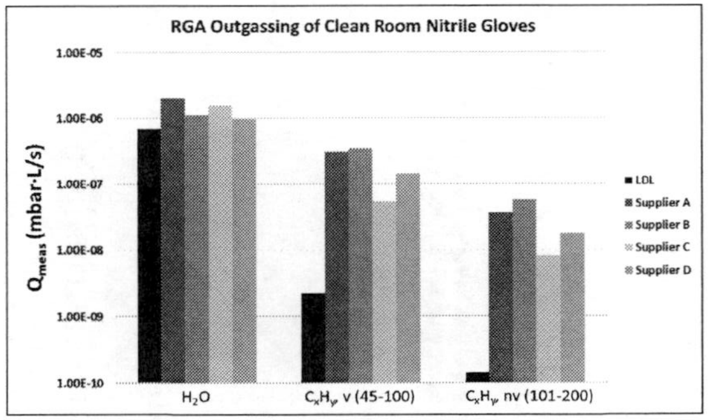

Figure 4: Glove Outgassing Results

Conclusion

QuantumClean® uses RGA to measure moisture and hydrocarbons while developing ACS™ critical cleaning processes for semiconductor equipment chamber components. Outgassing data is used specifically to recommend final bake processes based on substrate material and surface coatings. RGA analysis is used in conjunction other analysis methods to establish critical part cleaning protocols and specify packaging and handling supplies.

References

Generic Standards of ASML (11/25/2008) Ver.1 GSA 07 2211

Contact

Matthew House
Matthew.House@quantumclean.com
Quantum Global Technologies, LLC
Phoenix-East Facility
3925 East Watkins Street, Suite 100
Phoenix, AZ 85034

QUANTUM GLOBAL® TECHNOLOGIES, LLC

QUANTUM CLEAN®

SPCC 2017
Surface Preparation and Cleaning Conference
March 28-29 Austin, TX

Sulfate ion removal by combined UV and Bake process

Eric Guo[1], Crystal Wang[1], Sandy Qian[1], Mars Wang[1], Harry Zhang[1], Keanu Wu[1], Fei Xu[2], Wei Jiang[2], Jun Jin[2], Jian Shen[2]

[1]Semiconductor Manufacturing International Corporation
[2]Changzhou Ruize Microelectronics Co., Ltd.

Surface Preparation and Cleaning Conference
March28~29, 2017 Austin, Texas

Outline

- Introduction
- Experimental & Result
- Summary & Future study

Surface Preparation and Cleaning Conference
March28~29, 2017 Austin, Texas

Outline

- Introduction
- Experimental & Result
- Summary & Future study

Photomask cleaning requirement	
3R (Remove)	**Organics:** *Photoresist, Fingerprint, Pellicle glue mark, Airborne organics*
	Particle
	Ions SO_4^{2-}、NH_4^+、$N(CH_3)_4^+$
2P (Preserve)	**Pattern integrity** *(SRAF)*
	Material integrity *(CrOx / MoSiON / Ru)*

Surface Preparation and Cleaning Conference
March28~29, 2017 Austin, Texas

SO_4^{2-} sources	
Photomask blank manufacture process	1、Quartz cleaning process 2、Chrome cleaning process 3、Rework process
Photomask manufacture process	1、Photoresist strip process 2、In-process cleaning process 3、Final clean process 4、Glue mark removal process

Surface Preparation and Cleaning Conference
March28~29, 2017 Austin, Texas

Outline

- Introduction
- Experimental & Result
- Summary & Future study

Surface Preparation and Cleaning Conference
March28~29, 2017 Austin, Texas

SMIC REC

SO_4^{2-} measurement methodology

Step1: Preheat 500ml DI to 90 degree for 2hours

Step2: Measure the SO_4^{2-} level as baseline

Step3: Soak sample into 90 degree DI for 2hours

Step4: Measure the SO_4^{2-} level as sample

Step5: Difference between sample and base as SO_4^{2-} level for sample

SO_4^{2-} measurement tool:
Thermo-Fischer ICS2100 with 0.05ppb detection limit

Surface Preparation and Cleaning Conference
March28~29, 2017 Austin, Texas

SMIC REC

SO_4^{2-} cleaning efficiency by different methods for SPM contaminated mask

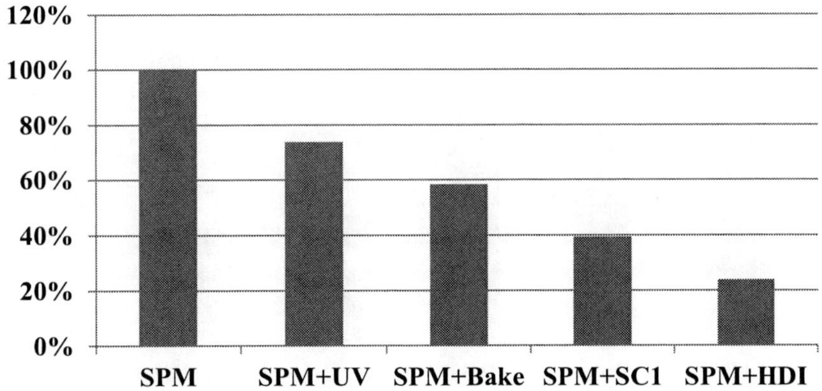

Surface Preparation and Cleaning Conference
March28~29, 2017 Austin, Texas

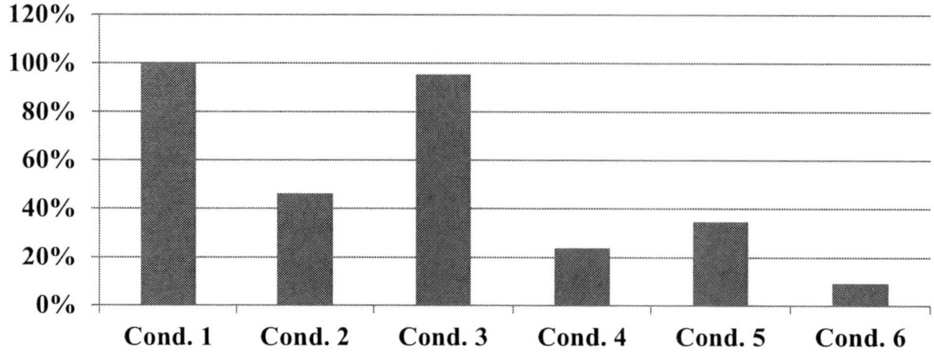

Cond. 1: Mask after regular cleaning process (SPM + 80° HDI + SC1)
Cond. 2: Cond. 1》 90° HDI soak 2hr》 80° HDI + SC1
Cond. 3: Cond. 2》 90° HDI soak 2hr》 105° bake + 80° HDI + SC1
Cond. 4: Cond. 3》 90° HDI soak 2hr》 80° HDI + SC1
Cond. 5: Cond. 4》 90° HDI soak 2hr》 105° bake + 80° HDI + SC1
Cond. 6: Cond. 5》 90° HDI soak 2hr》 80° HDI + SC1

Surface Preparation and Cleaning Conference
March28~29, 2017 Austin, Texas

SO_4^{2-} residue after different cleaning methods for mask treated by traditional cleaning process

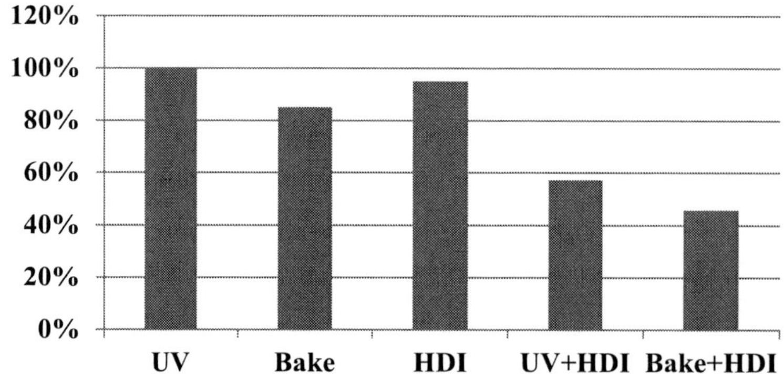

Surface Preparation and Cleaning Conference
March28~29, 2017 Austin, Texas

SMIC REC

SO_4^{2-} residue after different cleaning methods for mask treated by different acid temperature

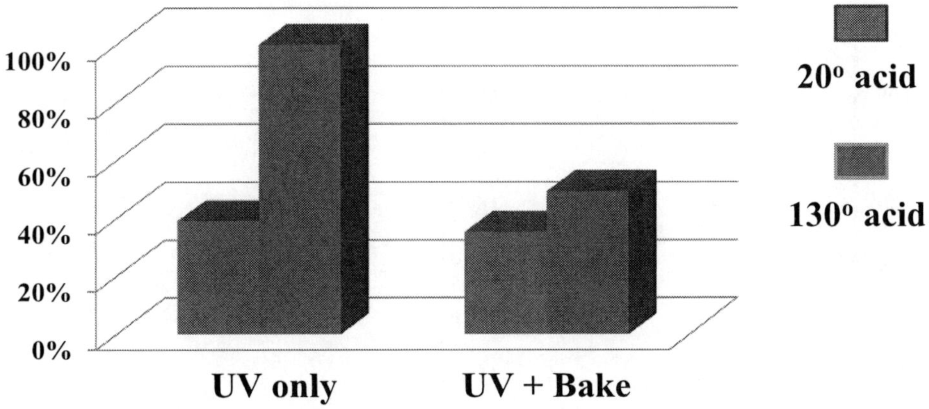

Surface Preparation and Cleaning Conference
March28~29, 2017 Austin, Texas

SMIC REC

SO_4^{2-} residue after new cleaning process for different masks

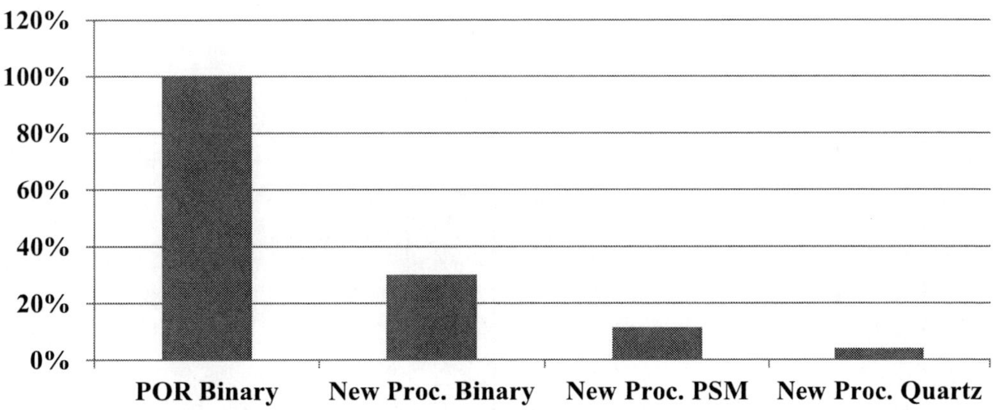

Surface Preparation and Cleaning Conference
March28~29, 2017 Austin, Texas

SMIC REC

130° SPM

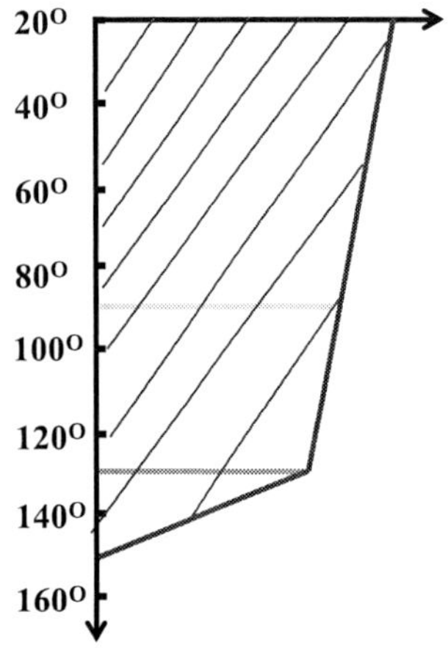

130° SPM
80° HDI
SC1

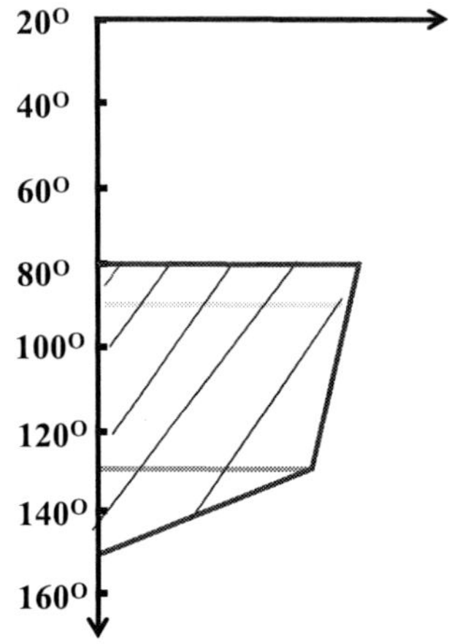

Surface Preparation and Cleaning Conference
March28~29, 2017 Austin, Texas

SMIC REC

130° SPM
80° HDI+SC1
130° Bake

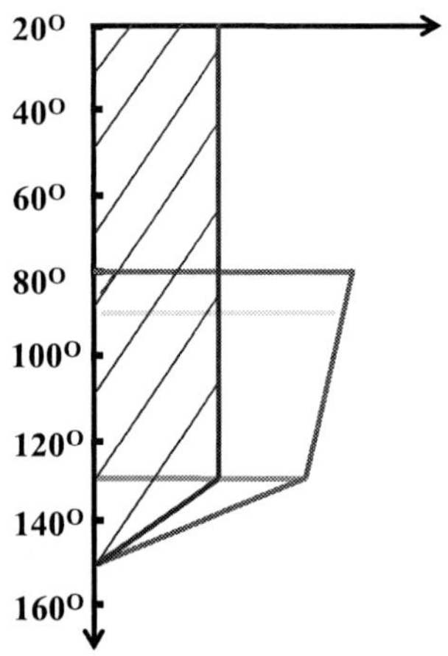

130° SPM
80° HDI+SC1
130° Bake
80° HDI+SC1

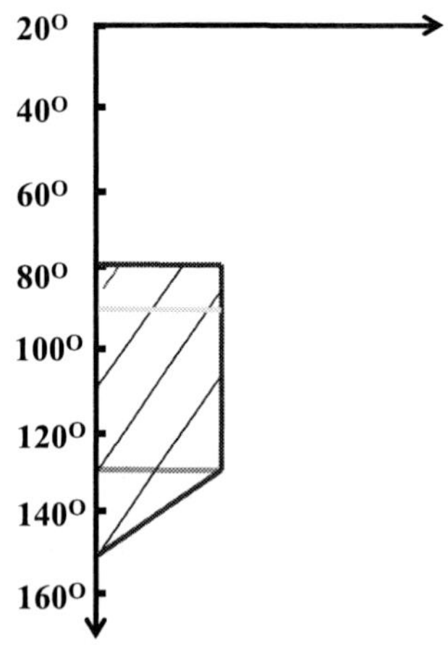

Surface Preparation and Cleaning Conference
March28~29, 2017 Austin, Texas

130° SPM
80° HDI+SC1
130° Bake
80° HDI+SC1
UV + 60° Bake

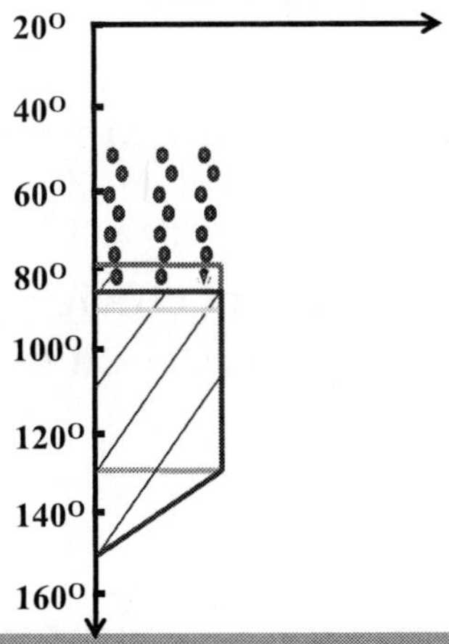

130° SPM
80° HDI+SC1
130° Bake
80° HDI+SC1
UV + 60° Bake
80° HDI+SC1

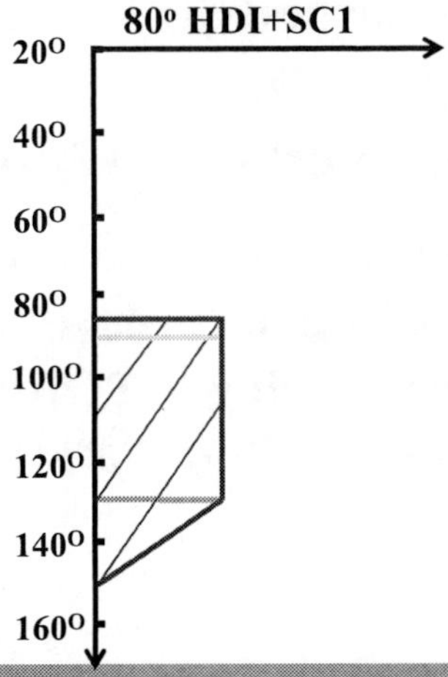

Surface Preparation and Cleaning Conference
March28~29, 2017 Austin, Texas

Outline

- ➢ Introduction
- ➢ Experimental & Result
- ➢ Summary & Future study

Surface Preparation and Cleaning Conference
March28~29, 2017 Austin, Texas

Summary

- Hot DI rinse has the best SO_4^{2-} removal efficiency comparing to UV, Bake, SC1
- Bake process can resurface SO_4^{2-} by diffusion mechanism
- Bake process can enhance SO_4^{2-} removal effect of UV process
- By combing UV and Bake process, the SO_4^{2-} removal efficiency improved 70% comparing to POR process. The SO_4^{2-} residue can reach <0.2ppb for PSM mask

Surface Preparation and Cleaning Conference
March28~29, 2017 Austin, Texas

Future Study

Further optimize the process parameters for better production control in terms of organics removal, particle removal, pattern damage prevention, phase change minimization, CD loss minimization, etc.

Surface Preparation and Cleaning Conference
March28~29, 2017 Austin, Texas

Acknowledgements

1. Sponsored by Project of Shanghai Municipal Commission of Economy and Informatization (2016 first breakthrough and demonstration applications of high-end smart equipment in Shanghai)
2. Management and technical support from Semiconductor Manufacturing International Corporation and Changzhou Ruize Microelectronics Co., Ltd.

Thank You!

Surface Preparation and Cleaning Conference

March28~29, 2017 Austin, Texas

Pall Corporation *Better Lives. Better Planet.* SM

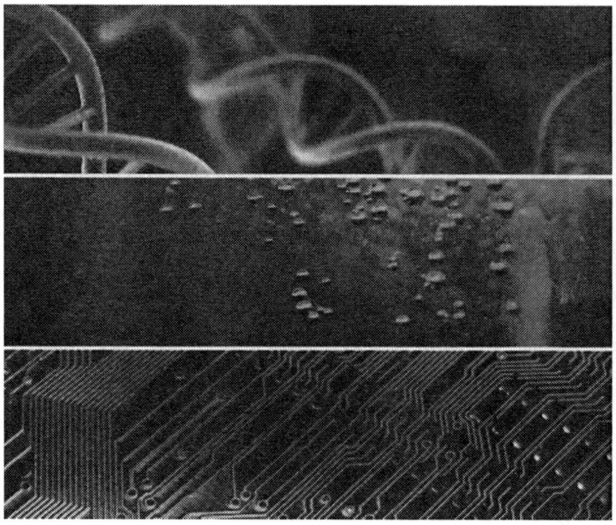

Particle Clean-Up of Various Filter Media in WEC Chemistry via 20 nm Particle Counting

Presented at SPCC 2017, Austin, TX
March 28, 2017

Patrick Connor, Lawrence Johnson, Rao Varanasi, Tony Shucosky
Pall Corporation

Rationale for Study

- Particle behavior of filters can be affected by the chemical in which they are used
  - Evaluation in actual chemical is needed to show actual performance
- Present-day light scattering particle counters are limited to a minimum size resolution of 30 nm
  - Particle sizes of interest for latest technology nodes are much finer
- Many process chemicals are prone to presence of light scattering bubbles that are counted as particles and skew assessment of behavior
  - Technologies that are less sensitive to bubbles would be preferred

Our study explores the use of an alternate counting technology enabling counting in chemicals down to 20 nm, with diminished impact of bubbles

> Can it show differences among different filter types?

Factors Impacting Particle Counting Operations

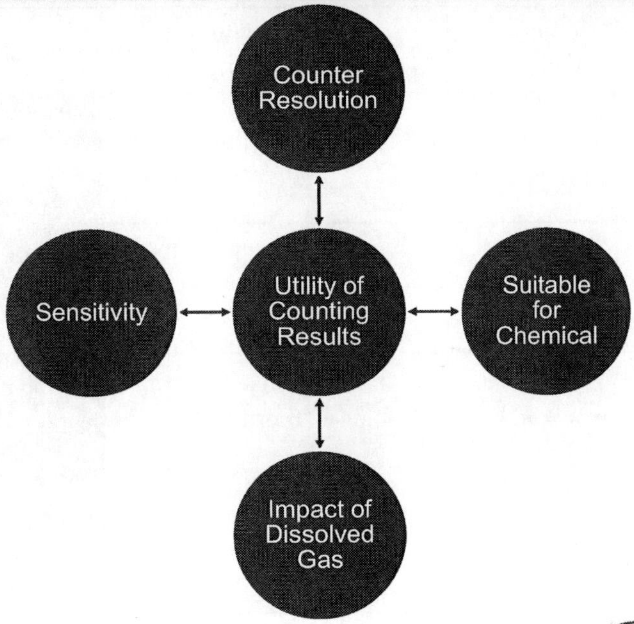

3

Pall Corporation

Background Information on Testing

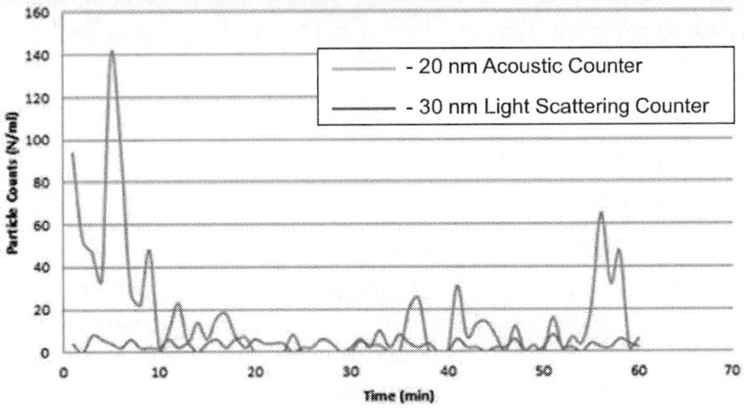

- Early testing with an acoustic counter in litho solvent suggests finer particle detection than previously possible in chemical
- This was viewed as an indication that this counter ought to be able to allow more discrimination among filters rated 20 nm and finer
- Current study explores chemical more broadly applied in the industry

4

Pall Corporation

Various Filter Media for Aqueous Chemicals

Material	Flow-ΔP	General compatibility	Wettability	Min. avail. rating	Appearance in SEM
HDPE	★★★	★★★★	-phobic	2 nm	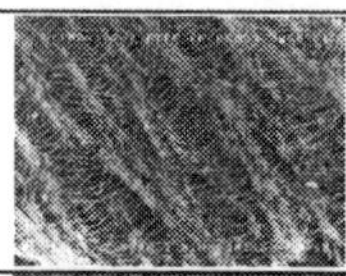
Nylon 6,6 asym.	★★★	★★	-philic	5 nm	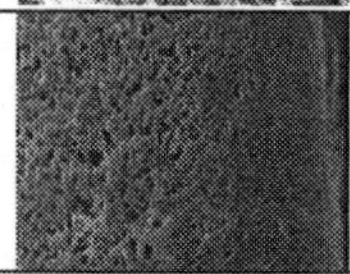
PTFE	★★★★	★★★★★	-phobic	5 nm	
HAPAS*	★★★★★	★★★	-phobic	5 nm	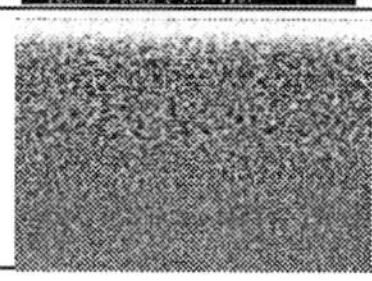

*HAPAS: highly asymmetric polyarylsulfone

Selection of Candidate Model Chemistry

- Tetramethylammonium hydroxide ("TMAH") selected as model chemical
  - Occurs widely for various applications
    - Cleaning chemistries
    - Silicon etchant
    - Photolithography developer
  - Also serves as a model for alkaline chemistries in general and alkylammonium type cleaning chemistries in particular
- TMAH is used and filtered under various concentrations in actual application
  - Moderate concentration of 2.5% was selected to model various applications and to allow safe handling
  - Diluted 25% TMAH was filtered for use in the work

Test Set-up and Operation

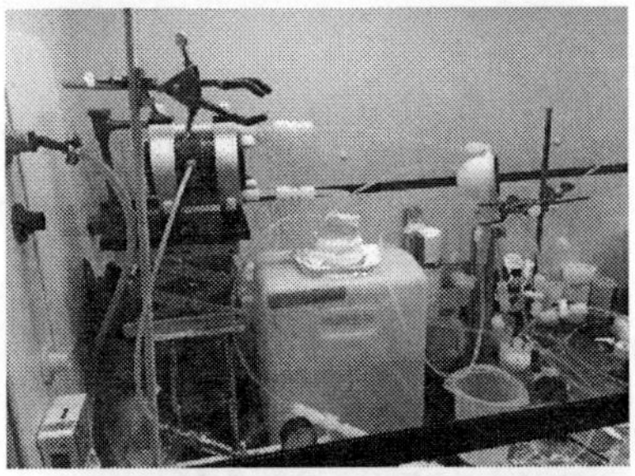

- Recirculative test system set-up with reservoir, pump, insertion point for filter, and side stream to particle counters
  - Effluent sidestream split to two counters
    - Acoustic (shown), 20 nm res.
    - Light scattering, 30 nm res.
  - Limited testing with UPW in single-pass, with UPW system providing pressure,
- Low backpressure used to mimic actual operational condition often occurring
- Filter capsules used for all tests
  - Flow rates selected based on size, typically in range 0.3 – 1 LPM
- Counts monitored over one hour period

7 PALL Pall Corporation

Relative Behavior Found for Various Filters Evaluated with Either Counter Type in TMAH

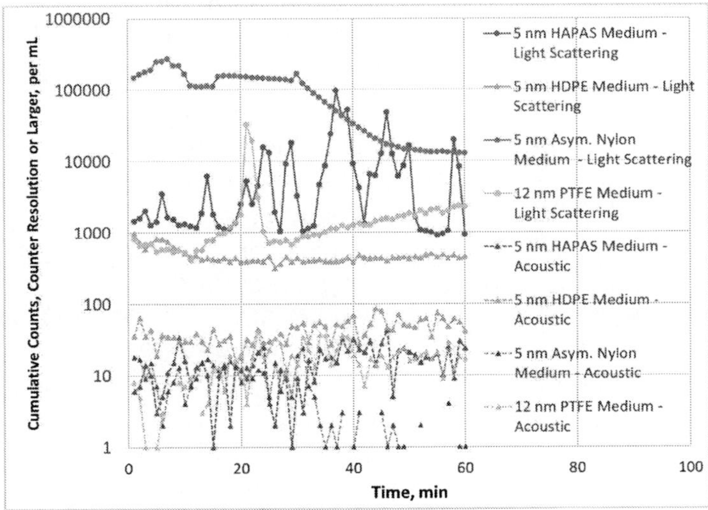

- Higher counts detected using 30 nm light scattering counter
  - Contrary to expected *lower* counts at larger size
  - High ≥30 nm counts suggest *extreme* presence of air bubbles
    - Wide variation among samples
- Generally more consistent counts detected with 20 nm acoustic counter
  - Differences detected among samples but in a narrower range
    - More in line with expectations for filters rated at 12 nm or finer

8 PALL Pall Corporation

Focus on HAPAS and HDPE Media Filters in TMAH via Acoustic Counting

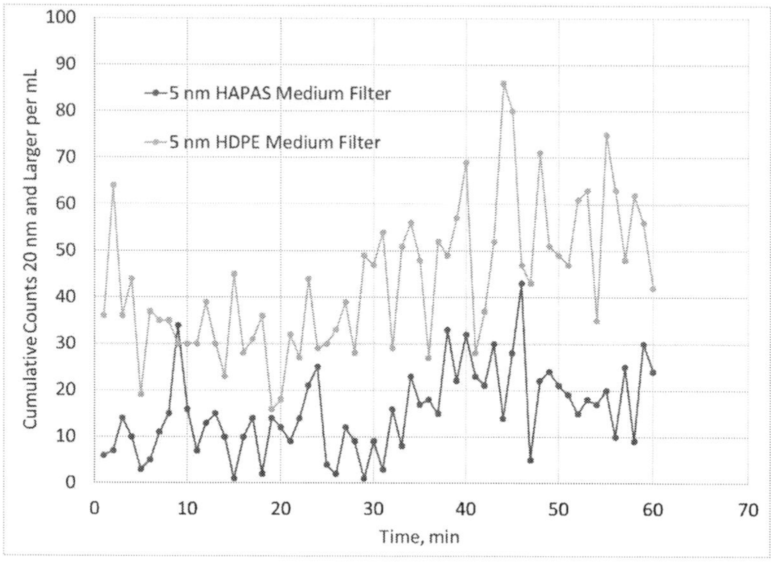

- Focus shifted to comparison via acoustic counting
- Clear difference seen in effluent particles detected
  - Lower levels found for filter using HAPAS medium

9

Pall Corporation

Acoustic Counting Comparison Including Filter with Asymmetric Nylon Medium

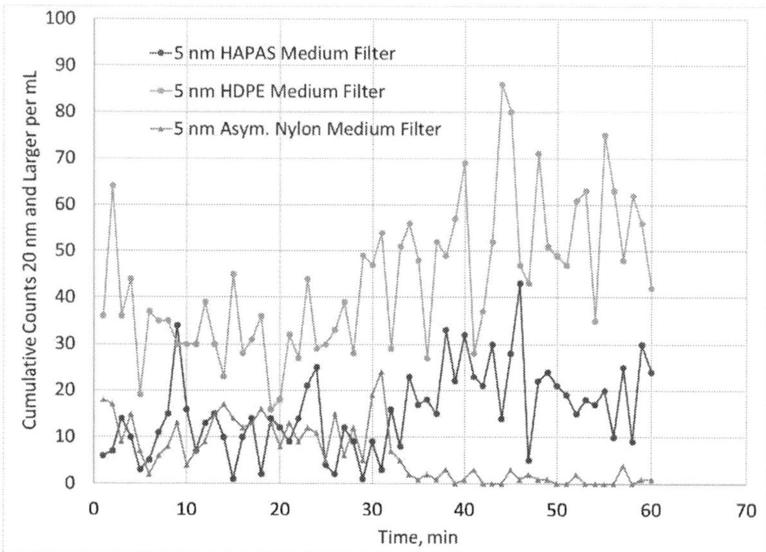

- Here the curve for the nylon medium filter is added
- This filter actually shows the lowest ultimate count level
  - Nylon medium filter exhibited *highest* counts with light scattering counter

(It is noted nylon not suitable for *extended* service in high pH)

10

Pall Corporation

Conclusions, Path Forward, Acknowledgments

- An acoustic particle counting methodology allows particle counting in chemical at smaller minimum size than for currently available light scattering counters
- Testing was conducted in 2.5% TMAH as a suitable model chemical, with four filter types evaluated
- Methodology was capable of showing differences among filters with different media
  - Results very different from those obtained via light scattering counting, likely due to impact of desorbed gas
- Further study using different WEC-relevant chemicals, different filters, and alternate test conditions appears warranted to validate the counter over a broader application space

The authors gratefully acknowledge the extensive test work conducted by Mai Nguyen and Ibrahim Mohamed-Ali of Pall

11

PALL Pall Corporation

Effective Point of Use Filtration of Ultrapure Water for Critical Tools and Processes

A new approach for reducing critical sized particles at the tool level

Problem

It's all about Particle Risk Management

Occurrence
- Filter performance
- Transient conditions

Detectability
- On wafer
- In UPW

Impact
- Yield
- Ability to remove

IRDS Input:

- The most advanced manufacturing processes are susceptible to particles as small as 10 nm
- Even the most advanced UPW final filters do not completely remove sub-50 nm particles
- Definition - critical concentration of "Killer" particles in UPW is 1,000 #/L (# = counts)

Page 3

Particle Occurrences from:

- Transient conditions (flow, pressure)
- Valves (on/off, throttling, regulating)
- Pumps (seals and shear)
- Rinse-up of new component
- Vibration
- Ineffective final filtration

Page 4

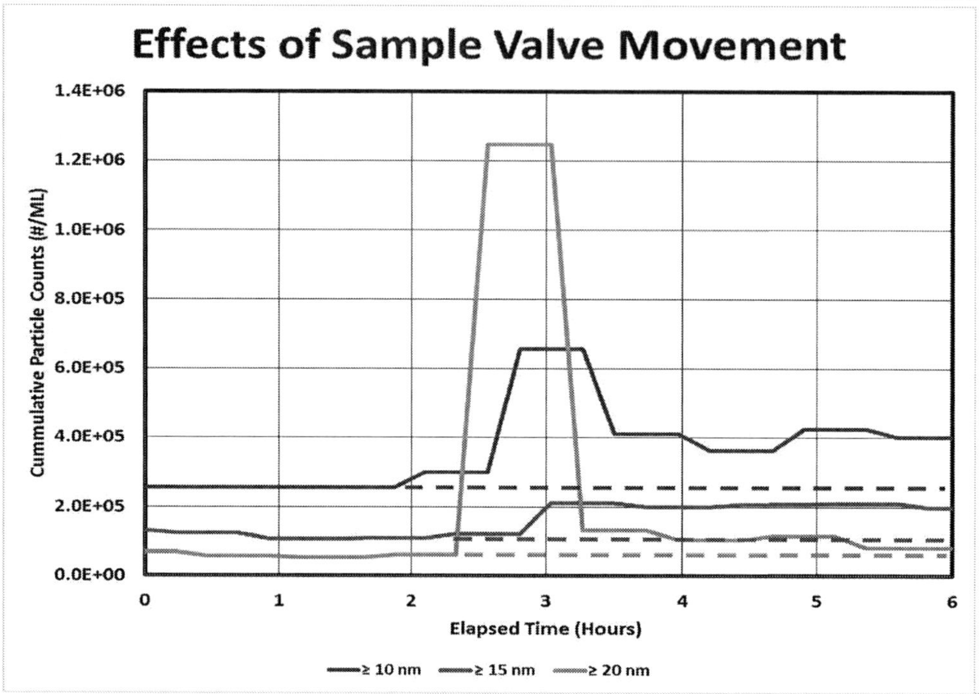

Page 5

Testing

SEMI C079 Test Method:

- 5E+9 #/mL challenge
- 12 nm Silica particles
- LNS, STPC and OPC
- 10 nm cartridge filter
- 4,000 MWCO UF

Page 6

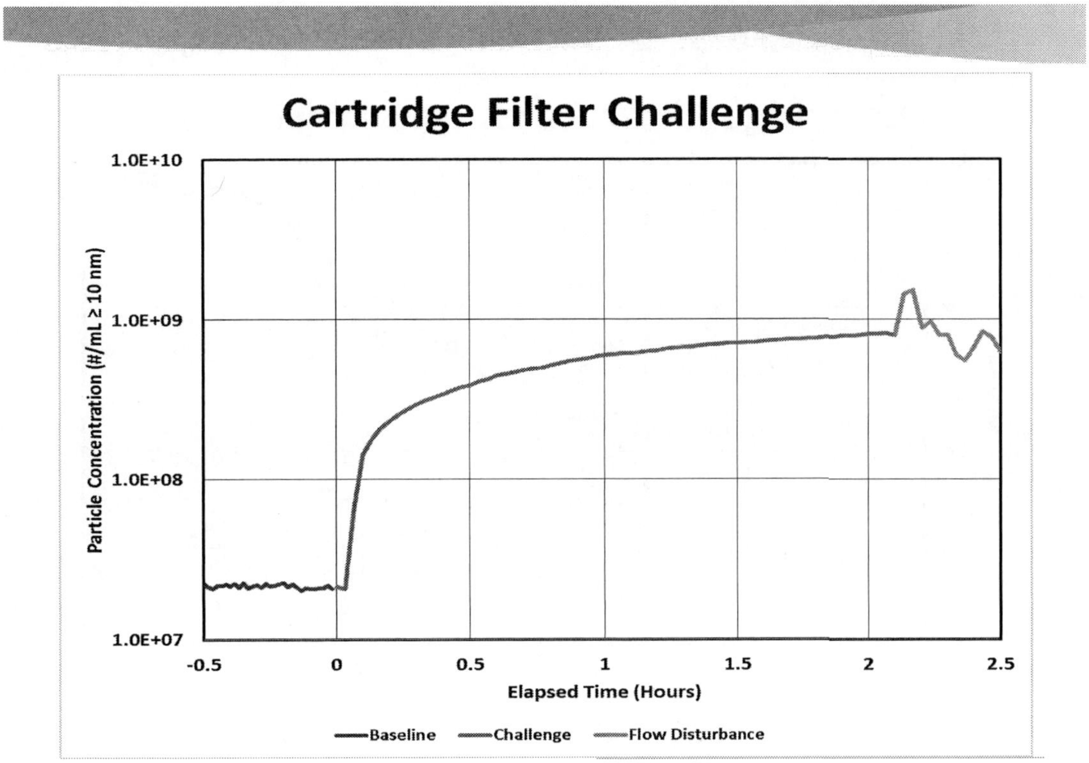
Cartridge Filter Challenge
1.0E+10
1.0E+09
1.0E+08
1.0E+07
Particle Concentration (#/mL ≥ 10 nm)
-0.5
0
0.5
1
1.5
2
2.5
Elapsed Time (Hours)
Baseline
Challenge
Flow Disturbance

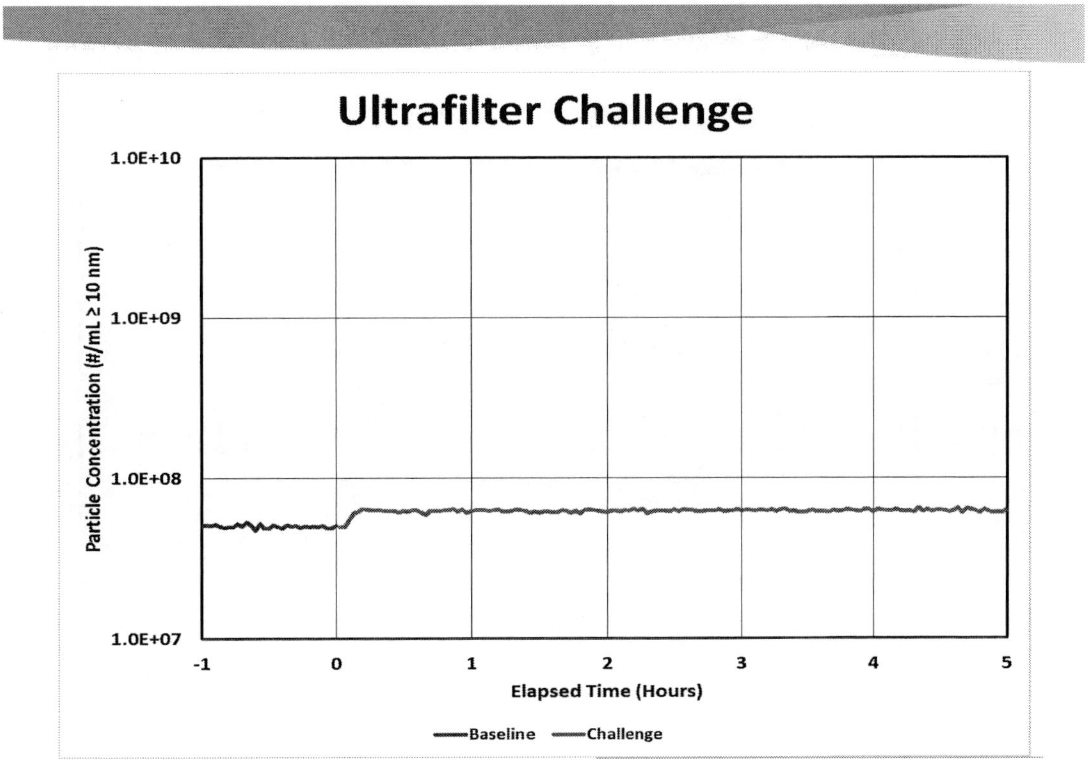
Ultrafilter Challenge
1.0E+10
1.0E+09
1.0E+08
1.0E+07
Particle Concentration (#/mL ≥ 10 nm)
-1
0
1
2
3
4
5
Elapsed Time (Hours)
Baseline
Challenge

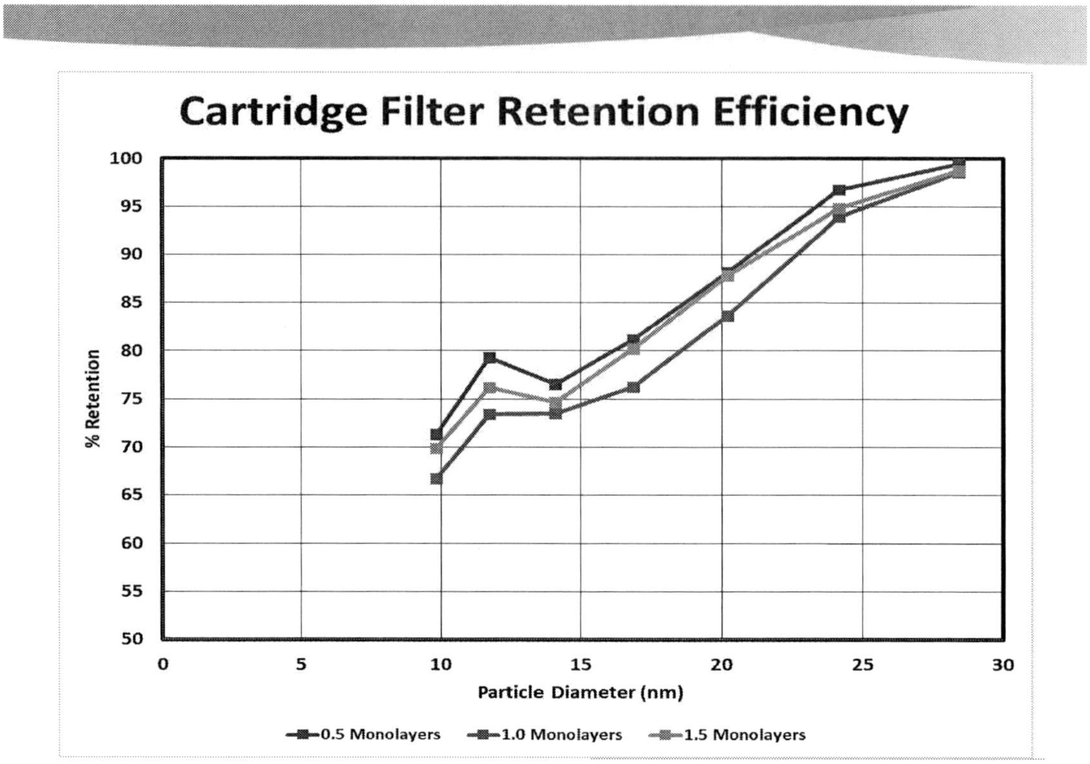

Page 9

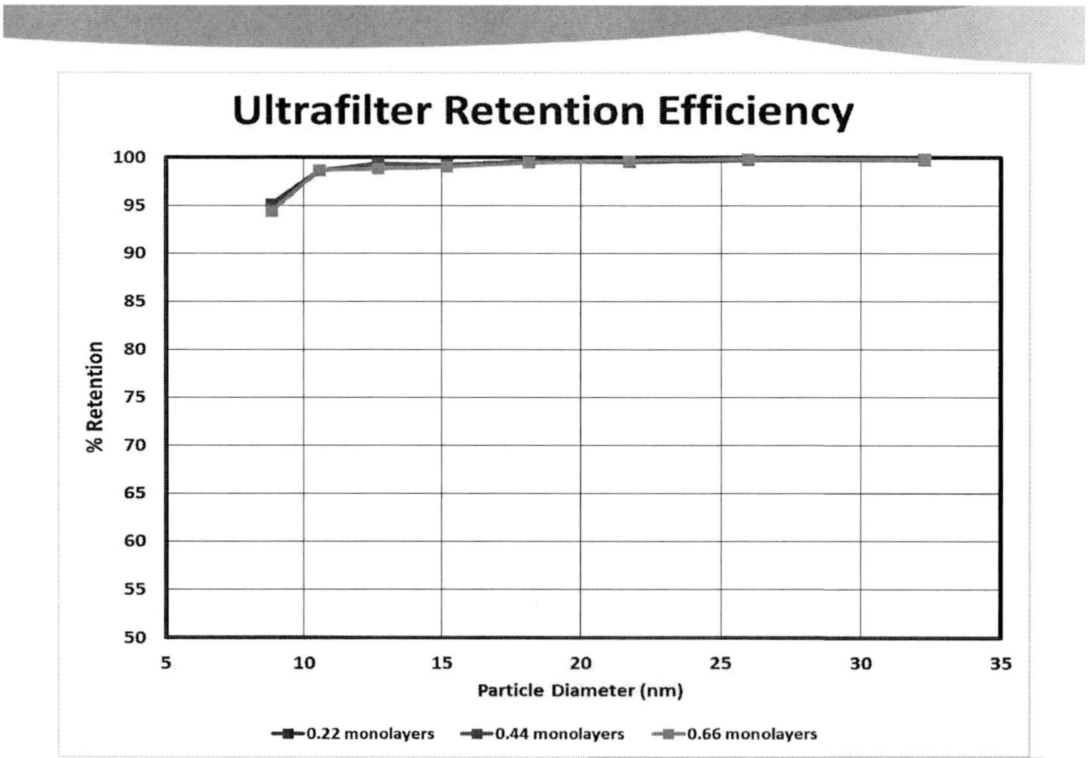

Page 10

Solution

- Bring particle control as close to the tool as possible
- Deliver filtered water at the same pressure as the UPW supply
- Reduce or eliminate fluctuations in velocities, especially through the filter

Page 11

Implementation of solution

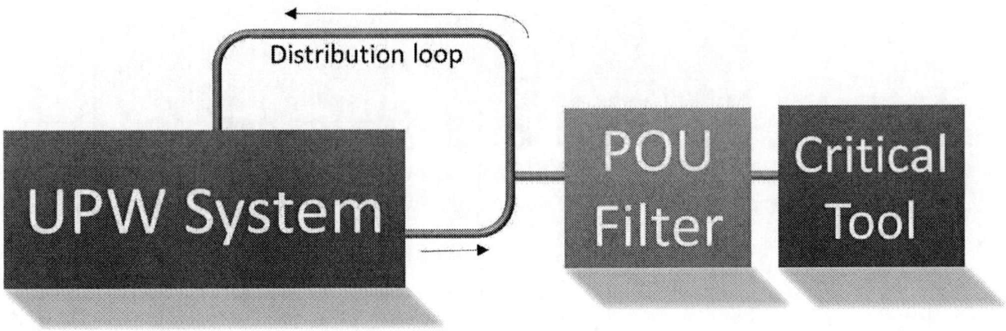

Page 12

VANOX® POU-F Self-Contained System:

- UF membrane filter
- 11 ft^2 of floor space
- 76 inches tall
- 25 gpm tool consumption

Page 13

Operation

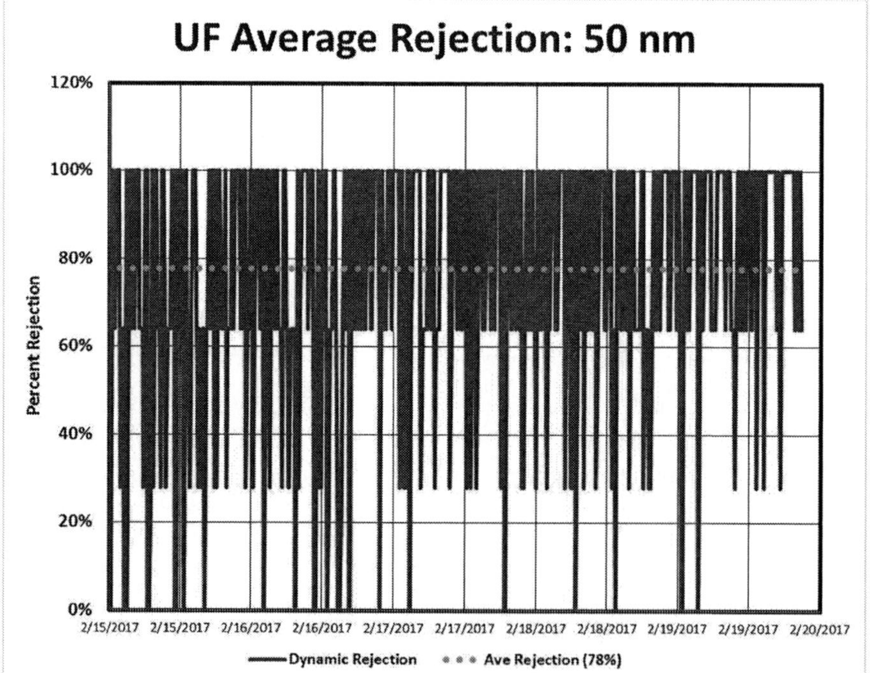

Page 14

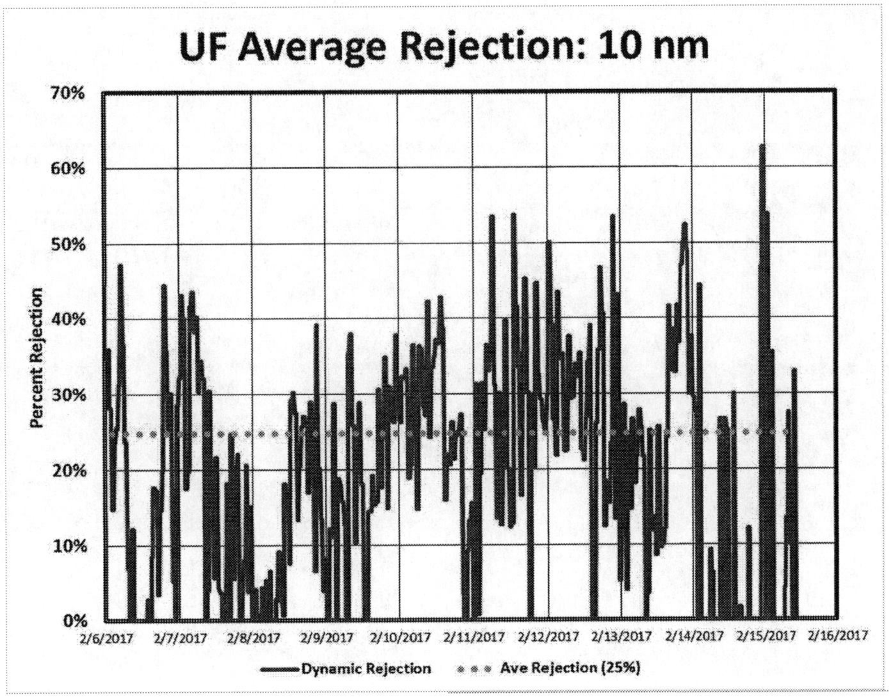

Page 15

Contacts

Glen Sundstrom, Evoqua Water Technologies
+1 815.921.8282
glen.sundstrom@Evoqua.com

Alan Knapp, Evoqua Water Technologies
+1 503.704.0366
alan.knapp@Evoqua.com

EVOQUA WATER TECHNOLOGIES

Page 16

PARTICLE MEASURING SYSTEMS | *Without measurement there is no control*

Continuous Monitoring of Particles at 20 nm in Critical Semiconductor Process Chemicals

Dan Rodier, Ph.D.

drodier@pmeasuring.com

Surface Preparation and Cleaning Conference
March 28, 2017

PARTICLE MEASURING SYSTEMS

Monitoring chemicals - SEMI & chemical operations

- Semiconductor Operations
  - Incoming quality measurement
  - On-line monitoring of distribution and blending systems
  - Monitor chemical filtration and recirculation
- Chemical Processing & Packaging
  - Qualification of chemical filters and components
  - In-line testing of packaging systems
  - Outgoing chemical qualification testing

Without measurement there is no control

20 nm particle measurement in chemicals

- React quickly to particle excursions before surface scan or yield data are available
- Improved sample population statistics for tighter process control
- Opportunity for statistical analysis
- Optimize chemical delivery systems from the loading dock to point-of-process

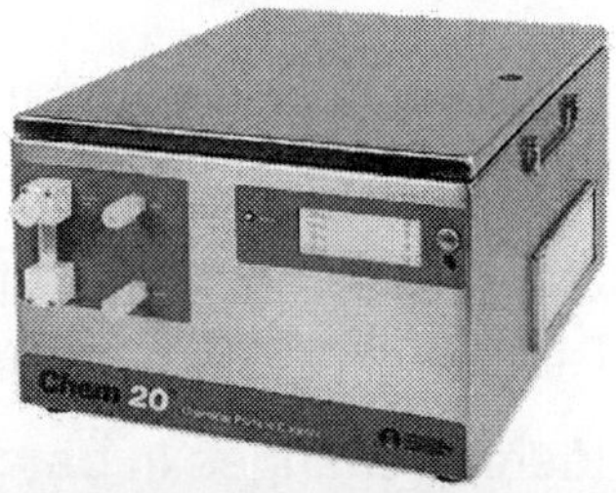

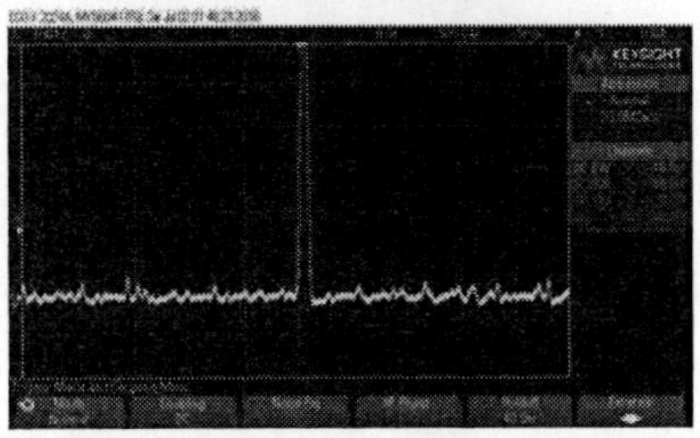

Confidential and proprietary

Without measurement there is no control

Information from a particle counter

- Level of particle contamination (#/ml)
- Particle contamination events
- Trends in particle contamination
- Particle size distribution
  - Indicate filtration performance
- Poisson's statistics
  - Indicate a systematic or non-random source of contamination
- Fourier transform analysis of cyclic contamination
- Contamination source isolated/addressed

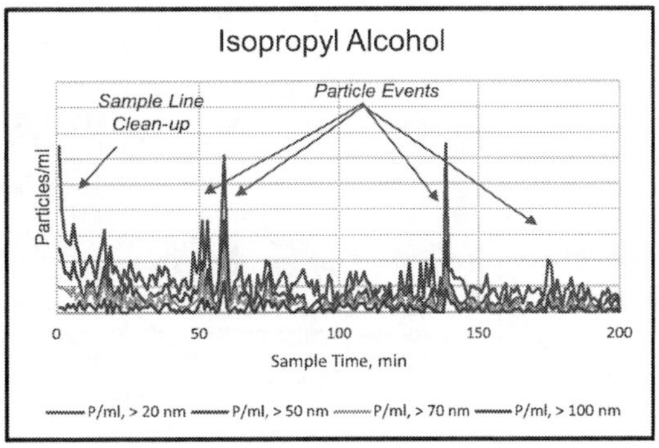

Confidential and proprietary

Without measurement there is no control

Why 20 nm?

- Detect particles you can't see with lower sensitivity
- Monitor effectiveness of advanced particle filtration
- Identify changes in batch-to-batch chemical purity
- Characterize and eliminate particle sources
- Protect wafer cleaning processes from upsets in chemical quality

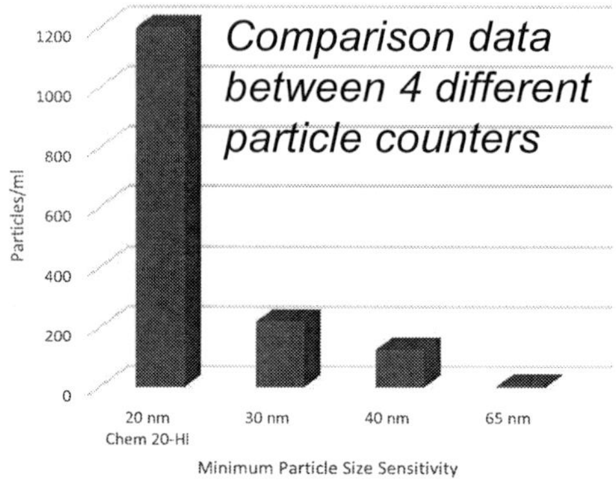

More particles detected at 20 nm—better process control

Impact from Recirculation/Filtration - 96% sulfuric acid

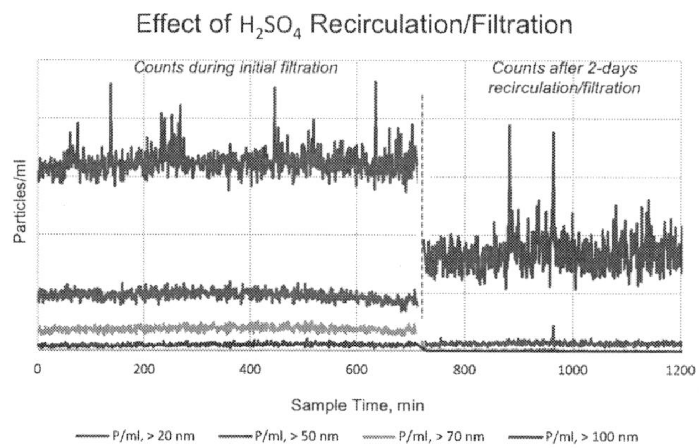

Size, nm	% Reduction
20	47.5%
50	85.6%
70	97.5%
100	98.0%

Particle Reduction after Recirculation

- Recirculated flow with filtration very effective at removing larger particles

Example: 29% Ammonium hydroxide

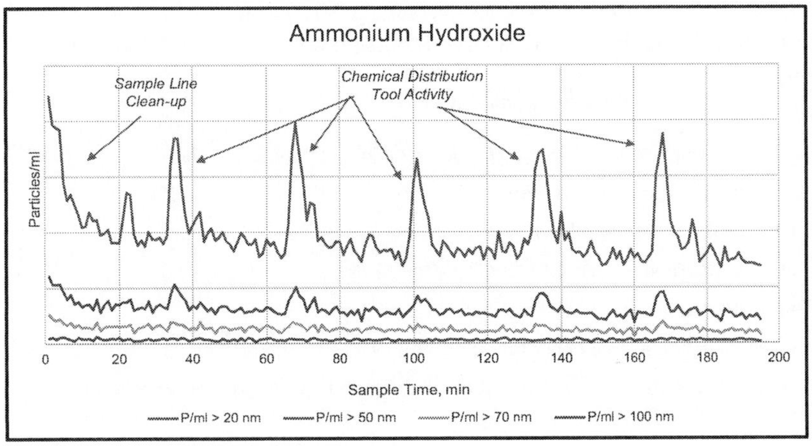

- Activities in the distribution system can cause significant surges in contamination

Particle size distribution

- Particle size distribution—plot on log-log and look at slope (-4.186)
- In H_2SO_4, PSD slope ranges from -2.5 to -4.5
- In filtered chemistry, filtration mainly determines PSD slope
    - Filter removal efficiency relative to particle detection efficiency

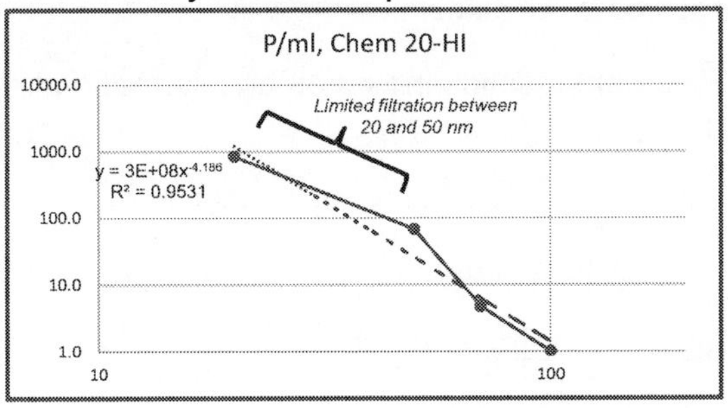

Poisson Statistics Review

- Useful for calculating the probability that a particle will be detected in a sample interval

$$P(k \text{ events in interval}) = \frac{\lambda^k e^{-\lambda}}{k!}$$

  - k is number of times a particle can be detected in an interval; 0, 1, 2, …
    - Applies to raw particle counts
  - λ is the average number of particles per sample interval
  - Standard deviation = square root of the mean, $\lambda^{1/2}$
- Key assumptions
  - Particle detection events occur randomly and independently
  - Events in one sample interval do not affect the probability of particle detection events in the next sample interval
  - The average rate of particle detection events is constant
  - Probability of detection in a sample interval is proportional to sample interval length

Confidential and proprietary

Without measurement there is no control

Implications of Poisson Statistics

- Ideal particle count rate occurs with Poisson distribution
- Actual particle rate deviates from Poisson distribution when:
  - Particle detection rate does not vary in a random manner
  - Contamination source is not random
- Deviations from Poisson statistics indicates a systematic contamination source
  - Identify & eliminate contamination sources in systems
  - Examples—valve switching, pressure pulses, changes in usage or flow.
- Quantify the deviation from Poisson as:
  - **DFP = $[(\sigma_{measured}/\lambda^{1/2}) - 1] \times 100\%$**

Confidential and proprietary

Without measurement there is no control

Ultrapure water system #1

- $\lambda = 353$
- $\sigma_{measured} = 40.8$
- $\lambda^{1/2} = 18.8$
- **DFP = 117.4%**
- System not very clean
  - No ultra-filter
- Definitely not Poisson behavior

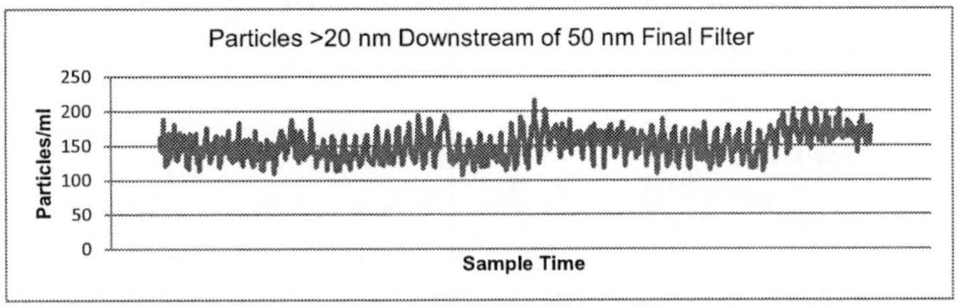

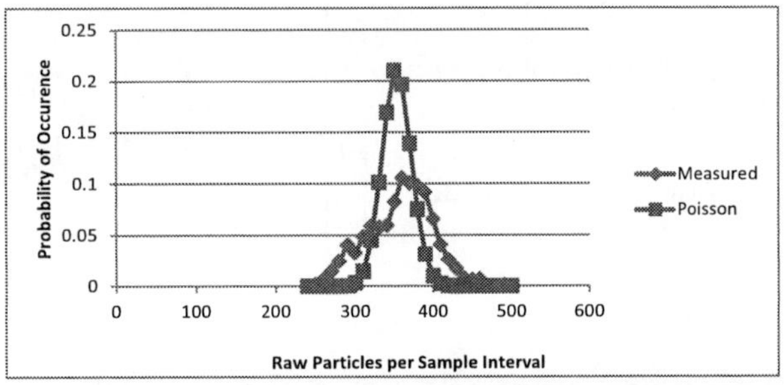

Confidential and proprietary

Without measurement there is no control

Ultrapure water system #2

- $\lambda = 1.34$
- $\sigma_{measured} = 1.25$
- $\lambda^{1/2} = 1.16$
- **DFP = 8%**
- Not the cleanest
- Very stable/well controlled
- Minimal non-random contamination

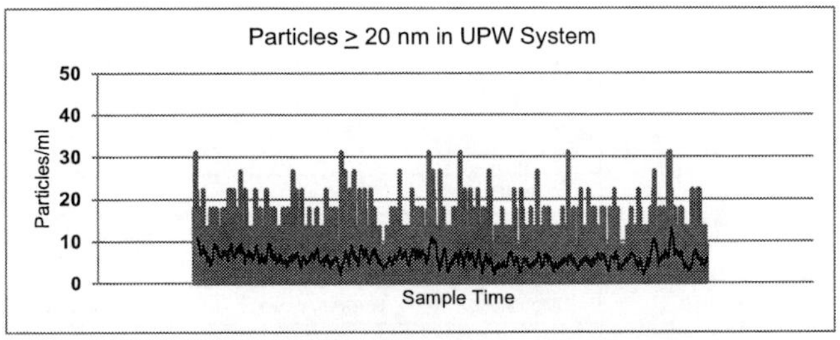

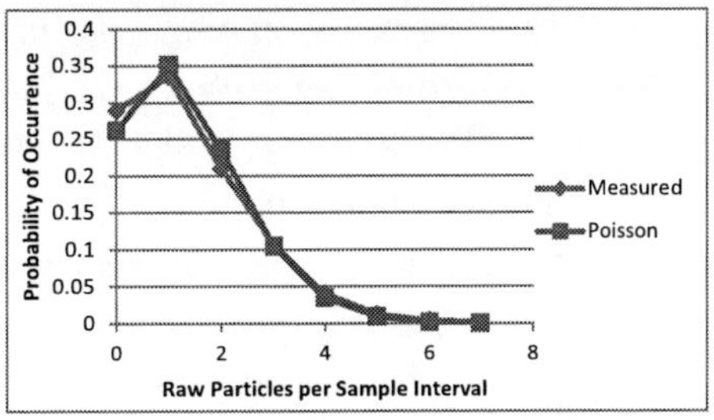

Confidential and proprietary

Without measurement there is no control

Example: Chemical System--31% H_2O_2

- λ = 1165
- $\sigma_{measured}$ = 768.9
- $\lambda^{1/2}$ = 34.13
- **DFP = 2153%**
- Definitely not Poisson behavior
- Indicates systematic contamination sources

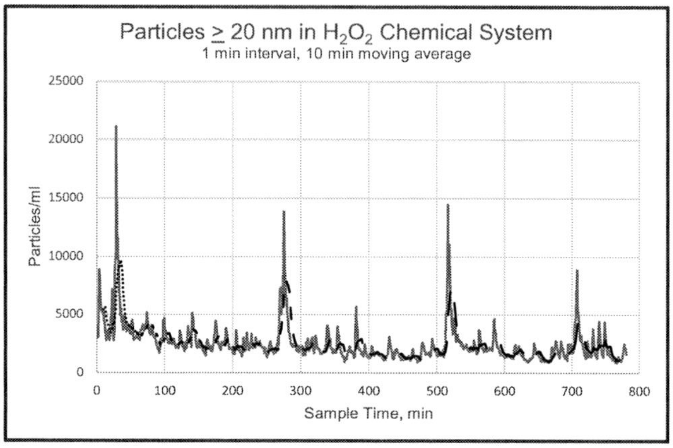

Confidential and proprietary

Without measurement there is no control

Summary of Poisson Analysis

Parameter	UPW System #1	UPW System #2	Chemical System-- H_2O_2
P/ml, ≥ 20 nm	157	5.96	1165
DFP	117.4%	7.8%	2153%
Configuration	Old 50 nm Filter, no UF	Electronic System, with Ultra Filtration	50 nm Filtered H_2O_2 Distribution System
Cleanliness	Not very clean	Nominal Cleanliness	Very Unclean
Stability	Very Unstable	Very Stable	Extremely Unstable

- Ultrapure water systems
  - Significantly cleaner and better-controlled than chemicals at 20 nm
- Chemical distribution systems
  - Comparatively higher particle levels
  - More systematic, non-random particle sources
    - Large opportunities for improvements!
  - Not as stable as ultrapure water systems

Confidential and proprietary

Without measurement there is no control

Next: Fourier Transform Analysis of Particle Contamination

- Use FFT to analyze frequency domain of contamination
- Isolate cyclical contamination sources

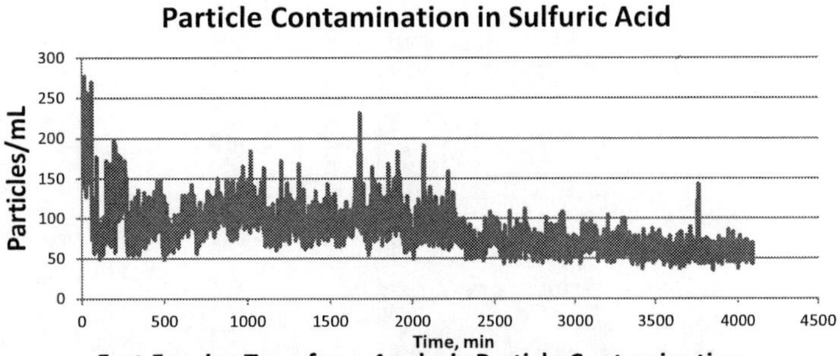

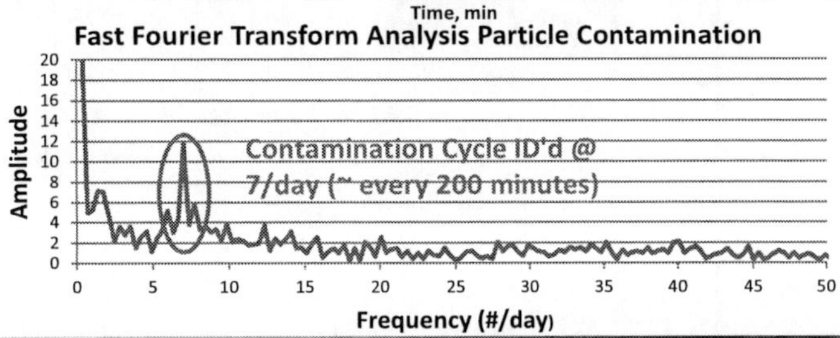

Summary

- Continuous, real-time data - react quickly to nanoparticle excursions
- Measure both the contamination level and the deviation from Poisson statistics
  - Quantify impact of system improvements (filtration, pumps, etc)
- Chemical distribution systems generally have large systematic contamination sources
  - Easy to see in 20 nm data
  - Fourier transform analysis to identify cyclic contamination
  - Measure it – control it

Implementing Chemical Cost Saving in 14 nm High Volume Manufacturing Line

Tejasvi Sadasivuni, Norberto DeOliveira
GLOBALFOUNDRIES, Malta, NY

Abstract
A continuous supply of chemistry with the right concentration and temperature is essential to ensure uninterrupted processing of wafers by wet cleans tools in an HVM environment. Chemical utilization must be optimized to reduce wastage and costs irrespective of the number of wafers in process. In this study, the utilization and chemical costs of Standard Clean 1 (SC-1) were analyzed over a period of three months. Excessive wastage of the chemistry was observed, with utilization levels at an average of only 23%. To reduce chemical wastage and costs, the input volumes of ammonium hydroxide, hydrogen peroxide and water were reduced by 50% per mixing batch. The overall chemical cost per move through this toolset was also reduced by 27%. Total utilization and cost savings before and after the changes have been compared.

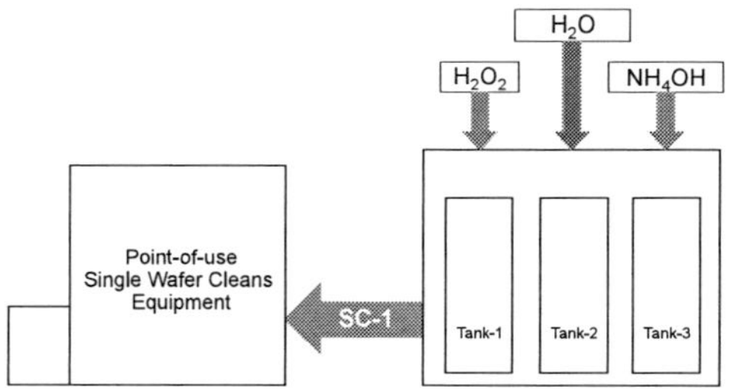

SC-1 supply schematic (3-tank configuration)

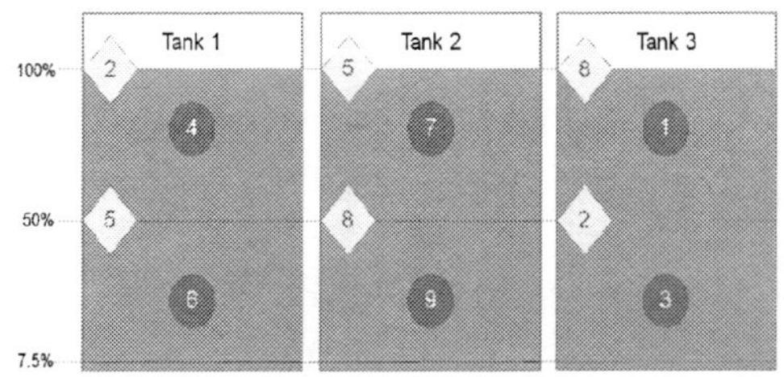

Tank Operating Sequence

	100% Volume	50% Volume
Principle	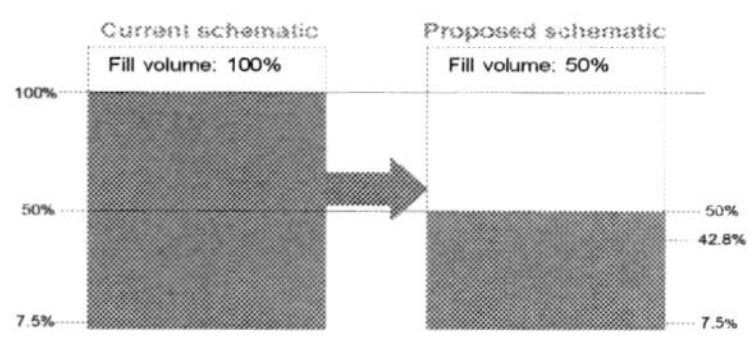	
Mechanism	➢ 100% working volume in each tank ➢ Mid level sensor at 50% volume ➢ Low level sensor at 7.5% volume	➢ 50% working volume in each tank ➢ Mid level sensor at 42.8% volume ➢ Low level sensor at 7.5% volume
Advantages	➢ None when compared to 50% volume ➢ Might lead to higher valve lifetime – not enough data to backup this point	➢ Very low chemical wastage due to 66% rise in utilization ➢ Tank ready time – reduced by 30%

Qualification tests run on new tank volume configuration are comparable to process of record

Particle count – test data compared to POR

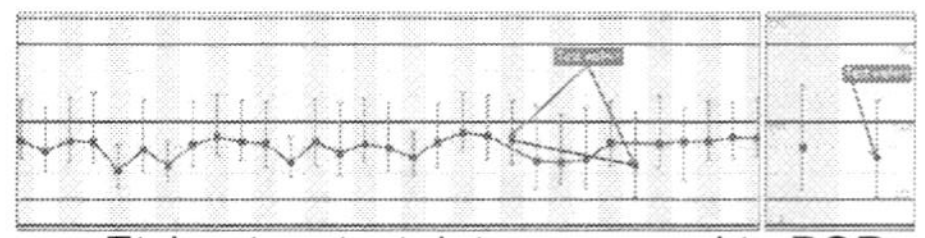

Etch rate – test data compared to POR

Benefit Analysis

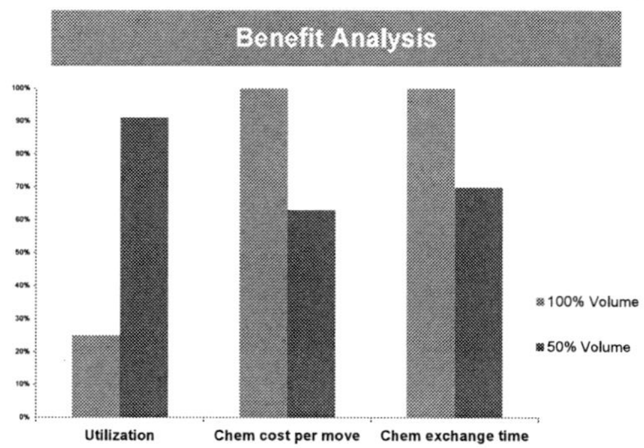

Summary

- ➢ Chemical cost savings by changing the working volume of mixing tanks must be executed with great care.
- ➢ No alarms due to chemical exchange of any sort have been generated after this change.
- ➢ When the WIP is low or the tool is idle, the chemical wastage is reduced by 67%.
- ➢ By reducing the working volumes in the mixing tanks, not only the costs of raw chemistries is reduced, but also the waste water treatment costs and the potential ecological impact[1] of ammonia is also greatly reduced.
- ➢ This method of chemical cost saving can be applied to other chemistries as well, as long as the tank configurations and chemical life times permit such a change.

Acknowledgement

- ➢ Kim, Taehyun (Process Engineering)

References

- ➢ [1] Pubchem-NCBI: https://pubchem.ncbi.nlm.nih.gov/compound/ammonium_hydroxide#section=UN-Number

Tejasvi Sadasivuni, MS
GLOBALFOUNDRIES INC., 400 Stone Break Extension, Malta, NY 12020
Email: tejasvi.sadasivuni@globalfoundries.com
Phone: (864)252-6740

Improved Missing Pattern Defects using Megasonic for Hard Mask Clean at 14 nm and Beyond

Sangita Kumari, SherJang Singh, FNU Devendar, Silas Scott
GLOBALFOUNDRIES, Malta, NY

Abstract

Scaling challenges in BEOL technology for 14nm and beyond makes lithography patterning extremely sensitive to particles generated at TiN deposition steps. These particles translate into missing pattern defects classified as multiple line opens (MLO) and single line opens (SLO). High MLOs and SLOs are one of the highest device yield detractors in BEOL technology.
Here we report improved HM cleaning with reduced MLOs and SLOs by adding Megasonic irradiation to conventional POR (process of record) cleaning recipe

Challenges

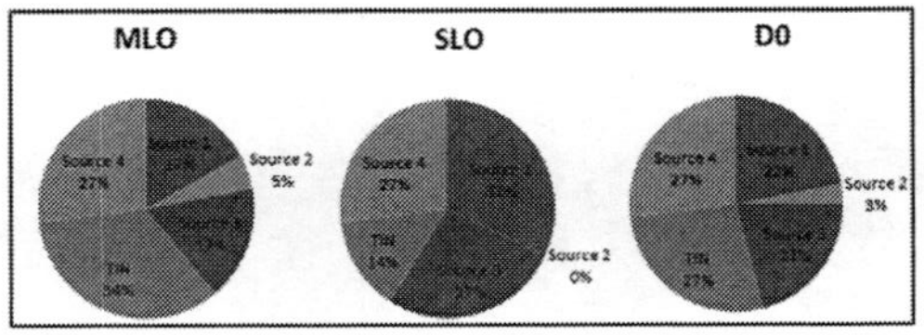

Fig above shows various sources that contributes to MLO SLO. ~27 % of the total defects (D0 = MLO + SLO #/cm2) are introduced during HM/TIN deposition steps

Defect Translation

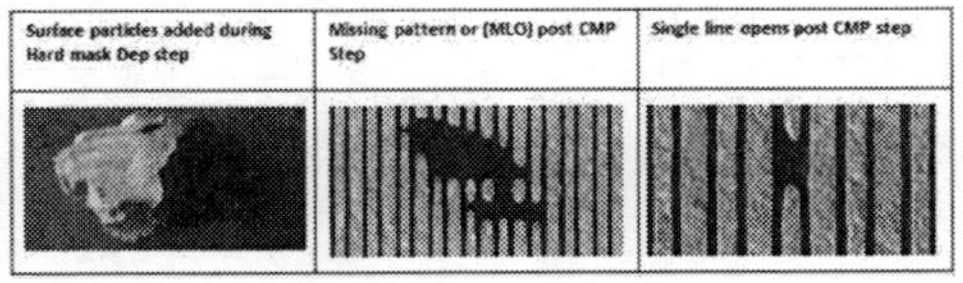

Fig. above shows translation of defects added during hard mask deposition into downstream missing pattern defects (MLO and SLO) post copper CMP (Chemical Mechanical Planarization). Line opens lead to electrical contact resistance fails and consequent die yield degradation.

Results

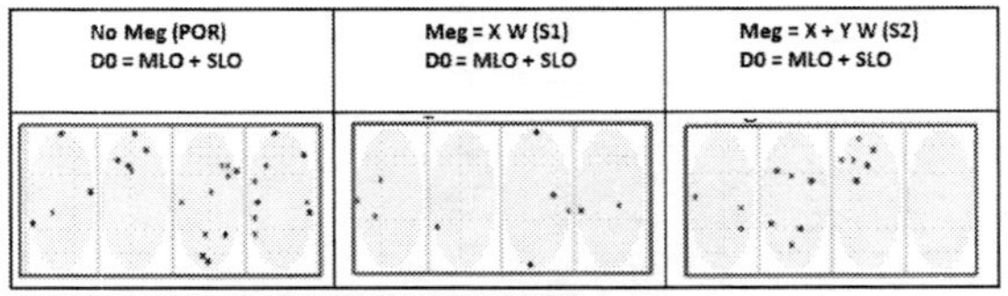

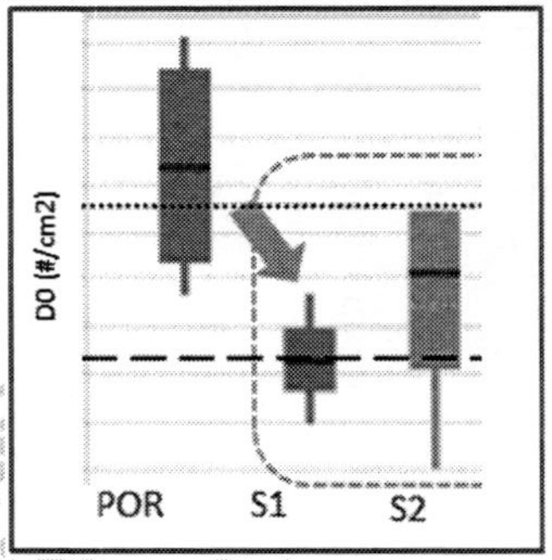

Fig shows Megasonic addition to POR recipe (S1), leads to ~ 65% total defect (D0) improvement

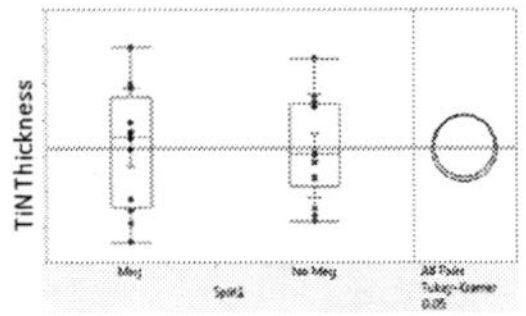

Fig above shows TIN thickness remains unchanged when Megasonic added

Summary

1. Litho patterning extremely sensitive to TIN dep clean steps for 14 nm and beyond Technology
2. ~ 27% of total defects (D0 = MLO + SLO #/cm2 are introduced during TiN deposition steps
3. Defects during TIN dep steps translate to single line and multiple line opens post CMP
4. Addition of Megasonics to conventional POR recipe improved baseline D0 performance by ~65 % with no change to inline TiN thickness.

References

1. Gale, G. W., and A. A. Busnaina. 1999. Roles of cavitation and acoustic streaming in megasonic cleaning.particul Sci Technol 17:229-238
2. .Karimi, P., T. Kim, J. Aceros, J. Park, and A. A. Busnaina. 2010. The removal of nanoparticles from sub-micron trenches using megasonics. Microelectron Eng87:1665-1668
3. S. Kumari, M. Keswani, S. Singh, M. Beck, E. Liebscher, P. Deymier and S. Raghavan, Control of Sonoluminescence Signal in Deionized Water Using Carbon Dioxide, Microelectronic Engineering, 88, pp. 3437-3441 (2011)

2017 SEMATECH Surface Preparation and Cleaning Conference

Sangita Kumari, PhD
GLOBALFOUNDRIES INC., 400 Stone Break Extension, Malta, NY 12020
Email: Sangita.kumari@GLOBALFOUNDRIES.COM
Phone: 520 548 9284

DUMMY POLY GATE ETCH RESIDUE REMOVAL

Wen Dar Liu*, [a], Yi Chia Lee[a], John G Langan[b], Tianniu Rick Chen[b], Aiping Wu[b], and, Andrew J. Adamczyk[c]

a. Versum Materials Taiwan LLC., 2F, No.25, Lane 62, Chung Ho Street, Chupei City, Hsinchu County 30267, Taiwan
b. Versum Materials, Tempe, AZ, USA
c. Computational Modeling, Versum Materials, Tempe, AZ, USA

E-mail address: Wen.Liu@versummaterials.com

Phone: (886) 978587956

VERSUM MATERIALS

Outline

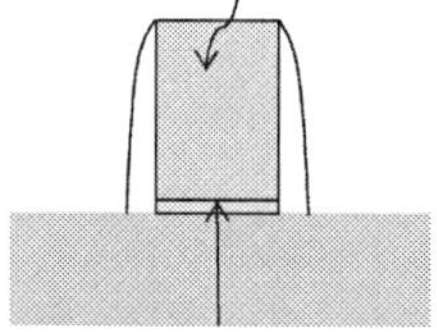

1. **Background introduction**
   - Technical challenge of conventional Si etchant
   - Anisotropic Si wet etching
2. **Design concept of novel Si etchant**
   - Etch rate requirement
   - Formulation design
   - Experimental trails
   - Computation modeling
   - Application in FEOL process
3. **Summary**

VERSUM MATERIALS

Technical challenge of conventional Si etchant

➢ Issue of conventional Si wet etchant

Due to the anisotropic etch characteristic, the pyramid shape etch residues will be generated for conventional Si etchant.

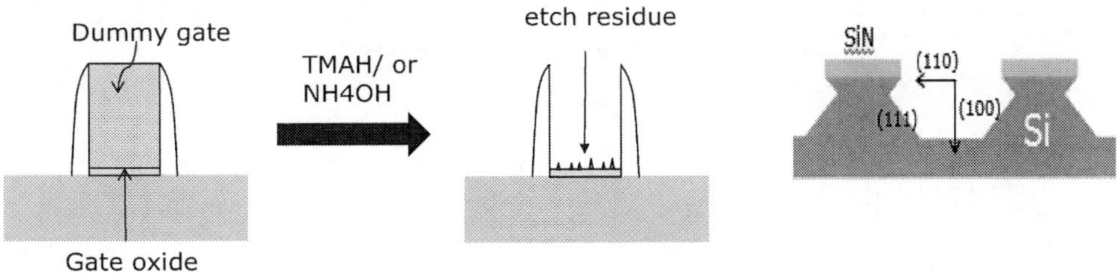

We need high Si etching power in Si(110) and (111) orientations.

VERSUM MATERIALS

Anisotropic Si wet etching

Etch rate of conventional Si wet etchant

	Etch rate (A/min)			Selectivity	
	Si (100)	Si (110)	Si (111)	Si(110)/ Si(100)	Si(111)/ Si(100)
conventional etchant 1	7117	2228	98	0.31	0.01
conventional etchant 2	4256	2075	870	0.49	0.20

For conventional Si wet etchant, the etch rate in Si(100) direction is much higher than in (111) which results in pyramid shape etching profile.

Bare Si (100) → Conventional Si etchant

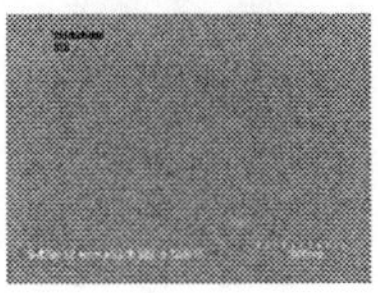

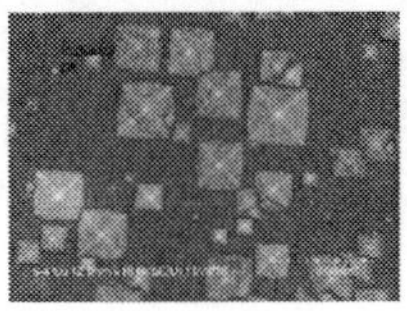

(Top view SEM)

pyramid shape etching profile

VERSUM MATERIALS

Etch rate requirements

To address the etch residue issue, etch rate in Si (111) and Si (110) directions should be increased.

	Etch rate (A/min)						Selectivity	
	Si (100)	Si (110)	Si (111)	a-Si	Poly Si	ALD oxide	Si(110)/ Si(100)	Si(111)/ Si(100)
Target	>1000	>1000	>500	>300	>600	<0.1	>1	>0.5

Bare Si (100) → New Si etchant

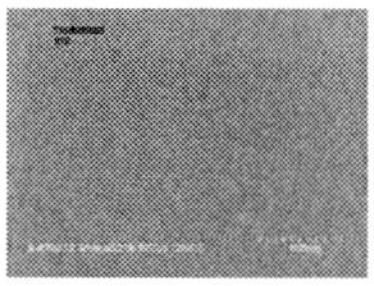

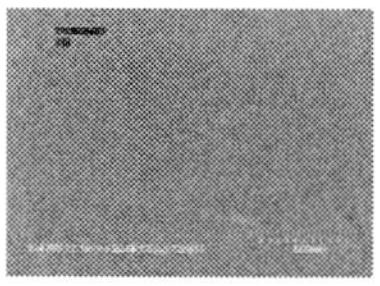

(Top view SEM)

No pyramid shape etching profile

VERSUM MATERIALS

Formulation design

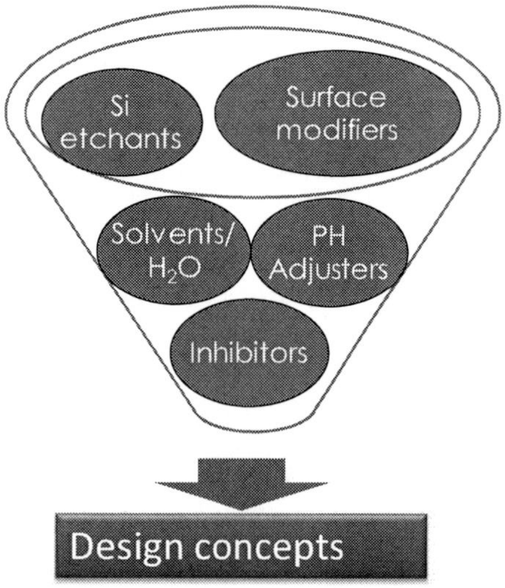

Etchants: Dissolves Mono-crystalline Si

Surface modifiers: Suppress the anisotropic Si etching; improve the wettability

Solvents/H_2O: Removes organic/inorganic residues Vary the viscosity of formulations

PH adjusters: maintains pHs

Inhibitors: Preserve SiOx, barrier liner and Cu

VERSUM MATERIALS

Computational modeling

➢ Impact of different quaternary ammonium hydroxide (QAH) in a-Si etch rate

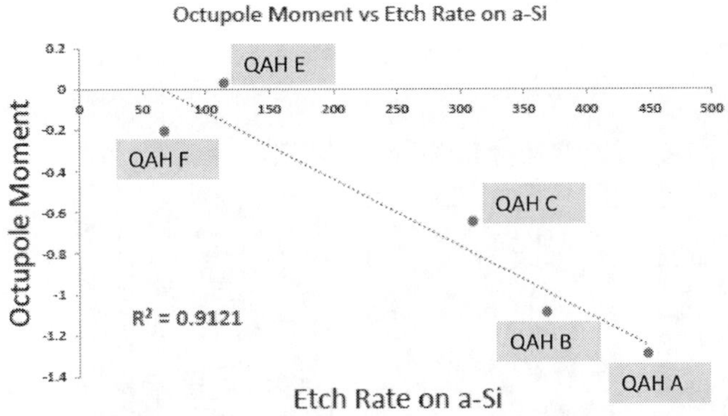

- Strong correlation between cation-only electrostatic octupole moment and experimental a-Si etch rate
- Etch rate increases with greater electrostatic octupole moment magnitude
- This result suggests that the charge distribution of our cation is playing a key role in the observed etch rate

VERSUM MATERIALS

Additive effect in Si etching selectivity

➢ An effective additive increases Si (110), (111) etch rates and maintains Si (100) etch rate at high level.

	Etch rate (A/min)			Selectivity	
	Si (100)	Si (110)	Si (111)	Si(110)/ Si(100)	Si(111)/ Si(100)
conventional etchant 2	4256	2075	870	0.49	0.20
conventional etchant 2 + additive	2821	3845	1155	1.36	0.41

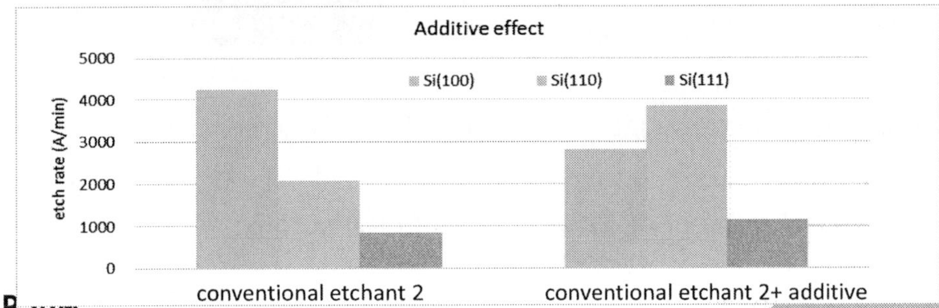

VERSUM MATERIALS

Computational modeling

➢ Si surface

Si (100) Si (110) Si (111)

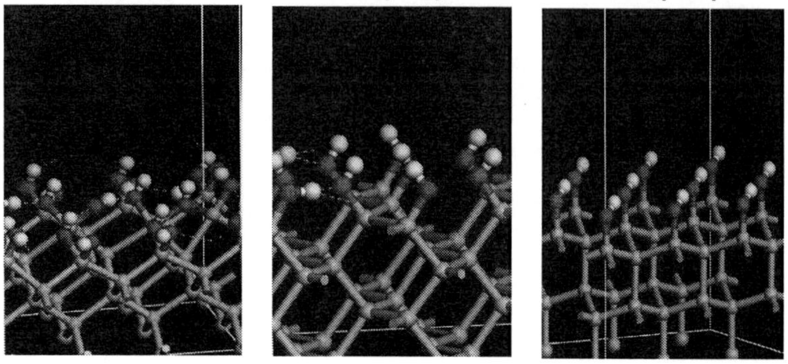

- Both Si (111) and Si (110) surfaces are Si(OH) terminated while the Si (100) surface is $Si(OH)_2$ terminated.
- The Si (110) surface has step sites (greater overall surface roughness).
- All surfaces showed some degree of H-bonding on surface with the greatest amount seen for the Si (100) surface.

VERSUM MATERIALS

Computational modeling

Effect of surface symmetries on additive binding

Average additive binding energies (kcal/mol)		
Si (110)	Si (100)	Si (111)
-4.4	-1.0	-1.8

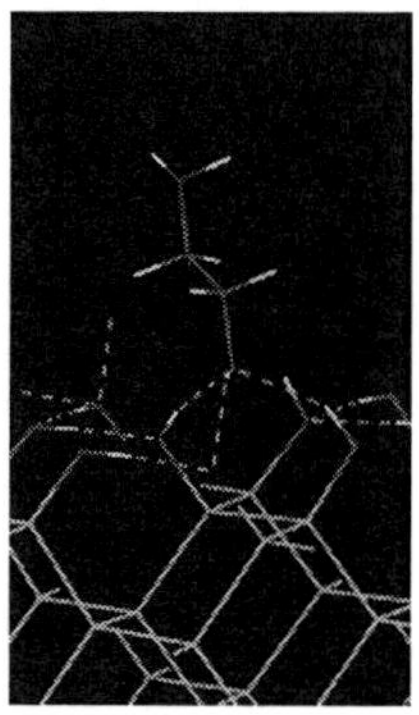

- All binding energies were negative and favorable on each of the surfaces.
- Highest surface coverage of additive is on Si (110) surface.

VERSUM MATERIALS

Our thoughts

- Si (110) : Si (100) : Si (111) surfaces have ratio of 4:1:2 additive surface coverages
  - Surface with greatest e/r in presence of additive also has the greatest surface coverage of additive which is consistent with experiments
- Previously, rate-determining step (RDS) for Si etching process in a basic environment was found to be the formation of the surface $-Si(OH)_3$group
  - additive which was found to interact strongly with surface OH groups can aid in reorganizing the active site to lower RDS activation barrier and increase e/r
- Also, once $Si(OH)_4$ is generated, additive has a ~-16 kcal/mol favorable interaction energy with it
  - additive can help to keep $Si(OH)_4$ from crashing out of solution because of high water solubility of additive
- Si (100) surface may undergo unfavorable water elimination at surface in the presence of additive, which is consistent with experiments showing a decrease in e/r in the presence of additive

VERSUM MATERIALS

Experimental trails

	Etch rate (A/min)						Selectivity	
	Si (100)	Si (110)	Si (111)	a-Si	Poly Si	ALD oxide	Si(110)/ Si(100)	Si(111)/ Si(100)
Target	>1000	>1000	>500	>300	>600	<0.1	>1	>0.5
conventional etchant 1	7117	2228	98	n.a	n.a	n.a	0.31	0.01
conventional etchant 2	4256	2075	870	n.a	n.a	n.a	0.49	0.20
Versum Material's new etchant	2674	3821	1561	586	1753	0.08	1.43	0.58

Note: All the etch rate were measured at 50C. Unit: A/min

VERSUM MATERIALS

Application in FEOL process

- **Sigma shape generation for S/D**

before etching

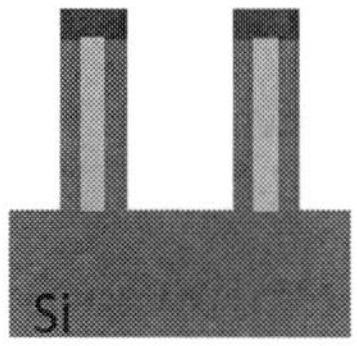

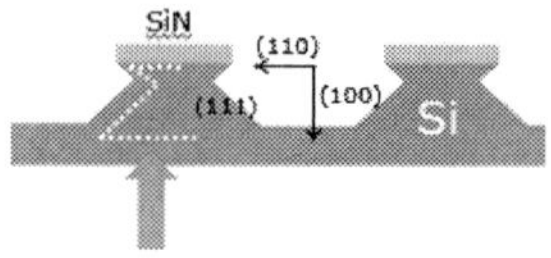

Sigma etching profile

after etching

Conventional

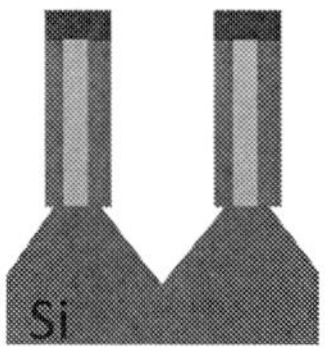

V shape

Target

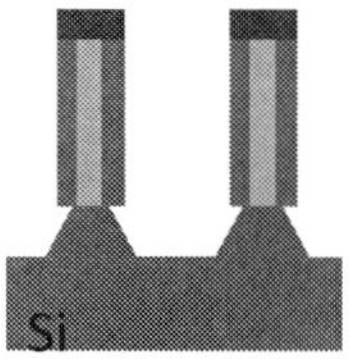

Flat bottom

Sigma shape generation by wet etching process is achievable by Versum's novel Si etchant.

VERSUM MATERIALS

Summary

For the studies of Versum Material's novel Si etchant, we demonstrate the following achievements:

✓ excellent dummy poly gate etch residue removal capability.

✓ etching selectivity of Si (110)/ (100) and Si (111)/ (100) reaches 1.5 and 0.5 respectively.

✓ compatible with HK, oxide, SiN and SiGe

✓ could be operated as recirculation mode, bath-life over 72hrs at 50C.

✓ applicable on single wafer tool

VERSUM MATERIALS

Indium bump liftoff using the MEI FluidJet

Scott Tice-Director Applications Engineering

Scott.Tice@meillc.com **Ph # 1-541-936-0844**

3838 Western Way NE, Albany, OR 97322

Ph: 541-917-3626 | Fx: 541-917-3623

www.meillc.com

Indium Bump Overview

MEI Built on Commitment

- High-performance infrared photodetectors operating in the range of 3–14 μm are highly needed for a variety of applications, such as security surveillance, chemical sensing, and industrial process monitoring. This has promoted increasing research interest in new materials and structures to improve detector performances, such as spectral sensitivity, leakage current level, and operating temperature

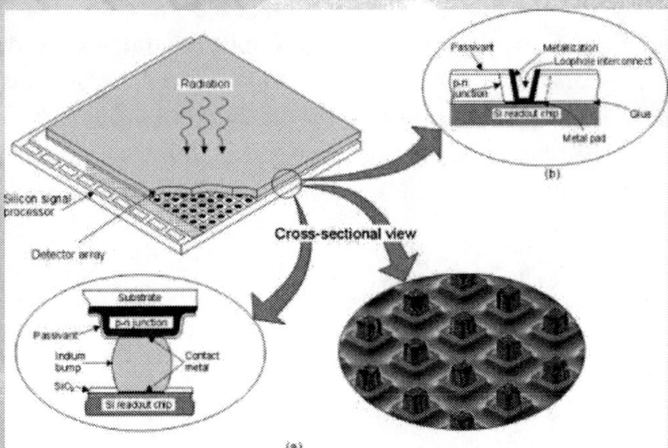

Liftoff Challenges

MEI Built on Commitment

- Indium bumps are extremely fragile
- Spray, Tape and Manual liftoff create severe damage
- Standard immersion systems leave metal between wafers which causes damage when removed
- Soaking and manual extraction are extremely slow with low device yield
- Limited budgets for equipment
  - Low volume production requires a lower cost system
- Production ramp is required
  - Driving inventive solutions to problems described above

www.meillc.com -3-

Metal Lift-Off with FluidJet™

FluidJet™ Design

MEI Wet Processing Built on Commitment

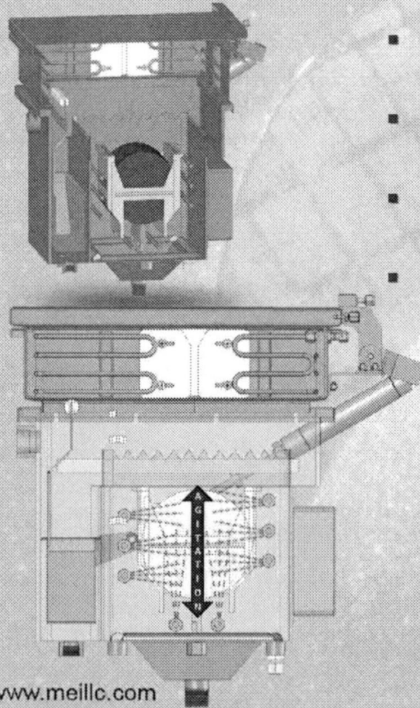

- Provides metal removal utilizing precise hydraulic forces with robot agitation to maximize coverage
- Oscillating fluid forces to optimize metal removal from cassette
- Tank fluid flow balanced to maximize metal removal and achieve maximum metal cleanout
- Fluid flow modeled to specific field proven velocity requirements for stripping metal

Adjustable High-pressure spray bars between wafers

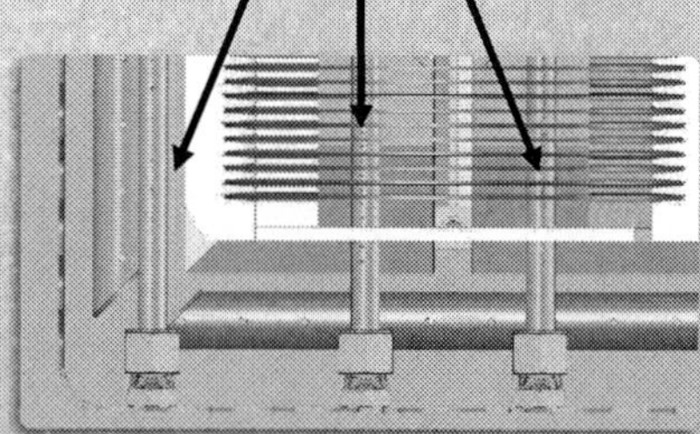

www.meillc.com

NMP-Acetone-IPA – Marangoni or SRD

MEI Built on Commitment

Objective: The test was conducted to evaluate if indium could be removed without causing damage to the sensitive structures below.

Test was performed using:

- ❖ MEI Metal lift tool for metal liftoff process w/NMP
- ❖ SS Recirculated filtered tank for Acetone
- ❖ SS Recirculated filtered tank for IPA rinse
- ❖ DIW Marangoni dry

❖This process testing was executed in the MEI LLC Applications Lab in Albany, Oregon in our Fluidjet system.

NMP-Acetone-IPA MLO sequence

MEI Built on Commitment

The Process:

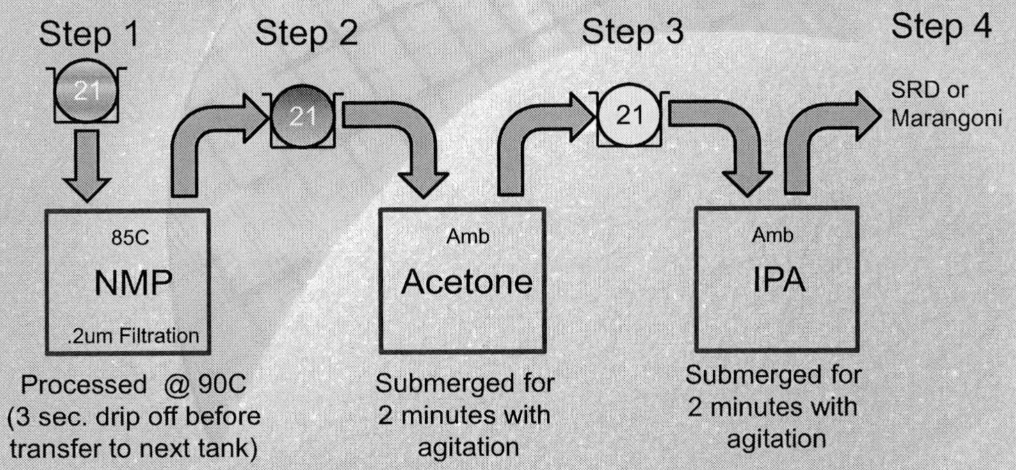

NMP-Acetone-IPA Indium liftoff

FluidJet Recipe

- NMP Tank temperature set 90C, ALL Steps with robot agitation on.
- Soak time 20 Minutes 25% recirculation pump power, Sonics 100% @ 72kHz.
- Gentle Lift "Bottom Center" 15 minutes 20% Fluid jet power, Recirc. 25% power, Sonics 100% @ 72kHz
- Gentle Strip 10 minutes 20% Fluid jet power, Recirc. 25% power, Sonics 100% @ 72kHz 60 Sec Cycle on strip pumps.
- Strong Lift "Bottom Center" 15 minutes 50% Fluid jet power, Recirc. 25% power, Sonics 100% @ 72kHz
- Final Clean and Strip 30 Minutes 50% Fluid jet power, Recirc. 25% power, Sonics 100% @ 72kHz 60 Sec Cycle on strip pumps
- Followed by 2 Min Acetone rinse
- Followed by 2 Min IPA Rinse and Dry

Indium Bump Damage

Indium Bumps have smeared due to rapid metal removal

Indium Bump Damage

MEI Built on Commitment

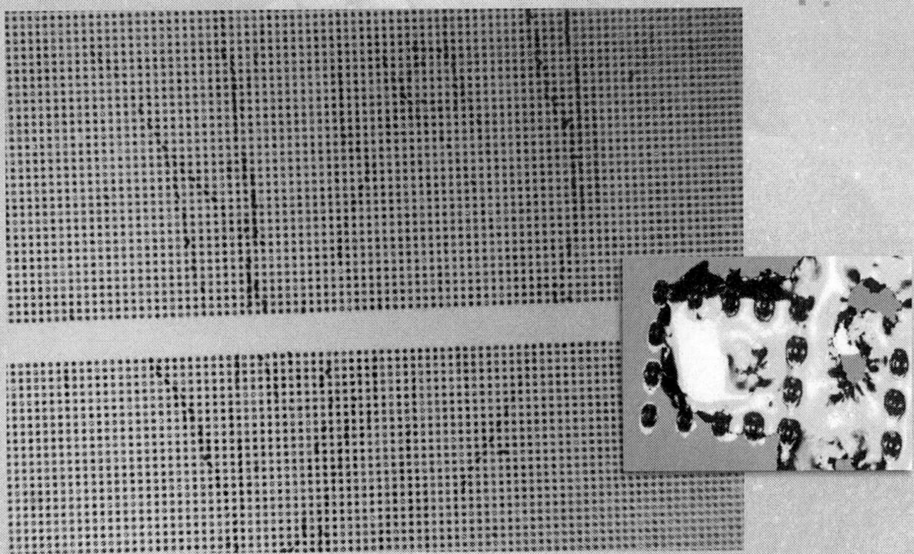

Damage due to rolling of metal as it was removed and rubbed the tops off the bumps

www.meillc.com -9-

Indium Liftoff High Flow

MEI Built on Commitment

Good cleaning but slight rubbing during removal creates shine on the tops of the bumps

www.meillc.com -10-

Undamaged Indium Bumps

MEI Built on Commitment

Clean Liftoff achieved at 1.5 hours with gentle liftoff followed by aggressive flow for final cleaning.

Conclusion

MEI Built on Commitment

- Indium bumps are very fragile and require gentle liftoff
- Long soaks and gentle strip forces are required to avoid smearing of bump formations during liftoff
- After liftoff is accomplished, final cleaning is performed
- Results on single wafer spray tools destroyed the indium bumps caused by aggressive striping
- The FluidJet is uniquely capable at producing a relatively fast liftoff of 25 wafer cassettes with no damage to features
- Automated liftoff operation with the Fluidjet provides dramatic productivity improvements compared to other methods

Thank you for your attention

AUTHOR INDEX

Adamczyk, Andrew J. 326
Asokan, Muthappan 15
Balan, V. 289
Balan, Viorel 275
Barclay, Joshua 50
Beitia, Carlos 121
Bender, Hugo 199
Berhe, Seare A. 15
Bhat, Talapady 191
Bhattacharyya, Dhiman 67, 102
Broussous, Lucile 59
Brown, Ian 15
Buisine, F. 59
Campo, Alain 112
Cano, Austin 227
Carolan, Patrick 199
Chen, Tianniu Rick 326
Chiou, N. 289
Choi, Eun-Suck 239
Chyan, Oliver 15
Connor, Patrick 302
Conrad, Mark 102
Croisy, Marion 112
Das, Atanu 256, 267
De Coster, Hanne 199
Deem, John 290, 291
Deng, Lu 50
DeOliveira, Norberto 102, 324
DeOliveria, Norberto 67
Devendar, F. 325
Du, Jincheng 50
DuMont, Jaime 227
Eynon, Ben 1
Faguet, Jacques 15
Fontaine, Herve 121
Friot, P.-Y. 289
Frye, Don 267
Fujimura, Yu 91
Gan, Nobuko 91
George, Steven 227
González-Aguirre, Paola 121
Gourvest, E. 289
Guiheux, Denis 112
Guo, Eric 292
Harris, H. Rusty 142
Heslop, Stacy 159
Holsteyns, F. 37, 170
Holsteyns, Frank 91, 199
Horiguchi, Naoto 199
House, Matthew 290, 291
Iordanov, Iordan 290
Ishikawa, H. 59
Jacob, G. 289
Jenny, Cecile 112
Jiang, Wei 292
Jin, Jun 292
Joblot, Sylvain 112
Johnson, Lawrence 302
Kesters, Els 37
Khalil, Osama 290, 291
Khatkhatay, Fauzia 102
Kim, Hyun-Tae 239
Knapp, Alan 308
Knotter, Martin 129
Kumari, Sangita 325
Lamaury, A. 59
Langan, John G. 326
Le, Q. 37
Lee, Dong-Hwan 239
Lee, Jae-Hwan 239
Lee, TaeHoon 252
Lee, Yi Chia 326
Li, Jian 191
Lieten, Ruben 267
Liu, Jun 267
Liu, Wen Dar 326
Lorant, C. 37
Lundgren, Jorgen 121
Manning, Hayley 102
Marquardt, Amy 227
Masaoka, Toru 91
McIntosh, Robert 77
Meldrum, Duncan 210
Mena, Alexander 102
Minato, Yasuharu 91
Miyazaki, Yoichi 91
Mukherjee, Tamal 15
Muscat, Anthony 159
Ogawa, Yuichi 91
Ohlsen, Jim 121
Okobiah, Oseoghaghare 50
Otsuji, M. 170
Pacco, Antoine 91
Pargon, Erwine 112
Park, Jin-Goo 239, 239
Parson, Thomas 256, 267
Patil, Sushil 102
Peckler, Lauren 159
Perrot, C. 289
Phely-Bobin, Thomas 77
Pierre-Edouard, Raynal 181
Posseme, Nicolas 112
Qian, B. 289
Qian, Sandy 292
Ragnarsson, Lars-Ake 199
Reidy, Rick 50
Richard, Claire 112
Rimal, Sirish 15
Rodier, Dan 316
Sadasivuni, Tejasvi 324

AUTHOR INDEX

Salvatore, F. 289
Sato, M. 170
Scott, Silas 325
Sebaai, F. 170
Sengphanlaya, Tina 50
Sengupta, Archita 77
Shen, Jian 292
Shirakawa, Hajime 170
Shucosky, Tony 302
Sidhwa, Ardeshir 290, 291
Sih, Vincent 67, 191
Singh, Kanwaljit 15
Singh, SherJang 325
Snow, J. 170
Sundstrom, Glen 308
Takahashi, H. 170
Thomas, Elizabeth 267
Tice, Scott 333
Trio, Paul 77
Valette, S. 289
Van Dorpe, Pol 199
Varanasi, Rao 302
Vereecke, Guy 199
Wang, Crystal 292
Wang, Mars 292
Wang, Volley 267
Weidner, Colin 67
White, Daniela 256, 267
White, Michael 256
White, Mike 267
Willems, Kherim 199
Wostyn, Kurt 91
Wu, Aiping 326
Wu, Keanu 292
Xu, Fei 292
Xu, X. 37
Yim, Ratanak 289
Yoshida, Y. 170
Yoshida, Yukifumi 91
Zhai, Hong-Ying 67
Zhang, Harry 292
Zoll, S. 59
Zuck, David 290, 291

Surface Preparation and Cleaning Conference (SPCC)
c/o Mark Thirsk
Linx Consulting-PO Box 384
Mendon, MA 01756-0384
USA

ISBN 978-1-5108-4814-6

CURRAN ASSOCIATES INC.
proceedings.com